Lineages and Advancements in Material Culture Studies

This volume comprises a curated conversation between members of the Material Culture Section of University College London Anthropology. In laying out the state of play in the field, it challenges how the anthropology of material culture is being done and argues for new directions of enquiry and new methods of investigation. The contributors consider the ramifications of specific research methods and explore new methodological frameworks to address areas of human experience that require a new analytical approach. The case studies draw from a range of contexts, including digital objects, infrastructure, data, extraterrestriality, ethnographic curation, and medical materiality. They include timely reappraisals of now-classical analytical models that have shaped the way we understand the object, the discipline, knowledge formation, and the artefact.

Timothy Carroll is principal research fellow in the Department of Anthropology at University College London, UK.

Antonia Walford is lecturer in Digital Anthropology in the Department of Anthropology at University College London, UK.

Shireen Walton is lecturer in Anthropology in the Department of Anthropology at Goldsmiths, University of London, UK.

Lineages and Advancements in Material Culture Studies

Perspectives from UCL Anthropology

Edited by Timothy Carroll, Antonia Walford, and Shireen Walton

Routledge
Taylor & Francis Group
LONDON AND NEW YORK

First published 2021
by Routledge
2 Park Square, Milton Park, Abingdon, Oxon OX14 4RN

and by Routledge
52 Vanderbilt Avenue, New York, NY 10017

Routledge is an imprint of the Taylor & Francis Group, an informa business

British Library Cataloguing-in-Publication Data
A catalogue record for this book is available from the British Library

Library of Congress Cataloging-in-Publication Data
Names: Carroll, Timothy (Timothy Andrew), editor. |
Walford, Antonia, editor. | Walton, Shireen Marion, 1986– editor.
Title: Lineages and advancements in material culture studies : perspectives from UCL anthropology / edited by Timothy Carroll, Antonia Walford and Shireen Walton.
Description: Abingdon, Oxon; New York, NY: Routledge, 2020. |
Includes bibliographical references and index.
Identifiers: LCCN 2020031267 (print) | LCCN 2020031268 (ebook) |
ISBN 9781350127487 (hardback) | ISBN 9781003085867 (ebook)
Subjects: LCSH: Material culture–Research–Methodology. |
Material culture–Case studies. |
University College, London. Anthropology Department.
Classification: LCC GN406 .L564 2020 (print) |
LCC GN406 (ebook) | DDC 306.072/1–dc23
LC record available at https://lccn.loc.gov/2020031267
LC ebook record available at https://lccn.loc.gov/2020031268

ISBN: 978-1-350-12748-7 (hbk)
ISBN: 978-1-003-08586-7 (ebk)

Typeset in Times New Roman
by Newgen Publishing UK

Contents

Figures

Tables

Contributors

Victor Buchli is professor of material culture at UCL and works on the material culture of 'Low Earth Orbit', architecture, domesticity, the archaeology of the recent past, and critical understandings of materiality and new technologies. Currently, he is principle investigator of the five-year European Research Council (ERC) funded research project, 'ETHNO-ISS: An Ethnography of an Extra-terrestrial Society: the International Space Station' (ERC Advanced Grant, no. 833135). His previous books include *An Archaeology of the Immaterial* (2015); *An Anthropology of Architecture* (2013); *An Archaeology of Socialism* (1999); and, with Gavin Lucas, *Archaeologies of the Contemporary Past* (2001). He has also edited *The Material Culture Reader* (2002), *Material Culture: Critical Concepts in the Social Sciences* for the Major Works Series (2004); and, with C. Alexander and C. Humphrey, *Urban life in Post Soviet Asia* (2007).

Timothy Carroll is a principal research fellow and a UK Research and Innovation Future Leaders Fellow in Anthropology at UCL. His research focuses on the role of material within the religious practice of Eastern Orthodox Christians, and the relationship between the body as a cultural artefact and the wider material ecology. His research works across religion, art, medicine, and ecological contexts. He is the author of *Orthodox Christian Material Culture: Of People and Things in the Making of Heaven* (2018), co-author of *A Return to the Object: Alfred Gell, Art and Agency* (2021), and co-editor of *Medical Materialities: Toward a Material Culture of Medical Anthropology* (2019) and *Material Culture of Failure: When Things Do Wrong* (2017).

Ludovic Coupaye is associate professor of anthropology at UCL and professor of history and anthropology of Pacific Arts at the École du Louvre, Paris. His research focuses and combines different strands of research, distributed over four interrelated topics: material and visual culture in Oceania; art and aesthetics among the Abulës-speaking communities ('Abelam') of East Sepik Province; anthropology of techniques, skills and materiality; and anthropology of technology and Modernity. His most recent strand of research deals with the wider question of the relationships between 'technology' and 'society', from the angle of technical activities, technical objects, and technical systems. He is the author of a monograph, *Growing Artefacts, Displaying Relationships: Yams, Art and Technology amongst the Nyamikum Abelam of Papua New Guinea* (2013), and co-edited a special issue in the journal *Oceania* on 'Interweaving of Vital and Technical Processes' in Oceania.

Adam Drazin is an associate professor at UCL, where he coordinates the MA in Material and Visual Culture and teaches design anthropology. He has worked in the past for HP Labs and Intel Ireland and taught anthropology courses in universities and design schools. His work has been published in books and journals, including *Ethnos, Social Anthropology* and the *Journal of Design History*. He is co-editor of the journal *Home Cultures*, and recently edited, with Susanne Küchler, *The Social Life of Materials* (2015), a volume about anthropological and ethnographic approaches to materials and materials innovation. He is a member of the Business of Fashion and Textiles research cluster, which received a major grant in 2018. His forthcoming book on *Design Anthropology in Context* will be published by Routledge.

Haidy Geismar is professor of anthropology at UCL; she is a curator of the Ethnography Collections and co-directs the Centre for Digital Anthropology. With fieldwork in Vanuatu and New Zealand dating back to 2000, her research focuses on museums and collections as sites of knowledge and value production. Her recent books include *Moving Images: John Layard, Fieldwork and Photography in Malakula since 1914* (2010); *Treasured Possessions: Indigenous Interventions into Intellectual and Cultural Property* (2013); *The Routledge Cultural Property Reader* (with Jane Anderson, 2017); and *Museum Object Lessons for the Digital Age* (2018). She is an active curator, working with the Cambridge Museum of Archaeology and Anthropology, the Metropolitan Museum of Art, and the East West Gallery, Honolulu. She is chair of the Royal Anthropological Institute Photography Committee, a founding editor of the open access series, *Anthropology and Photography*, and is associate editor of the *International Journal of Cultural Property*.

Laura Haapio-Kirk's research examines the experience of ageing in Japan, in particular how new communication technologies impact older adults. Between 2018 and 2019 she conducted 16 months' ethnographic fieldwork in Kyoto city and rural Kōchi prefecture. Laura is a researcher on the ERC-funded 'Anthropology of Smartphones and Smart Ageing' project at UCL Anthropology. In 2018–2019 she was a visiting researcher at the Graduate School of Human Sciences, Osaka University. Laura was a Leach/Royal Anthropological Institute Fellow in Public Anthropology for 2020, working to improve the profile of anthropology in public debate. Prior to starting her PhD, Laura was a research assistant and public engagement fellow on the 'Why We Post' project at UCL Anthropology. Laura's training in Visual Anthropology (MSc University of Oxford, 2011) informs her ethnographic approach, which combines participant observation with drawing and film as research methods. She is currently writing a book about her research, which includes reportage illustration.

David Jeevendrampillai, Julia Burton, and Eva Sanglante are members of the *Objects of Desire* (OoD) collective. OoD is a collaborative project between artists, anthropologists, and sexworkers. The collective is sexworker-led and OoD writing is done in a collaborative fashion as per the ethics of the collective. OoD preserve sexworkers' stories through archiving and exhibiting the artefacts of sexworkers in order to highlight the empirical details of sexwork in the gallery spaces and wider public discourse. The collective has held two major shows in London (2016) and Berlin (2019). OoD's activity has been supported by the generous collaboration of the sexworker community in London and Berlin, The Open Society Public Health Program Grant and the Schwules Museum, Berlin.

Hannah Knox is associate professor of anthropology at UCL and director of the UCL Centre for Digital Anthropology. Her work focuses on the study of technical projects as sites of social and cultural change, and she has conducted ethnographic research in the UK and Latin America. Recent books include *Roads: An Anthropology of Infrastructure and Expertise* (co-authored with Penny Harvey) and *Ethnography for a Data Saturated World* (co-edited with Dawn Nafus). Her latest book, *Thinking like a Climate: Governing a City in Times of Environmental Change*, was published in 2020 with Duke University Press.

Susanne Küchler is professor of anthropology and material culture at University College London. She has conducted ethnographic fieldwork in island Papua New Guinea and Eastern Polynesia over the past 25 years, studying from a comparative perspective the modular, composite image in its relation to political economies of knowledge. Her work on the history of the take-up, in the Pacific, of cloth and clothing as 'new' material and 'new technology' has focused on social memory and material translation, and on the epistemic nature of pattern. The question of why images prevail, set within a re-examination of Oceania's flourishing image-based polities, is the central theme of her current work, which follows publications on *Malanggan: Art, Memory and Sacrifice* (2002); *Pacific Pattern* (2005) and *Tivaivai: The Social Fabric of the Cook Islands* (2009).

Delphine Mercier is collections curator of the UCL Ethnography Collection. In the past, she has taught art history, iconography, fashion history and textile techniques at École du Louvre and other schools in Paris while researching textile techniques during the Middle Ages. She has also worked in various French institutions focusing on preservation of cultural heritage and developing projects in partnership with local communities. As a curator in one of the UCL Study Collections, she not only works with students but also with various publics, such as schools and source communities. She is currently a research postgraduate student in UCL Science and Technology Studies department focusing her work on non-specialised publics in museums contexts.

Daniel Miller is professor of anthropology at UCL and principle investigator of the ERC Advanced Grant project (no. 740472) 'Anthropology of Smartphones and Smart Ageing'. Recent books include *The Comfort of People* (2017); *Social Media in an English Village* (2016); *Webcam* (2014), with Jolynna Sinanan; *Migration and New Media* (2012), with Mirca Madianou; *Digital Anthropology* (2012), edited with Heather Horst, *Visualising Facebook* (2017), with Jolynna Sinanan; *How the World Changed Social Media* (2016), co-authored with eight others; and *Blue Jeans* (2012), with Sophie Woodward.

Aaron Parkhurst is a lecturer in biosocial and medical anthropology at UCL, with a focus on the anthropology of the human body, bioethics, cyborg and health technology, urban anthropology, and outer-space studies. His work addresses men's health issues, chronic illness, and the effects of urban life on both physical and mental health. Prior to his work in London, he conducted research in Dubai and Abu Dhabi to study local understandings of health in the Arabian Gulf. His current work is focused on biosocial understandings of biomedicine, genetics and mental health; humanity's struggles and anxieties with 'modernity'; the anthropology of sport, medical materiality, and the human body in outer space. Some of his major publications focus on the intersection of medical anthropology and material culture studies, as well as the relationship between the human body and rapidly growing cities.

Christopher Pinney is professor of anthropology and visual culture at University College London. He has held visiting positions at Australian National University, University of Chicago, University of Cape Town, Northwestern University, Boğaziçi University (Istanbul), and Jagiellonian University (Krakow). In 2019 he was a visiting professor at EHESS Paris and delivered a series of Global Initiative of Academic Networks lectures at Jawaharlal Nehru University, Delhi. His research interests cover the art and visual culture of South Asia, with a particular focus on the history of photography and chromolithography in India. He has also worked on industrial labour and Dalit goddess possession. Amongst his publications are *Camera Indica* (1997), *'Photos of the Gods'* (2004), *The Coming of Photography in India* (2008), *Photography and Anthropology* (2011), *The Waterless Sea* (2018), and *Lessons from Hell* (2018). He is currently leading the ERC Advanced Grant project (no. 695283) 'Citizens of Photography: The Camera and the Political Imagination'.

Jill Reese is a senior teaching fellow in visual culture and anthropology writing tutor at University College London. Through ethnographic work with artists and lower-level members of five political parties in Madurai, Tamilnadu, she examines the materiality of images, the productivity of spectacle in assertions of power, and the bodily performance of visibility and patronage in the making of local politicians. Her research also analyses the aesthetics, processes of making, display and demise of political imagery installed within contested urban spaces. She currently teaches on material politics, photography, writing and visual methods. Prior to an academic career, she produced documentary programs for broadcast television and taught video production, editing, photography, and documentary history courses in India and the United States.

Rafael Schacter is an anthropologist and curator working on both public and global art. A lecturer in the Department of Anthropology at University College London, Schacter received his PhD in the department in 2011 and was a British Academy Postdoctoral Fellow at the same institution from 2014–2017. Schacter has published three books: *Street to Studio* (2018); the award-winning *World Atlas of Street Art and Graffiti* (2013); and *Ornament and Order: Graffiti, Street Art and the Parergon* (2014). He also has an active curatorial practice. Schacter co-curated the Tate Modern's *Street Art* exhibition in 2008; he curated *Futurismo Ancestral, Mapping the City* and *Venturing Beyond* at Somerset House (between 2014 and 2016); *Silver Sehnsucht* in London's Silvertown in 2017, and *Motions of this Kind* at the Brunei Gallery, SOAS, in 2019. He is also currently curating an ongoing, four-year public art project in Heerlen, the Netherlands.

Antonia Walford is a lecturer in digital anthropology at UCL. Her research explores emerging practices of digital quantification and calculation and how they affect political, social, and cultural imaginaries, with a focus on data, the environmental sciences, and Brazil. She has co-edited the following collections 'Is There an Ontology to the Digital?' (2016), with Hannah Knox; *A World Laid Waste? Responding to the Social, Cultural and Political Consequences of Globalisation* (2018), with Francis Dodsworth; *Environmental Alterities* (Forthcoming), with Cristóbal Bonelli; and 'Towards an Anthropology of Data' (Forthcoming 2021), with Rachel Douglas-Jones and Nick Seaver. She also has published several articles and book chapters on critical anthropological approaches to new knowledge objects in data-driven science. She is working on a new book provisionally entitled *Data Nature: The Digital Worlds of Environmental Science.*

Shireen Walton is a lecturer in anthropology at Goldsmiths, University of London. Her research focuses on digital and visual anthropology/ethnography, and she has carried out fieldwork in Iran, the UK, Italy, and online. She was a postdoctoral research fellow at UCL Anthropology between 2017–2020 as part of the European Research Council funded project,

'The Anthropology of Smartphones and Smart Ageing' (ASSA). Prior to this, she was a teaching fellow in material and visual culture at UCL Anthropology (2016–2017). She received her PhD in 2015 from the School of Anthropology and Museum Ethnography, University of Oxford, with a thesis on popular digital photography in/of Iran, about which she has published several articles and book chapters, alongside publications on digital and visual research methods in/for anthropology. Her forthcoming monograph, *Ageing with Smartphones in Urban Italy*, will be published with UCL Press as part of the ASSA book series.

Preface

This volume has emerged as a conversation, one that has happened in the halls, offices and meeting rooms of the material culture research group at UCL Anthropology. It is not a definitive statement, but rather a multivoiced exploration. It includes those colleagues present in the group during 2017 and 2018. The process of rendering this conversation as a volume began in response to a felt need to have some updated version of *The Material Culture Reader* (Buchli, ed. 2002) and, to a lesser extent, the *Handbook of Material Culture* (Tilley, et al. eds 2006). We felt this need particularly in terms of new directions that have developed in material culture studies, and the availability of an easy resource to use in teaching and research training. So, while this volume does not replace either of these earlier works, it is specifically designed to sit alongside these volumes. Unlike the *Reader*, however, the chapters in this volume are each original pieces, written in 2018 and 2019; and, unlike the *Handbook*, it is not a comprehensive survey of the whole field of study, but rather a reflection of new developments and directions of possible research enquiry.

Contributors were invited to submit chapters that considered where material culture studies is, currently, and what directions it is, or should be, developing. This means that some of the chapters are somewhat speculative, casting forward the shape of research very much still in development. This is exciting and new, and therefore could run the risk of aging poorly – so we also asked contributors to situate their positions within the longer (and, in some cases, forgotten or minor) reading of material culture studies and anthropology.

In total, this volume has fifteen chapters, each coming from a different perspective and addressing different, albeit overlapping, research areas. While this is not our agenda, we are very aware that there is a 'school' of thought that has developed here in the Material Culture Section of UCL Anthropology. As a research cluster, and as colleagues sharing teaching and administrative duties, this is to some degree inevitable. It is also a result of the legacy left by Daryll Forde in how he founded the department at the intersection of British Social Anthropology, American Cultural Anthropology, and Archaeology. As such, even within the multiple perspectives and voices of this volume – where many disagreements exist between the authors (some of which are discussed more fully in the Introduction) – there is a sense that material culture studies (MCS) at UCL Anthropology has taken up a new analytical edge. As a research group, the authors represented in this volume have increasingly come to privilege the material as a methodological and analytical imperative within the project of social anthropology, in a way that is not necessarily seen in MCS in other schools and other disciplines. As an imperative, the material can be seen to direct research – not only being a subject of enquiry but an ongoing source of research questions and continual source of ethnographic data – in a way similar to how local concepts (such as *mana* or *hau*) are understood to shape ethnographic enquiry and anthropological analysis via 'indigenous exegesis'. However, the specifics of *how* material shapes this enquiry is distinct within each voice in this discussion.

In offering the present volume, we invite the reader to participate in this conversation. There is no specific order to the chapters, and they are not arranged into sections based around specific themes. Instead, we as editors highlight in our Introduction central themes that emerge across the contributions. These themes came to the fore in a daylong workshop held in June 2018, and following this workshop contributors were encouraged to consider these issues as they finalised their chapters.

Acknowledgements

We would like to thank our colleagues for participating in this conversation, and also to thank the editorial board of the *Journal of Material Culture* for covering the fee to make this volume available through Open Access.

1 Introduction

Timothy Carroll, Antonia Walford, and Shireen Walton

Taking insight from the wisdom of Marie-Jeanne Rose Bertin, who – as Marie Antoinette's dressmaker – cautioned, 'There is nothing new except what has been forgotten,' this volume is in part a project in highlighting how the lineages of thought and methodological approach in Material Culture Studies (henceforth MCS) produce the anthropology of material culture as it stands today. These advancements in MCS, represented in the chapters, take a number of forms. Some of the chapters are explicitly position pieces, challenging how anthropology of material culture is currently being done, and arguing for new directions of enquiry or new methods of investigation. Other chapters advance new typologies of objects or take old theories into new areas. Many of the chapters explore the ramifications of specific methods and offer new methodological frameworks to address areas of human experience that demand a new or reimagined analytical approach. While the scale of investigation, and the types of object in question, vary widely across the collection, five key themes emerge around a reconsideration of what the object is. These five themes – concerning the self and personhood, temporality, scales and topology, representation, and participation – draw the object into sharp relief, as an anthropological imperative, and allow us to explore the role the object plays as both a topic of study and an ongoing source of research questions within the anthropological project.

However, the advancements that this book proposes must be understood in relation to what has come before. In framing this book as a stage in a long conversation that is happening within the research group at University College London Anthropology, it is helpful to have a brief account of how that conversation has been shaped so far. There are many ways to retell this history and lineage of MCS, and important contributions to this genealogy have been given by the editors of the *Journal of Material Culture* (Miller and Tilley 1996), Mike Rowlands (1983), Victor Buchli (2002a), Daniel Miller (2005b), Christopher Tilley (2006), Dan Hicks and Mary Beaudry (2010), Paul Basu (2013) and Haidy Geismar et al. (2014), amongst many others. Our intention in this volume is not, however, to produce another genealogy of MCS for introductory context-setting purposes. Rather, the volume illustrates how the historical foundations of MCS – or what we, the editors, call 'lineages' – are central to current and nascent work in the field. By making explicit how these lineages have developed within UCL, we are situating the 'advancements' in theory and method put forward in this volume, not in terms of wheels re-invented, but by recognising that the epistemological foundations of MCS allow us to explore very old discourses in light of new phenomena, and to reconsider classic frameworks within new contexts. Each of the chapters actively engage with an epistemology of the lineages and, through this, advancements in various directions are made. In this Introduction, we first offer an overview (albeit brief) of how MCS has developed at UCL and then turn to the five themes that the volume addresses.

MCS at UCL, its foundations and threads

In establishing the *Journal of Material Culture* in 1996, the editors opened the first issue with an editorial that made the case that material culture studies is an undisciplined field of study. In framing it in this manner, they highlighted the intellectual freedom gained by drawing from multiple disciplinary insights and methodological approaches. There were no certain ancestors to whom homage must be paid, and no need to guard the borders of disciplinary territory. Nonetheless, in this Introduction we argue that there are strong lines of influence that have shaped how the research group at UCL Anthropology (who, notably, are the managing editors of the *Journal of Material Culture*) has come to claim, and maintain, that undisciplined nature (see also Hicks and Beaudry 2010 and Basu 2013 in this regard).

In founding the Anthropology Department at UCL in 1947,[1] Cyril Daryll Forde drew upon his own multidisciplinary training – first in Geography and then in Ethnology – as well as important work being done at the time at the University of London on physical Anthropology (in Anatomy) and material culture as technology (in various collections). Being trained in Ethnology in the United States, under Alfred Kroeber and Robert Lowie, Forde's ethnographic method was also heavily influenced by Linguistic Anthropology. While Forde rejected the social theories associated with the anatomical studies at the time, he insisted on the study of physical anthropology, and the human relation to material technology, as core to the anthropological project. This broad, interdisciplinary position set him at odds with the dominant school of Social Anthropology at the time, following Bronisław Malinowski, at the London School of Economics (LSE).

Malinowski's interest in the object was limited to its social function and role within the broader context of meaning (Young 2000; Bell and Geismar 2009), and was marked by an 'indifference to structural problems,' giving too little detail to 'significant structure' (Lévi-Strauss 1963: 132; cf. Mosko 2013, though see Basu 2013 for a defence of Malinowski's influence in material culture studies). For his part, Forde maintained that the study of the material basis for humanity was vital. In his book *Habitat, Economy and Society*, he is concerned with cultural difference and social change, investigating both diffusion and the 'functional relations' that 'any element [of civilisation] plays in the life of a people'; he argues that, 'These active cultural factors operate on the relatively static materials of race and physical environment,' but require 'a fairly full and balanced picture of actual peoples' to be understood (Forde 1963[1934]: 8).

In 1962, Forde appointed Peter Ucko to help develop the study of technology within the department. For his part, Ucko framed this primarily in terms of art, having studied Near Eastern anthropomorphic figurines during his doctoral research, carried out at UCL. This emphasis on the visual and formal elements of art objects was part of wider interest at the time, such as by those such as Anthony Forge at the LSE. It was Ucko who started using 'material culture' in its present sense, that is, the study of the relationship between people and things irrespective of time and space. This was a movement away from specific contexts of historical study in museums or by archaeologists of the distant past and critiquing the ethnographic study of material culture for its overarching concern for categorisation of morphological classification. Concurrently, different groups at other research institutions developed different emphases within MCS, some – such as that at the School of Anthropology and Museum Ethnography at the University of Oxford – maintaining and developing the centrality of museums in MCS. For his part, Ucko broadens the remit of MCS, and argues that 'the study of human artefacts can act as a bridge between most other aspects of anthropology'; moving outside the museum, he sees the study of material culture as central to 'the future development of anthropology as an integrated academic subject' (1969: 28). In the comparative project that he undertakes, in his 1969 Curl Lecture, Ucko highlights the fact that as material culture data is manufactured artefacts, comparison can be done regardless of the specific time of origin (ibid.: 29).

Students of Anthony Forge and Peter Ucko – notably Francis and Howard Morphy, Robert Layton, and Alfred Gell – took forward their interest in material culture as related to the anthropology of art. This group of scholars was also heavily influenced by French structuralism and American symbolic anthropology and, during this period, in the 1960s, the core elements of material culture studies at UCL were established. This can be characterised by a central focus on the object, especially the importance of visual art – and later the image more broadly, an interest in technology, the environment, and archaeological insight into landscape and the contemporary past.

It is also in this period – and especially into the 1970s and 1980s (Hicks 2010) – that a growing interest in structuralism and Marxism shaped the kind of ethnographic work being done. Students of the department, such as Mike Rowlands (1984), responded to the wider Marxist interest in social anthropology, moving emphasis away from the functionalist interest on the object and its uses in favour of wider examination of the processes and modes of production by which they were formed. This gave rise to sustained attention to the means by which objects are made, and the social and historical influences of production, but has been fundamentally shaped by the longer history, especially in francophonic, ethnography and theory on technique (e.g. Lemonnier 1986, 1992; Leroi-Gourhan 1971[1943], 1973[1945]; Mauss 1973[1935]).

In the 1980s, Daniel Miller, trained in Archaeology and Anthropology at Cambridge, joined the research group and brought with him an interest in the work of Georg Wilhelm Friedrich Hegel, Karl Marx, Nancy Munn, and Georg Simmel. In his reading of Hegelian dialectics, Miller (1987) was able to help frame a theory of objectification that sought to overcome the subject/object dualism and open material culture studies to examination of how 'things make people as much as people make things' (2009). Miller's interest in the 'humility of the thing' (1987:85ff) also helped bring attention to mundane objects, at the same time as a similar move in culture studies (e.g. Hebdige 1988), that maintained the comparative empirical method of ethnographic enquiry.

In the 1990s, the research group was expanded, incorporating more of an archaeological influence from Cambridge, with the addition of Chris Tilley and Victor Buchli – both students of Ian Hodder. This expanded the already important work on landscape and the built environment by the likes of Barbara Bender (1998). Tilley's work drew upon the post-processual school of archaeology, with a strong emphasis on phenomenology in dialogue with structuralism (1974, 1996). With his interest in architectural forms and archaeology of the recent past, Buchli's work opened up new attention to the home (2013), as well as immateriality and decay (2017).

In this same period, two students of Alfred Gell at the LSE, first Susanne Küchler and then later Christopher Pinney, brought new approaches to the anthropology art and visual culture. Küchler's work on *malanggan* mortuary statues, and later textiles, proved an important means to rethink the relations between persons and objects, specifically in terms of the role of object as extensions of thought (Küchler 2002; Küchler and Eimke 2009). Pinney's interest in the image, and the bodily responses they evoke, has brought new perspectives concerning the phenomenology of aesthetics and the localisation of global practices, such as photography (1997, 2004).

The shape of the research group, has had, as *one* sees when reviewing this history, strong influence from the perspectives of Anthropology of Art at the LSE and Archaeology from Cambridge, as well as the close institutional affiliation, shared teaching and use of the Ethnographic Collections with Archaeology at UCL. In this light, the undisciplinedness of material culture studies, as expressed by Daniel Miller and Christopher Tilley (1996), may be seen in contrast to the stronger (and older, more well-established) schools of thought represented at the institutions from which many of the core members of the research group had come. For its strengths, the Malinowskian tradition of anthropology has, nonetheless, replicated some of his own shortsightedness (Mosko 2013), and, as Tilley has shown (1991, 1994, 1999), MCS affords a much greater interdisciplinary reach than is common to archaeology. Whereas Miller and Tilley made the comparison to language which, while a broad phenomenon has found a home in its own dedicated discipline, may be the better comparison is to Conservation which, drawing from a range of discrete disciplines, benefits from a broad and unifying umbrella. As the Department of Anthropology and Conservation at Kent (or indeed the Human Ecology Research Group at UCL) demonstrates, the holism of sociocultural anthropology allows for this multi- or undisciplined area of study to flourish.

In the breadth of attention to various kinds of objects and technology, the emergence of new genres of objects – such as seen in digital devices and digital objects – gave rise to a new area of research focus, in the establishment of the digital anthropology research group as part of wider MCS. In establishing the new area of study, Miller was joined by a series of people working on various aspects of digital media, e-communications, and earth observation, who have each brought widening perspectives and interdisciplinary backgrounds. At present, this group includes Haidy Geismar, Hannah Knox and Antonia Walford. While 'the digital' in its broad sense can be read as a problematisation of material culture, in its claim to 'virtuality' and 'immateriality,' the fact of the matter is the digital is simply another genre of material culture and, while at times marked by ephemerality, it is also deeply reliant upon established infrastructures of the built environment and articulated within human practice as a material entity with which to engage. For example, drawing on a background in critical museum studies, intellectual property rights, and photography in anthropology, Geismar's exploration of the practices of digitalisation in museum and archive settings interrogates the continuity of the normativities built into digital architecture within the virtual spaces of digital collections (2018). Knox's work on roads and hard infrastructure has led to new research pathways in digital infrastructure and 'smart' technology developed out of an interest in the politics of material – such as roads and concrete – and implicitly carries forward Forde's interest in the economy of the material environment and the technological adaptation within a given ecology. Similarly, Walford's training in the intersection of STS and anthropology of science has led to ongoing research in technology and observation from a relational perspective, shifting from the visual object to rather consider the emergence of new knowledge economies of digital metrics and data, and their social and political efficacy.

Taken in this light, the material culture research group at UCL has had a consistent focus on the object, particularly within the framework of visual art and technology, broadly conceived. This tradition set down by Daryll Forde, has continued through the subsequent generations of the research group. The original engagement around art between Ucko and Forge has continued through students of Forge and Gell, and the original close relationship between archaeology and the research group exemplified in the work of Ucko has been a constant element as well.

In the recent years, as new members of the group have been added – most of whom are represented as contributors in this volume – the central interest in the social roles and implications of the object, the relational capacity of the artwork and the importance of technology and material environment, has continued as important influences. Several different research agendas and pedagogical initiatives have emerged over the last decade that demonstrate the enduring importance of these themes for MCS more broadly. Ludovic Coupaye has developed a research programme around the role of technology drawing on French theorists, often overlooked in anglophone academia, and draws upon his formal training in art history and archaeology to examine the practices and techniques enacted upon and demanded by the object in order to deconstruct the category of 'technology' and its sociopolitical role in both public and academic discourse. Adam Drazin's focus on Design Anthropology brings together classic interests of MCS, such as attention to the form of the object or engagement with the aesthetic qualities of materials, critically examining how academic scholarship is instrumentalised within design contexts, and examining how dialogue with public and private sector institutions around issues such as aging, mobility, and the home can bring new anthropological insight.

It is also worth highlighting that alongside the permanent members of the research group mentioned so far, there is a group of early-career research and teaching fellows. In many cases these have been brought into the group as part of European Research Council grants led by Miller, Pinney, or Buchli, or they have been hired to help support the teaching of MCS within the department. Those included in this volume (Carroll, Jeevendrampillai, Reese, Schacter, and Walton) were in the group as of summer 2018, when the collection began to be collated; alongside these, Delphine Mercier, who is the Collections Curator of the UCL Ethnography Collection, also supports teaching and the intellectual project of object-oriented study within the group. Within this new generation (four of whom completed their PhDs in the research group), the idea of 'un-disciplined' MCS was a core defining mark of the kind of anthropology to pursue. This has become something of a self-fulfilling prophecy, as the research trajectories being developed (most explicitly in this volume seen in Schacter's contribution) push in directions and develop methods of that undisciplined nature. It is an undisciplinedness, however, that is still marked by an interest in collections, museums, visual media, and technology.

Whether this continuity within the tradition established by Forde and Ucko is an intentional act of design or an accident of interest within the wider field is obviously debatable, but it is our contention that this consistency is at the heart of the research group's capacity to generate innovative research within the broader movement back to 'materiality,' seen across many disciplines within the humanities and the social sciences. This tension between the foundational approach to objects in the work of Forde and Ucko, and the new forms of material culture that confront us as scholars of MCS, gives the contributions to this volume a specific inflection. The chapters in this book demonstrate the importance of a relational approach to the object, and how placing objects at the centre of our analyses allows the reimagining of a range of fundamental aspects of social life, such as personhood, temporalities, scales, representation, and politics. The challenge of re-theorising the object and materiality in this way lies in staking out new conceptual territory that does not return us to the deterministic and reductionist perspectives of evolutionists, but draws on lineages of MCS that allow objects, artefacts and materials their full range of social and cultural possibilities and efficacies.

Over the last few decades, the 'material turn,' broadly conceived, has converged around a move away from representationalism and semiotics and to a rejuvenated interest in teasing out the affordances of materials and materialities.[2] This has been done in different conceptual languages and using different intellectual coordinates, from tracing out object biographies to attending to the physical attributes of images, to a focus on the political materiality of infrastructural systems. However, their convergence indicates a shared commitment to re-theorising the constitutive role of the object in social life in a way that does not reduce objects to inert vessels in human semiotic systems. The chapters in this book all take on this challenge in different ways and, in so doing, push the discipline of MCS in new and exciting directions. Unlike other approaches in MCS that have also taken on this challenge, such as 'new materialism' (see for example Barad 2007; Bennett 2010; Braidotti 2013), the authors assembled in this volume do not seek to redefine the material world in terms of vitalism, so much as work through rich ethnographic material in order to propose a framework for how the object is active within social practice in often pre-social ways. That is, the properties and affordances of objects means that the possible sociality is there, at least in some sense, before the socialisation of the material form. Thus, there is a drive to understand, through empirical observation of the matter of society, what the role of material is in culture.

Bringing the object back (again) – from lineages to advancements

This volume's analytical focus rests centrally on the object and what it is as an analytical point of access to, and as an ethnographic element within, wider social and cultural phenomena. This has always been a point of research and debate within MCS, so it should be no surprise that it comes to the fore in this volume, where we are explicitly thinking about the lineages of thought in MCS. Across the collection, the authors each deal with object-ness differently. There is no dominant agenda or definition, and – while some contributing authors present specific cases for articulating what an object, or kind of object, is (e.g. Walton on 'place-object' and Coupaye on 'technical object') or reimagining what it might be or do within society (e.g. Jeevendrampillai's argument on objects as labour, or Drazin's insight on the object biography), or suggesting what the epistemic benefit of an object might be (e.g. Mercier, Schacter, Knox, Pinney) – in this Introduction we are using it simply in a broad sense.

While we are cognisant of the analytical burden terms such as *object* (or *artefact*, *thing*, *stuff*, *substance*, *material*, etc.) carry, we draw upon the empirical benefit of the object in Peircean semiotics – as the source or vehicle of the sign – as a way to tie the semiotic entanglement of culture to the concrete certainty of an observable and material world, away from the arbitrariness of the Saussurian sign. While the shift from symbolic to semiotic anthropology did recognise the importance of the object as the anchor and producer of the sign, it is still a significant step to the kind of MCS that investigates the capacity or affordance of that object, and how this shapes the form and social ability of the sign. So, in emphasising the object in the Peircean sign, we also, conversely, are able to link the empirically observable lived-in world to the relationality of social form. In doing so, we come to trouble the subject–object dyad by highlighting how objects relate within interartefactual domains or within various milieu of relational and sequential actions. In this manner, this empirical object orientation is also a driving force within the analytical project of anthropological theorisation (e.g. Küchler) and application (see especially Miller and Haapio-Kirk, Chapter 11).

So, while MCS, as it was (re)born in the 1990s, was an 'undisciplined' project of enquiry, it can be seen to have developed, at least as an anthropological project within this institution, to be, rather, an anthropology of materials that is emphatically ethnographic in its method, and relational rather than semiotic in its epistemology. If, as Hallowell argued (1955), human society is characterised by an 'object orientation,' then the investigation of these objects and orientations is significant to the broader anthropological project, and it is in this sense that the emphasis on the object comes to fruition. In distinct ways across the chapters, this collection forefronts new perspectives on the material object as a place for knowledge production, and it reimagines what exactly the capacity of the object is in society.

Within this broader conversation, five dominant themes are important to highlight. These include issues of selfhood and post- /humanism; temporality; scales and topologies; representation; and politics.

Self, personhood, (post-)humanism

In *The Soul of the Primitive* (1966[1928]), Lucien Lévy-Bruhl presents the idea that the person is distributed through their belongings, and that this allows for an understanding of person–object relations that is not simply representational, but facilitates the intersubjective continuity of persons, even past death. The mentality of 'primitive' versus 'civilised' peoples, and how their cognitive and logical (as both evolutionary and cultural) apparatus worked was, for Lévy-Bruhl, intimately connected to the role that artefacts played in the social, and especially ritual, capacity of a culture group. In thinking through what the person is, and what it means to be human, the relationship between humans and objects, and specifically tools, is often articulated as central to the human project. Debates about the evolutionary development or the psychic unity of (hu)mankind rested, to a great degree, on the material culture used (or left behind) by various societies. And for its part, recent anthropological debates have considered the interconnectedness of people and materials at a global (and beyond) scale (Povinelli 2016; Moore 2016; Olson and Messeri 2015). While the scale (see below) of this can be disorientating, the simplicity of the object, as Drazin suggests, is such that investigating the object's biography allows new avenues into interrogating the specific narratives of contemporary global crisis, as well as wider politics and norms of identity and selfhood.

If the definition of humanity rests upon some articulation of technology, then it is fair to consider the possibility that a new frontier in technology – be it cybernetics, or AI – could frame a new stage of humanity, or indeed posthumanity. However, Coupaye's methodological emphasis on the autonomous sequences of action within the internal milieu of technical devices, acting apart from, but impinging upon, the actions of the human user, calls into heightened relief the assumptions about society and anthropocentric bias in anthropology. In Walford's chapter, we see how scientific subjectivities are themselves derived through the aesthetic effects of knowledge objects. The scientific, technological frontier is itself deeply inculcated in the ethical bias and political landscape of human society. As German media theory has shown, there is an aspect of everything 'new' wherein we have always already been doing it; the question is not what is new, but rather, 'How was the human always already historically mixed with the nonhuman?' (Siegert 2015: 6). In this vein, the contribution by Carroll and Parkhurst asks us to be critical of the (ethnographic and analytical) claims to 'posthumanity' and the novelty of cyborgian enhancements of/to the twenty-first century body, mindful of the long tradition in religion and philosophy of human aspiration to something more-than-human.

Time: futures, histories, presentness

Within the anthropological attention to material culture, there has always been the critical issue of temporality as it is inscribed in the object. As Pinney reminds us (this volume), the Durkheimian tradition approaches objects as a historical record of society. However, as Hallowell argues, the 'object orientation' of society means that objects 'are sources of power to human beings through the "blessings" they bestow' (2002 [1960]: 21). The possibility of futurity, not historicity, being the primary temporal framework of objects has important implications, as a social anthropology of materials should, then, be examining objects not as a record of society, but as key players within society, and in the processes by which futures are imagined and brought into being. Objects have been understood to play constitutive roles in mediating the temporality of social practices, as ecological approaches to anthropology demonstrate, wherein the objects and material resources are understood as part of, especially the timing and ritual of, social custom (Rappaport 1999). However, the way in which objects can actively produce futures – and in so doing, inscribe pasts – is developed in several chapters of the volume. The futurity of the object is most explicitly addressed in Pinney's consideration of both prophetic illustrations and photography, where it is the indeterminacy of the image that lends it so readily to becoming an index of the future, and where the photographic is a form of disturbance rather than illustration that anticipates what is to come rather than records what is past. Similarly, Reese's discussion of the animated luminescent *gopurams* in Indian political festivals highlights the futurity of care and obligation between the divine political persona and the publics drawn into the warm glow of their incandescent atmosphere.

In a different manner, Coupaye's elucidation of the technical object points to the processual ontogenesis of the device, and the lineages of social, material, and political influences that come to produce any given artefact. This historicity, however, is shown to have a critical decisiveness in shaping the future, as the automatisation of the technical life impinges on the person's quotidian movements, knowledge, and expectations. The relation between historicity and futurity is also productively problematic in Schacter's discussion of belatedness as it appears as a theme in the co-curated gallery exhibition *Motions of this Kind.* In this latter setting, the curatorial as method brings together various histories and presentnesses – often taut with contention and politics – and makes explicit the benefit of considering the project of data elicitation in fieldwork, and the museological display of objects, to be both ongoing and iterative projects directed at future potentialities.

It is clear here how the question of the critical moment of the contemporary period of post-modern (colonial, human, etc.) is also implicated in these discussions of the temporality of the object. The putative 'post'-colonial in Schacter's chapter, and 'post'-human in Carroll and Parkhurst's, are both challenged by a consideration of how objects hold together different temporal frames, be it the way that colonial pasts (and presents) linger on in ways that can be creatively reworked in contemporary artistic production and collaboration, or how objects from any given historical time period have always been enrolled in human extensions of themselves into the future. This dynamism of objectual temporality is mirrored in Drazin's attention to how objects have biographies, moving simultaneously through time as well as space, and how in this their futures are tangled up in their present. This vision of possible futures, of ideal futures, shapes not only the production of objects now, but also, as Drazin shows, shapes how objects are moved, exchanged, kept, or discarded within their individual lives. Such an emphasis on the futurity of objects is also a challenge to MCS to engage in new modes of objects, and both Knox and Coupaye make a case for approaching certain contemporary technical objects with an eye to revealing their unexplored social and political potential.

Scales, space(s), topologies

Re-theorising spatiality has been a crucial element of several branches of MCS. One particularly influential re-theorisation came out of the archaeological and phenomenological approach of Chris Tilley, who has argued consistently for the importance of a sense of place rather than space – or 'human space' over 'abstract space' (Tilley 1994). This has resonated with other direct challenges put to a specific form of spatialisation, variously called 'cartographic' (Turnbull 1989), 'euclidean' (Küchler 2001), 'mathematical' or, at times, simply 'western' (Ingold 2000). Space and place has been a contentious re-visited issue for anthropology more broadly, founded as it is on a method that privileges a certain sort of colonial spatial imagination (or 'Christian,' see Boyarin 1991), that of the exotic field site. But within MCS this problematisation has been specifically attuned to the role of the material world in generating the coordinates for social action, coordinates that are often shown to escape the confines of a cartographic world view. In his paper on means of navigation, Gell (1985) makes a distinction between token indexical and non-token indexical images to frame how a subject's perspective – a view of a landmark, for example – works in tandem with a general, abstracted view of the landscape. In the movement through space, the situated view of the landmark, the token indexical image, relates to other token indexical images within a sequence that allows the navigation in relation to the non-token indexical view of space. In this sense, a map, as a non-token indexical artefact, allows the person to anticipate their position within the world. Similarly, Küchler's work on the *malanggan* of New Ireland (2002) and the *tivaivai* of the Cook Islands (2007) both demonstrate the non-Euclidean ways in which the object, like a knot, is a topological artefact that allows the folding in of space, capturing it in the concretised artefactual form. Like the token indexical map, objects hold within themselves the information of complex systems, as an abstract geometry, allowing this conceptual space to be held and contemplated in manageable forms (Küchler and Carroll 2021b). Alternatively, Tim Ingold's work (2000) on the emergence of places as a result of human interactions with their environment emphasises the extent to which people always move through, and dwell, in space, and it is these relations with the material world that surround them, that generate places.

In a similar way, and often linked to the issue of space, the concept of scale has also undergone a thorough interrogation within the broader field of anthropological MCS. Influential here has been the work of Bruno Latour, who in developing Actor-Network Theory (ANT) argued for a 'flat' approach to studying social worlds in which there is no presumption of scalar differentials before the research has been conducted (2005). Different 'scales' – such as local and global, or micro and macro – emerge from practice and cannot exist independently of the specific networks of persons and things that constitute any social setting; that is, events or things can only become local or global. It is Latour's problematisation of the scalar concept of a 'society' or 'the social' that perhaps resonates most clearly with Marilyn Strathern's widely-adopted position on scale which she developed over the 1990s, in which she questions the analytical usefulness of concepts such as 'society' and 'individual' and the relations thought to inhere between them. Drawing on her work in Melanesia, and putting it into generative relation with mathematical ideas from the 1980s, Strathern proposed a scalar image different to that of parts encompassed by a whole, that of Mandelbrot's fractal, a non-Euclidean mathematical form that conserves complexity at every scale (Strathern 1991). In this she was in creative conversation with the ideas of other Melanesianists, such as Roy Wagner's notion of a 'holographic worldview' (2001). This Melanesianist perspective, also seen in the work of Gregory Bateson (1972, 1979), was influential in Gell's formulation of the role of the art-like object and 'index' within his theorisation of the social agency of things and in the role of objects as part of the extended mind (1998).

The chapters in this book build on this previous body of work, starting from the realisation that both cartographic or abstract space and Euclidean dimensionality are of questionable usefulness for MCS. But, rather than pushing back against the constraints of any one particular spatial or scalar imaginary, the chapters seek to explore the new geometries that present themselves when the enquiry is explicitly object-led. The object is, as Küchler highlights, something in sequence, both temporally and spatially, knowing its predecessor and anticipating its successor, within the abstract geometry and spatiality of how objects relate to each other. In her telling, certain objects – like the machinist's model – 'show off' complex social systems under perpetual construction that extend well beyond the object itself and allow us a vision of the scope of object relations beyond the human. This extensive capacity of objects to work within a number of different geometries is also picked up on in a very different register in Buchli's chapter (Chapter 2). Asking what sort of ethnographic object the International Space Station (ISS) is, Buchli points to the challenges of an extraterrestrial ethnography for traditional anthropological fieldwork. Not only does being in a Low Earth Orbit confound the physical and material expectations of terrestrial settings, in terms of gravity and, indeed, diurnal and nocturnal rhythms, but also the ISS itself is distributed between the terrestrial and the extraterrestrial, the human and the cosmic, requiring theorisation of the attunement and worldings between these distributed scales of the ISS.

The spatial presumptions of method itself are here brought into stark relief, and, indeed, attention to the problems that different scales of objects pose for traditional MCS is a theme that runs through various of the other chapters, including that of Knox's on the agency of digital devices; Coupaye's on the efficacy of technical objects; Walford's on the aesthetics of scientific data; and Mercier's on the wide reaching implications of the object in solitary observation or in a locally produced set. In Walton's chapter (Chapter 16), the potential for digital objects to have spatialising effects – for objects to also be 'places' – is explicitly drawn out in order to argue for a theoretical position and methodological approach that can grapple with the types of movement and reproduction afforded by digital photographs, and attend to the kinds of dispersed, but highly sensual and affective, social configurations made possible because of the image as place-object.

Representation: aesthetics, signs, semiotics

While scholarly work in the first decades of the twentieth century held in high regard the role of the physical and environmental factors within the human society (e.g. Forde 1963[1934]; Boas 1955[1927]; Hallowell 1955), the broad trend in anthropology was away from object orientation. In its place, anthropology focused on symbolic interpretation, seen for example in the Geertzian tradition, wherein material culture was increasingly read as representational, following the semiology of Ferdinand de Saussure, thus holding the object to be valued socially only for its capacity to hold attributed meaning. By taking culture as 'a system of inherited conceptions expressed in symbolic forms' (1973: 89), and religion as 'a system of symbols' (2004[1966]), Geertz frames the anthropological project as one principally concerned with the interpretation of symbols. With the shift from symbolic to semiotic anthropology, the increased preference for the semiotics of Charles Sanders Peirce, over Saussurean semiology, afforded a greater importance to objects, as Peirce's 'triadic relation of sign, object, and interpretant includes an ontology of objects as well as an epistemology of subjects who conceive or know the objects through mediation of the sign' (Singer 1980: 491). As such, semiotic anthropology granted a greater focus to the context of representation and the qualities of the objects (Mertz 2007).

In other object-oriented disciplines, such as art history and archaeology, the latter half of the century saw an increased focus on the aesthetics and cultural value of the artefact as a means to shift attention away from the problematic aspects of the politics of representation (Fowles 2016; Rubin 1984; Küchler and Carroll 2021b). The return in the 1990s towards objects in anthropology reflected what Christopher Pinney and Nicolas Thomas (2001) signalled as a need to look 'beyond aesthetics' in our interests in the visual (and indeed language), to wider material forms, registers and contexts wherein objects circulate. In a similar move, scholars like Birgit Meyer (2009, 2012) and Webb Keane (2003, 2013) have sought to articulate the role of objects, specifically, and material more broadly, within the cultural and semiotic worlds of human society. Meyer's interest in the mediating capacity of objects, whereby they are instilled with an authority that allows them to hold key positions in social, and especially religious, settings has made important contributions to the ways objects work within the social, and deeply intersubjective, relationality of human society. However, in her emphasis on mediation, the object is quickly lost, and appears manifestly present only when it does not work correctly (Meyer 2011). In this sense, the approach of media studies is all too often beguiled by the 'humility of things,' such that the social ability of the object to hide in plain sight is successful, not only in the ethnographic context, but also in the analytical arena. By contrast, as seen in Drazin's chapter (Chapter 5), there is great analytical insight to be gained via an almost absurd attention to the object, not only as it mediates human relations, but as it exists as an entity outside of those relations as well.

Similarly, the Peircean influence in Keane's work has had immense benefit in terms of critically engaging how signs work within society, but the intellectual burden of his work rests on a linguistic assumption that constrains the possibilities of what the object might do. If, as Küchler suggests, the object – and specifically its sequenciality – is able to hold in its relations knowledge that cannot be articulated in human language (and indeed may only be partially grasped in human cognition), then the place of objects as signs far exceeds what may be achieved within a linguistic analytical model.

Turning back to objects in this way is not, then, simply a question of eschewing representationalism, but of allowing the object a fuller range of analytical affordances which, in turn, pushes us to once again reconsider not only the relation between objects and representations, but also the relationality of objects per se. This is clear in Knox's chapter (Chapter 8), in which she asks us to consider how knowledge is itself an object, and how the 'epistemology' of digital devices as 'empirical technologies' permits new ways of knowing or, in Walford's chapter (Chapter 15), where 'objectivity' is recast as a form of relational capacity inhering in knowledge objects like scientific data. Across several of the chapters, objects emerge as relational and dynamic generators of knowledge, rather than static vessels of meaning, as we see clearly in Schacter's analysis of the capacity for art objects to challenge and question concepts and ideas and thus produce new paradigms of thought.

From this perspective, representation and linguistic analyses in MCS need not be relegated to the concerns of the past or rendered outmoded for the sake of new research sites, questions, and approaches. What images show, tell, and represent, remains fundamental to understanding what certain kinds of visual objects are and how they have e/affect in a range of material forms, registers, and contexts; this can be seen in the 'prophetic' futurity of almanacs in nineteenth-century England (Pinney), in the camera phone digital photography and the visual (geo-)politics of place (Walton), and the *gopurum* light towers in South India, which manifest the form of deities or individuals that are often rendered from print photographs or illustrations (Reese). Despite the many moves beyond representation, aesthetics, language, and semiotics, the present volume highlights a need to maintain these analytical registers in exploring material cultural phenomena, while also exploring theoretical frontiers that bring us to areas such as temporality, affect, place, and digitality.

Participation, politics, people

Moving beyond representation, however, is more than simply a theoretical motivation animating this collection. In thinking about what objects might represent, Jeevendrampillai, Burton, and Sanglante demonstrate that the representational capacity of the object is not just to stand in for people, but is also formative of labour and labour practices. In their work, and the wider *Objects of Desire* collective, the possibility of the object carries with it a certain political potential. Objects here are not merely representative of sex work, but mediators of relations that constitute the labour of sex work, and within this framework of possibility, the subject may find the political efficacy to assume or resist a particular subject position. Similarly, in Reese's chapter (Chapter 13), it is the infrastructures and politics in which the object – in her case a *gopurum* electric light tower – is embedded along with the materiality of the object that a/effect both the human eye and specific bodily responses, such as squinting, as a politics of the ways of seeing.

Questioning the subject/object binaries on which representational strategies rest also implies questioning the social role and status of MCS, asking how it comes to matter outside of elite academic circles. Several contributors to the volume examine the contemporary role(s) of the anthropologist. While MCS has been traditionally directed at what objects are and do in the world, several chapters in the volume actively explore and advance the role of the anthropologist as participant in the politics of presence, participation, and visibility. Geismar explores how anthropologists can be social commentators, investigating how the characteristic inside-outside stance of the ethnographer can be employed to advance the social and political participation of academics as cultural critics. Geismar advocates for an engaged social responsibility, but one that challenges a certain contemporary instrumentality of academic research into neoliberal regimes of knowledge production. Similarly, Miller and Haapio-Kirk take up the issue of anthropologists participating in the world as part of a broader position piece about making anthropological research, and MCS specifically – as an object of knowledge: matter in the world. Public education through anthropology, or a 'public anthropology,' is hereby advocated as something that 'matters.' As with Geismar (Chapter 6), Miller and Haapio-Kirk (Chapter 11) push back against the neoliberalisation of academic knowledge, and in effect both chapters highlight the political significance of anthropological knowledge itself as a kind of 'epistemic object' or 'knowledge-producing entities' (see Schacter and Knox, respectively, Chapters 14 and 8).

The role of the anthropologist also comes to the fore in Schacter's chapter (Chapter 14), which delves into the curatorial pursuit in the context of exhibiting art. Here, co-curation, collaboration, and art objects are implicated in the wider postcolonial politics of participation. In this way, the object itself becomes generative of new theoretical positions and has methodological implications for further research. In Schacter, the emphasis is on co-curation, while in Mercier's piece, the researcher is inculcated within an intimate, exploratory project of analysis, via 'be with' the object, in order for its qualities to be brought forth to the researcher.

In the multiple voices of this conversation, and the themes that cut across the chapters, it is important that even while framed around the object, these themes start with questions of humanity and end with people. Ultimately, even in the un-disciplinedness of MCS, it is – or at least has come to be in this research group – a deeply anthropological project that takes the anthropological imperative of the object as a source of analytical and methodological insight. To adapt Robert Farris Thompson's insight about aesthetic objects more broadly, considering the object turns 'ancient objects of thought into fresh sources of guidance and illumination' (1973: 67).

Notes

1 We are indebted to the authors of Geismar et al. (2014) for providing an English working draft of their German-language publication. While the interpretation and errors are our own, the broad narrative of the research group relies on this paper, drafted in 2013, as a key source.

2 For an overview of the debate around materiality, see Miller (2005b), Ingold (2007), and his respondents, or Carroll and Parkhurst (2019) in considering how these affordances may engage with wider analytical domains.

2 Extraterrestrial methods

Towards an ethnography of the ISS

Victor Buchli

The anthropological study of extraterrestrial settings might seem novel and exotic. However, this chapter proposes that the methodological tool kit available to social scientists for the empirical study of such contexts has a well-established genealogy, informed and developed by recent innovations in the ethnographic study of social media and their attendant communities. These studies in social media have challenged what is traditionally known within anthropology as the 'field' (Miller and Manadiou 2012; Beualieu 2010; Zhao 2003), and the methodological challenges of an extraterrestrial ethnography allow us to examine what happens to some of our fundamental categories of analysis in the social sciences. To date, our understanding of fundamental concepts that underpin the social sciences – such as transcendence, kinship, materiality, architecture, sovereignty, and the body – have been mostly conceived in terrestrially bound terms of Earth's gravity (see Gorman 2009c; Gorman and Walsh [forthcoming]), and even the concept of 'fieldwork' as it has been enshrined in the discipline is a distinctive artefact of terrestrial geometries of time and space. However, the International Space Station (ISS) has, for twenty years at the time of writing, been a home for humans that orbits the planet every ninety minutes in Low Earth Orbit (LEO) (Figure 2.1). This unique extraterrestrial society has received little attention within mainstream anthropology and material culture studies with notable exceptions such as Gorman's work on the archaeology of gravity (Gorman 2009c) and Gorman's recent and innovative archaeological approaches for the study of the ISS proposed with Justin Walsh (Walsh and Gorman 2020 and Gorman and Walsh forthcoming).[1] The ISS is a vibrant nexus of constant processes of calibration, coordination, and attunement that brings realms of experience into novel relations of entanglement (Hodder 2012 and Reno 2018) at unexpected scales and degrees of intimacy that have yet to be fully understood. This chapter outlines what the ISS is as a nexus and field of ethnographic study, and it proposes a methodological approach that draws upon the model of 'armchair anthropology,' reimagining it in a multi-sited, distributed, and space-age research context.

The ISS

The International Space Station (ISS) is the longest-lasting extant extraterrestrial society in Low Earth Orbit. As it is a place of dwelling, not just scientific discovery, any anthropological study of the ISS must focus on the quotidian and material dimensions of the ISS and its bodily and material techniques, re-examining traditional empirical assumptions within the innovative conditions of the new polymedia environments (as informed by the work of Madianou and Miller 2012) in which the ISS is situated. These polymedia environments include the respective Mission Controls (in Moscow, Munich, Houston, and Tsukuba) and their wider communities. These wider communities may comprise, for example, Orthodox Christians tracking the movement of relics to and from the ISS, or various groups on social media who follow and engage with the ISS mission and crew. These communities, I argue, are coterminous with the ISS site, simultaneously constituting and occupying the same 'field' of co-presence.

Co-presence here is used in distinction to co-location, which has characterised traditional ethnographic research, following Beaulieu (2010) and Zhao (2003). As Beaulieu and Zhao have both noted the rise of virtual reality, the Internet, telephony, and social media have suggested a new importance for the understanding of co-presence as the condition of contemporary ethnographic work and social life. As Beaulieu highlights: 'The effort needed to sustain co-presence should not be underestimated. Co-presence is a very active form of "field making." The "field" is constituted in the interaction' (Beaulieu 2010: 463). As such, the media of co-presence are productive of the conditions of social life and the conditions of ethnographic knowledge and the structures of its production. The proposition here is that identifying the wider polymedia field of co-presence and participating in them is integral to the very conditions by which 'worlding' (Heidegger 1993) takes place and produces the nexus of terrestrial and extraterrestrial sites that comprise the 'field' of the International Space Station – not just simply the structure itself in Low Earth Orbit but the various communities and mission controls on Earth that constitute this ever-expanding nexus that is at once the 'field' one 'enters' and are the media of its constitution – this study among them (following Beaulieu 2010).

This wider field can be examined as a complex nexus of inhabitation, encompassing both terrestrial and extraterrestrial realms in a novel configuration that is dynamic and expanding. Such a method must involve research on each of the governments and space programmes that constitute the ISS: Russia, United States, Europe, and Japan. This expanded and mediated 'field,' constituted by the quotidian aspects of habitation, the wider communities, and the governmental institutions in both the terrestrial and extraterrestrial spheres, is one united field of ethnographic enquiry, held together by various registers of co-presence rather than the traditional registers of co-location (see Beaulieu 2010; Gorman 2009c). This field is unified under the general theme of 'worlding' (Heidegger 1993) through the one common object: the ISS in its expanded and combined terrestrial and extraterrestrial nexus.

The modular architecture of the ISS represents a unique and highly complex habitat that has transcended the bounds of Earth's gravity, following a long-standing desire to transcend the Earth, a common impulse prominent in many known cosmologies historically and archaeologically (Milbrath 2009). In the early part of the twentieth century the Modernist avant-garde's predilection to dematerialise architectural form and transcend Earth's gravity was most eloquently expressed in utopian schemes such as Krutikov's floating cities of the 1920s (Khan-Magomedov 2015) and Malevich's suprematist arkhitektons from the same period, which were envisioned to orbit Earth (Kovtun and Douglas 1981; Malevich 1920). An extraterrestrial anthropology expands studies of such utopian schemes towards understanding the workings of the ISS. It is the Soviet space programme that, in addition to putting the first man and woman in space, also produced the earliest extraterrestrial habitats in Earth's orbit on the Salyut and Mir Space Stations. The current ISS is based on this earlier modular architecture first established by the Soviets, and the oldest modules of the ISS were produced by the post-Soviet Russian Space Agency (Nixon 2017; see also Chladek 2017).

Quotidian attunement

A central element in the question of an extraterrestrial methodology is the means by which the various forms of terrestrial and extraterrestrial attunement converge to produce this novel realm of human habitation and its expanded and expanding 'field' of co-presence. In this respect, such a method is informed by the spirit of recent work by Stewart (2011) on 'attunement' and 'atmosphere' (see also Pérez-Gómez 2016, Reno 2018, Ingold 2015) where

> incommensurate elements hang together in a scene that bodies labor to be in or to get through [...] bodies labor to literally fall into step with the pacing, the habits, the lines of attachment, the responsibilities shouldered, the sentience of a worlding.
>
> (Stewart 2011: 452)

The inhabitants of the ISS and its participants within the wider nexus of ISS terrestrially and extraterrestrially are 'attuned' to each other in surprising, unprecedented, and mutually constitutive ways. This goes back to the very beginning of space exploration, when Sputnik's 'beep' in 1957 was deliberately calibrated so that amateur radio enthusiasts on Earth could literally tune in (Miller 1991: 17). This practise continues with various ISS missions and extends into the present day via the wider complex polymedia environments of the ISS that Jakubowski (2016) observes as part of the 'expandable space aesthetic' he describes, and which brings the terrestrially quotidian into a new relationship with the extraterrestrial – a process that has been observed historically by Maher (2017) in his discussion of the history of the American space programme. These processes of 'attunement' and innovative 'worlding' can be seen in a wide range of examples, including: 9/11 commemorations (Catchpole 2008); the astronauts Tim Peake and Sunita Williams running marathons on the ISS in sync with marathons on Earth; national holidays such as the Russian 'Immortal Regiment' celebrations held on Earth and in the ISS (see Buchli [forthcoming]); as well as religious holidays, family rituals, and the literal attunement produced through periodic radio contact and various forms of social media within the wider 'field' of co-presence.

Following Turner (2013) regarding the political aesthetics of the 'democratic surround' and Messeri's work (2011; 2016; 2017) on extraterrestrial placemaking, we see that the material conditions of the ISS and the extension of human habitation extraterrestrially expand and reconfigure the conditions of the terrestrial in new and unexpected ways (see also Maher 2017). The quotidian data of the ISS provides a rich trove of information that evidences the ways in which 'worlding' emerges simultaneously intimately and cosmologically (see also Allen and Holbraad 2014). The ISS is a vibrant nexus of constant processes of calibration, coordination, and attunement that brings realms of experience into novel relations of entanglement (Hodder 2012) at unexpected scales and degrees of intimacy that have yet to be fully understood.

Humans have always already been going to space

Despite its novelty, an extraterrestrial methodology participates in a long-standing tradition within anthropology and archaeology that has dealt with the transcendent in terrestrially based cosmologies and the material cultures that constitute them. In fact, the sort of distinction the term extraterrestrial implies, and the binaries that it manifests, can be quite foreign to many non-EuroAmerican contexts where the transcendent and celestial are engaged in various forms of co-presence in everyday life (see also Gorman 2009c).

For instance, Ye'cuana traditional dwellings of Northern Brazil and Venezuela often serve as analogues for the celestial sphere (Rivière 1995). The conventionally empirical and material present realms of the day to day are seen to be the faint reflections of enduring ancestral and divine relations in a transcendent dimension. Similarly, Batammaliba dwellings (Blier 1987) track the passage of celestial ancestors through various light apertures in the structure linking the living and the dead through the practices of daily life – the terrestrial and the extraterrestrial converge into one common 'field' of inhabitation. As Milbrath notes, humans frequently structure ritual life and daily activities in relation to celestial cycles and, more importantly, both celestial and terrestrial cycles are often linked and require one another to continue (Millbrath 2009: 158). As Reno (2018) also notes, the conventional Euro-American clock-time of capitalist expansion and regulation is facilitated by attunement in relation to given celestial coordinates and their resulting daylight hours, which are challenged when considering the context of Low Earth Orbit, where sunrise and sunset are experienced 15 or 16 times a day. The dwelling is the nexus whereby these terrestrial, transcendent and extraterrestrial cycles are regulated and experienced in embodied form in our everyday material culture and architecture (Carsten and Hugh-Jones 1995; Douglas 1991).

In another vein, Messeri highlights how Inuit shamans have been going to the moon long before NASA, and Native American traditions have often reckoned kinship in celestial terms (Messeri 2011: 15). As Messeri (2011) notes, a multiplicity emerges between the terrestrial and extraterrestrial, where relationships are uncovered between Earth and extraterrestrial worlds with a particular emphasis on the capacity of language and visuality to render these realms into perceivable habitats. Projects such as the mapping of Mars democratise the planet in relation to American neoliberal ideals, making it that much more 'habitable' through democratic mapping processes such as Google Mars (ibid.). These technologies transpose and merge the technologies for mapping and inhabiting Earth in order to do the same on Mars, and in doing so, they create new intimacies with other worlds. As Messeri states, 'Simulation does not imitate; it generates' (Messeri 2011: 250; see also Maher 2017). The mirroring of these extraterrestrial and transcendent realms in whatever register, serves to generate these wider sets of relationships rather than represent them. It does this as part of a wider process of 'worlding' (Heidegger 1993) and the continuous embodied microprocesses of 'attunement' (Stewart 2011; see also Ingold 2015) that expands these realms and brings them into being.

Dialogic worlding

'Worlding,' as used here, relates to Heidegger's neologism describing the dynamic, mutually constitutive, and continuously unfolding processes whereby that which is intimate (on a bodily level) as well as that which is far-reaching (in terms of relationships with the cosmos) are dynamically and mutually configured in relation to one another in terms of empirically describable constantly expanding worlds (Heidegger 1993 and following Battaglia et al. 2012). This concept informs an extraterrestrial methodology, helping to understand the ways in which the fundamental characteristics of human life, sociality, and material culture are reconfigured through the expansion of our habitat extraterrestrially (see also Gorman 2009c; Gorman and Walsh forthcoming). Scholarship on space exploration has focused on organisational studies and the anthropology of extreme habitats and environments (Olson 2010, 2018; Stuster 2011; Kintz and Palinkas 2016). An extraterrestrial methodology builds on this work and on work such as Zabusky on the European Space Agency (ESA) (Zabusky 1995), to examine how the body and the cosmos are attuned to one another within space habitats. Olson's account of NASA describes how the body is radically manipulated to reinterpret what its very potential is in relation to space environments. By tracing space analogues, developed, researched, and inhabited at NASA, she describes how one universal totalising paradigm is produced. Earth and Space are not dichotomised, as her informant claims: 'There's only one paradigm, it's all the same' (Olson 2010: 225). Such a methodology proposed here helps ask what forms of human 'being' (Heidegger 1993) emerge from such constellations of body, technology, architecture and data?

However, that 'same' paradigm identified by Olson, is not immediately realised or apparent; Battaglia et al. (2012), in turn, focus on the 'extreme' characterisation of the extraterrestrial and how its 'extraness' produces the effect of the sublime in relation to the ordinary and extraordinary. This productive capacity questions the most basic assumptions of what constitutes life, habitability, time, and the Earth itself. More importantly, the trope of 'extraness' is seen to offer a new way of examining and reconstituting the human in terms that are not exhausted by terrestrial failures (see Carroll et al. 2017) but rather by the extreme conditions of the extraterrestrial itself. The 'extreme' becomes a form of dynamic worlding and, more importantly, a realm that is novel, tentative, emergent, unstable, and unknown. It becomes a site of radically redemptive alterity in the 'subjunctive' mode (Battaglia et al. 2012: 1011) in which a future for humanity could unfold. And, as argued here, this 'extraness' is also understood in terms of the novel micropowers at work within novel material nexus that bring forth this emergent worlding. As Valentine et al. note (2012), activities such as the commercialisation of Space, the expansion of neoliberalism off-world in the post-Soviet era and the fascination with the extraterrestrial and 'extreme' situations represent a radical new field for anthropological study with methodological challenges.

The constantly invoked themes of new forms of kinship, neo-unilineal evolution and common humanity echo earlier nineteenth-century notions such as the 'Psychic Unity of Mankind,' as can be seen in Farman's (2012) examination of 'singularitarians.' Singularitarians recall earlier nineteenth-century Russian Cosmist philosophy, notably Fedorov, which is at the heart of Soviet and Russian space intellectual histories (Siddiqi 2010; Young 2012). As these studies suggest, the study of the extraterrestrial reconfigures our understanding of terrestrial realms in a profound way. In this vein, the ethnographies of Olson and Messeri both look at the various scales in which extraterrestrial realms are inhabited: how extraterrestrial worlds are made Earth-like, and how Earth is made extraterrestrial-like (see also Maher 2017), through an examination of the processes of 'worlding' that brings these incompatible realms together to extend the notion of human habitability. Zabusky's (1995) ethnography of satellite launches, notes, anticipating Messeri, that the activities surrounding such launches are generative, not simply of the ESA satellite programmes themselves, but also of the continuing political, cultural, and social project that is the European Union itself: a very terrestrial consequence of these extraterrestrial activities. To date, ethnographies of extraterrestrial activities and Space offer critical insights into our terrestrial institutions and societies, describing as much the institutions and science these activities take place in as the conditions under which these institutions and their socialites are reproduced, extended, and augmented. An extraterrestrial methodology proposes to build on and extend this field of study in a cross-cultural, multi-sited and empirically focused ethnography of the ISS and its nexus, both terrestrially and extraterrestrially.

Alice Gorman's earlier archaeological work on the material culture of extraterrestrial settings (2007, 2009a, b) focuses on the heritage of space sites (such as the first lunar landing), historically significant satellites, equipment in orbit and innovative proposals for how the cultural heritage and archaeology of space can be preserved. The generative capacity of these extraterrestrial activities reconfigures and reproduces terrestrial arrangements and relations in novel and unexpected ways. The very terrestrially based concept of heritage at the heart of nation building extends beyond the Earth here to critically question what, in fact, such a common heritage might be, both on Earth and in Space. It is in this vein that an extraterrestrial methodology might examine the wider nexus of ISS activities in LEO and on Earth materially examining how they are mutually constituted within their expanding terms of interaction and how national territory might be understood when there is no terrain.

Space from the armchair

Despite the apparent novelty of an extraterrestrial site, this chapter proposes that in order to develop such an extraterrestrial methodology, it is necessary to excavate and update a tradition in anthropological thought and method that emerges from its nineteenth-century origins: the so called armchair anthropology that formed the basis of our comparative discipline before the advent of a 'modernist' social anthropology characterised by co-locational field work (see Strathern 1987a for a wider intellectual history and discussion of this tension between pre-modernist, modernist, and post-modern methodological traditions within anthropology).

Traditional empirical notions of physical co-location (see Beaulieu 2010 and in particular Zhao 2003 for a relevant taxonomy) are not possible in the Low Earth Orbit of the ISS – however, it is argued here (following Beaulieu and Zhao) that we are most emphatically co-present in terms of the polymedia networks sustaining it, from its various mission controls to social media robots and live feeds linking the ISS to terrestrial communities within expanding nexus that allow any smart phone to be a virtual mission control. This is 'armchair anthropology' of a particular magnitude that is co-present ethnographically within the innovative registers of co-presence at the ISS but not within the conventional material registers of co-location that inform terrestrial ethnography as it is traditionally conceived in its 'modernist' vein (Strathern 1987a). Further, Willerslev (2011) offers an analogous and instructive distinction whereby the 'armchair' is the site of knowledge both intellectually and empirically, where 'actual' empirical observations, produced via conventional co-location, are constitutive of the 'virtual' – in its more archaic sense of 'inherent virtues or power' (Willerslev 2011: 506) rather than the contemporary sense of 'virtual reality.' As such, the 'armchair' is constitutive of anthropological knowledge more fully: 'the virtual in an important sense is more real than its actual manifestations' (Willerslev 2011: 506). Yet, I would like to suggest that such polymedia nexus are an even more intensive and extensive form of sociality – more 'virtual' in both the archaic and contemporary sense and, hence, more real, as Willerslev might suggest. The ISS and Low Earth Orbit are a new part of Earth even though as such it is arguably not part of the planet and the traditional home of *Homo sapiens*. Yet the method by which such an investigation takes place challenges not only what our notions of territorial space and habitat are, and with that notions of presence and community, but also what is the status of nature and the human when occupying the decidedly hostile environment of Low Earth Orbit and the innovative 'virtual' conditions of human being, following Willerslev.

In fact, the ISS's nexus – both terrestrially and extraterrestrially – troubles the traditional terrestrially based methods that have characterised the discipline (see also Gorman 2009c; Gorman and Walsh forthcoming). The methodological investigation itself participates in and reconfigures deeper historical shifts related to the pre- versus post-Copernican views of the world. This is understood here as the tension between Earth-centric (pre-Copernican) views of the cosmos and post-Copernican views that displace the Earth as centre (Oliver 2015). In fact, much of the processes of attunement discussed here participate in this wider accretive and productive context of the vacillating characteristics of the relationship between the Earth and the cosmos. At the heart of this issue is the vacillating placement and displacement of the human as the centre of a cosmology, and the new models of society with the human alongside shifting material registers of attenuated degrees of materiality and immateriality that shape our continuously expanding relations into novel configurations and concerns.

An extraterrestrial site such as the ISS inverts the established terrestrially based geometry of ethnographic work. The object itself is not directly accessible, yet it exists within a highly accessible nexus of several terrestrial sites and in polymedia. Armchair anthropology of the nineteenth century relied on reports and surveys (Stocking 1992; Urry 1972) to collect data from the imperial peripheries along traditionally seafaring routes and bring that data into the imperial centres for study. The spatial geometry of this method reinforced a strict binary between anthropologists and their subjects (Strathern 1987a) with a very particular methodological configuration of distance and surveillance constituting the primacy of imperial centres and knowledge. This geometry is bound by the terrestrial conditions of time and space, and the historically contingent dynamics of power, technological advancement and mobility that animate this geometry.

Josh Reno (2018), observes that the time/space of extraterrestrial objects is radically distinct from conventional Euro-American clock time, based as it is on Earth's rotation and orbit around the sun. The ISS experiences 15 or

16 sunrises and sunsets in a day producing a radically distinctive time/space from that on Earth despite the coordination of time on the ISS with GMT. The GMT standard was established by the British Empire to regulate imperial time/space, and its use extraterrestrially represents an echo and refiguring of that imperial order as a compromise between two other competing orders on the ISS, the Russian (Soviet) and American. The two realms – the extraterrestrial ISS and the terrestrial realms of its mission controls – are kept in tune with each through the extension of this vestige of British imperial seafaring chronometry.

On another scale of investigation, the terrestrial viewer (re)produces a novel encounter through their micro-bodily techniques of observing, creating a new – personal and democratically individuated – embodied intimacy with the extraterrestrial after the original Copernican event. As Paddy Edgley (personal communication) notes in this regard in his current doctoral research, amateur astronomical observers learn to delicately squint their eyes to produce the fixed image of an astronomical object through the telescope's lens – the strained eye of the body and the astronomical object are literally held and attuned in relation to one another – through this subtle gesture. It is within these microprocesses that a particular methodological approach is suggested that allows us to understand what new embodied relationships are formed by the terrestrial observer and the extraterrestrial object – via the material affordances of the technology of observation (Ihde 1998 and see Reno 2018) – that produces this novel moment of worlding that is attuned through various micro-gestures and materialities.

New material cultures

Methodologically, this forces us to consider material culture in a new light, namely in terms of the effects of micro-gravity on our conceptualisations of material culture premised by the conditions of Earth's gravity as noted by Gorman (2009c). Consider how a focus on techniques, following Mauss (2006), might bring attention to certain kinds of artefacts of attunement such as 3-D printed clips or containers that serve as 'surrogates,' as Gorman has suggested, for Earth's gravity under the conditions of microgravity in Low Earth Orbit. Consider, for instance, Gorman's observation regarding straps, Velcro, and other means of anchoring objects and people, as 'surrogates' for Earth's gravity, literally strapping things down in simulation of gravitational effects (Gorman 2017). Such novel forms of material culture serve as true 'semiophores' (following Pomian 1990) and zones of contact between two incompatible realms brought into a novel relation. This might be posited to be analogous to Willerslev's distinction between 'actuality' and 'virtuality' where the 'actual' is the empirical realm and the 'virtual' is the unseen set of conditions constituting the 'actual' (Willerslev 2011: 506). These artefacts are not so much a point of contact (when considering museums and artefacts as 'contact zones') but a point of co-constitution – properly 'intra-active,' to use Barad's expression (2003, 2007). They are constitutive of a novel and dynamic process of 'worlding,' what I would like to consider as 'artefacts of attunement' following Ingold (2015) and, not unlike the fetish (Pietz 1985), the product of a radical incommensurability.

In this vein, the question of materiality and transcendence takes centre stage when considering the question of immateriality and the latest technologies of the immaterial, namely 3-D printing (Buchli 2010, 2016). The first extraterrestrially printed artefacts were printed on the ISS in 2014. Here the coherent status of the material object and the conventional effects of terrestrial gravity give way to the immaterial stability of digital code shaping 3-D printed objects in almost any 'gravity-defying' configuration. This disruptive new technology has yet to be understood in terms of the emerging moral and material orders that are unfolding at the ISS.

For instance, take into consideration the manufacture of fibre-optic cables for telecommunications and their more perfect shape microscopically when manufactured in microgravity. When exported back to Earth, such fibre optic cables provide near perfect means of information transmission, radically changing terrestrial modes of communication (Made in Space 2020). It is these subtle changes in material culture at the microscopic level, which are at the heart of the process of worlding and attunement as the terrestrial and extraterrestrial are brought into a tighter nexus representing just such an artefact of attunement. The extraterrestrial artefact facilitates novel processes of convergent 'worlding' through the various 'exo-surprises' (Battaglia 2012b) that constitute this novel emergent nexus.

Spatiotemporal distantiation

There is an archaeological analogy (following Reno 2018) that is apt here in terms of methodological origins that can be imaginatively extended to Pitt-Rivers' famous image regarding the study of material culture and the origins and evolution of human societies through his chart of 1868 (Figure 2.2). The artefactual 'semiophores' of Pitt-Rivers' chart radiate inwardly towards an originary mythical 'primitive' source (that is 'virtual') that reveals human origins, whereas the astronomer/cosmologist, as Reno suggests, extends this chart in the opposite direction outwards reminiscent of a decidedly pre-Copernican terra-centric arrangement (Figure 2.3) that, nonetheless, places the Earth at the perspectival centre of extended radiating spheres. As Reno notes, looking from Earth into Space, one sees the images of ancient cosmological forms; the further one looks upwards and deeper into Space, the deeper back into time one sees (consider also Allen and Holbraad 2014).

What this extended analogy here suggests is that the methods for engaging with such physically removed time/spaces is not a recent methodological dilemma as the challenge of studying ethnographically the ISS might suggest, and certainly not exotic in terms of the conditions of mediated social life and the conditions of co-presence in the present day. But the methodological

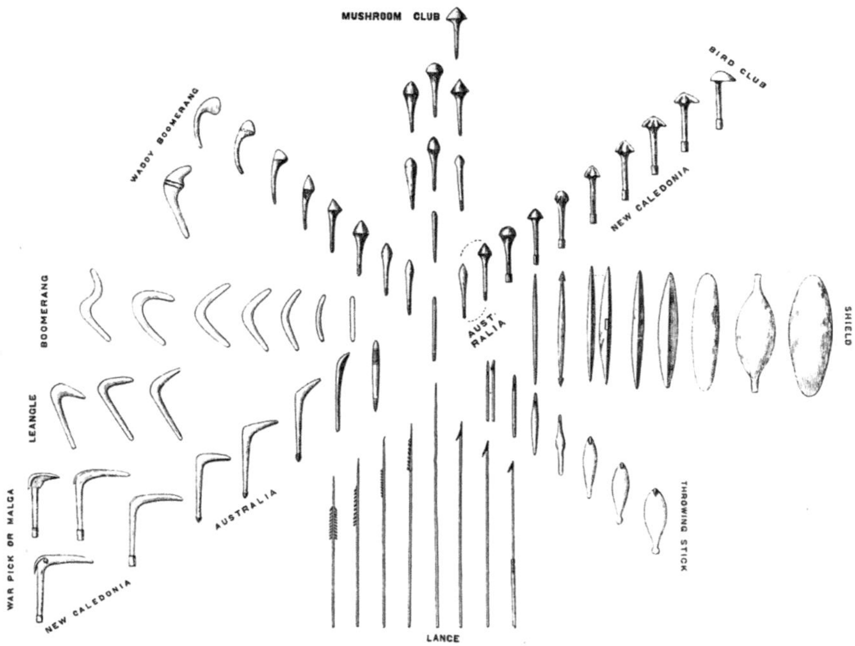

Figure 2.1 Primitive Warfare II, Pitt-Rivers 1868.

challenges in studying ethnographically the ISS is as old as the discipline of anthropology itself. Just as in the beginnings of the discipline where it was impossible to travel in time (and still is), the artefacts of the archaeological record were and still are arguably the 'semiophores' of such realms (on Tylor's 'object lessons' see Buchli 2002a). Similarly, as the far-flung imperial peripheries of the EuroAmerican centres where anthropology emerged were similarly distant – apprehensible through limited reports by missionaries, explorers, sailors, and adventurers and the artefacts they brought back. These reports and objects served as the 'encounters' within the imperial centres from which a nineteenth-century comparative science developed. Not until the systematic missions of the nineteenth century, such as the Torres Straits expeditions (Herle and Rouse 1998; Pickles 2009; Stocking 1992) and the emergence of a fully fledged ethnographic method performed *in situ* as epitomised in the method and research of Malinowski, Boas, and others, was a co-locational field established as the norm. Such ethnographic encounters and methods were a distinct product of a terrestrial geometry, and its attendant political economies. Here, at the time, was a methodological innovation that produced a radically different time/space that was coextensive and coeval with its object of study. Although such co-location, as decades of critical research have since

Figure 2.2 Cosmographia, Peter Apian 1539.

demonstrated, is epistemologically and ethically fraught, the reader was theoretically made virtually present (co-present) through the rhetorical tropes employed by the ethnographer. The reader is co-present in the register of a textual ekphrasis, and, even though not co-local, is able to experience a given society (The 'full flavor of native life' Malinowski 1988: 48).

The illusion of an adequate method here is long-rehearsed (see Clifford and Marcus 1986). However, the deeper history of such encounters and the methods they engendered as regards evidently and radically distinctive material time/spaces suggest the conditions under which novel forms of relationality and materiality begin to emerge. An extraterrestrial anthropology speaks to the seemingly incompatible realms of material time/space and evidences the distinctive ways such realms 'world' distinctive social and material life into

Figure 2.3 Artist's rendering of ISS above earth, passing over the Straits of Gibraltar. NASA ID 9802668.

being; to use Heideggerian language – they serve to 'gather' seemingly incommensurate realms within an expanded field of 'worlding.'

As material culture studies itself emerged from a reconsideration of these artefactual 'semiophores' the 'object lessons,' described by Tylor, which formed the core of the revaluation of the study of material culture in the wake of British Social Anthropology and the rise of ethnoarchaeology within archaeology in the post-war period (Buchli 2002a) – so, too, here, so-called armchair anthropology with its distinctive and divergent time/spaces mediated by artefactual 'attunement' provides a useful point of analogical comparison (cf. Strathern 1987a) and development in terms of the distinctive, though by no means exotic, conditions of the ISS, as I have argued.

The methodological approach to the ISS is informed and advanced by two main methodological developments within anthropology, the establishment of multi-sited methodological approaches and the emergence of online ethnography in digital realms (Boellstorff 2008; Boellstorff et al. 2012; Madianou and Miller 2012, Beaulieu 2010, Zhao 2003). In something of a historic irony, the advent of multi-sited ethnography – in this case, sited in the four mission controls globally – allows for a return to armchair anthropology, as these sited ethnographic accounts will return, in our current ERC-funded ETHNO-ISS project, to London as a centre for the comparative analysis emblematic of the 'armchair.' The arbitrary, yet, as highlighted above, historically contingent choice of the Imperial British standard of GMT as 'local time' in the ISS, echoes this now-outdated relation between London and various global sites as one of the centre and periphery, and simultaneously troubles this history. Taken alongside the polymedia constantly being recorded 'locally' in the ISS, an observer in GMT time is able to attune their daily cycles in line with those off Earth. Within both approaches spatially distinctive realms and long-held presumptions regarding virtual and non-virtual realms are challenged and are methodologically brought into alignment to understand these novel conditions.

Conclusion

The overall theoretical and methodological frame for studying an extra-terrestrial nexus such as the ISS, despite the seeming novelty of the site, as suggested earlier, harkens back to historical conditions in the discipline when sustained direct contact and embedded field work were not feasible or even desirable (Strathern 1987a). We are, however, co-present in terms of the media and networks sustaining the ISS. These include the Mission Controls in Houston, Munich, Moscow, and Tsukuba as well as sophisticated polymedia environments (Madianou and Miller 2012; Jakubowski 2016; Walton 2017) in which every human being on the planet with Internet access can be linked directly to the activities of the ISS site. Here, an extraterrestrial methodology has to consider polymedia such as the ISS live feed,[2] where at any one time at least five hundred people across the planet are co-present with the ISS and its daily interactions between mission controls and their communities (see Jakubowski 2016).

Such a method requires the sampling and study of the extensive video and audio collections held at the Johnson Space Center in Houston – similar to Gorman and Walsh's (Gorman and Walsh forthcoming; Walsh and Gorman 2020) proposed archaeological endeavour, which is a continuous recording of the habitation of the ISS since its inception. This invites 'armchair anthropology' of a particular magnitude, which is co-present ethnographically within the innovative registers of co-presence at the ISS but not within the conventional registers of physical co-location. Such polymedia on Earth, as Madianou and Miller (2012) have shown, are an even more intensive form of social interaction, despite the lack of physical co-locational presence.

More broadly, anthropological research has to date heralded the wider social and cultural parameters in which extraterrestrial activities take place, placing the disciplinary study of the extraterrestrial on sound foundations regarding its critical social and political economic consequences (see Battaglia et al. 2015; Battaglia 2017; Valentine et al. 2012; Valentine 2016; Messeri 2011, 2016, 2017; Olson 2010, 2018). Similarly, in material culture studies, scholars such as Gorman (2007, 2009a, b) have been ground-breaking in terms of outerspace and the heritage of extraterrestrial realms. My argument here in relation to the material culture of extraterrestrial contexts suggests a more intensive and thoroughly empirical engagement with a focus on quotidian material culture and techniques of the body that are manifest within the day-to-day activities and lives surrounding the ISS and, more significantly, its wider distributed terrestrial nexus with their attendant communities.

This focus goes back to an earlier tradition in material culture studies exemplified by Mauss (2006) at the beginning of the twentieth century, which emphasises the seemingly banal aspects of daily life and embodiment in order to understand the most intimate and enduring aspects of human social life. This emphasis on body techniques and quotidian material culture needs to be updated in relation to the new challenges that the ethnographic study of material culture, daily-life, and embodiment holds under the conditions of microgravity in both its terrestrial and extraterrestrial contexts where the new materialities and bodily techniques that bring forth mutual attunement facilitate a novel and dynamic nexus of worlding. To date, anthropology has mostly only understood these fundamental aspects of human and social life under the conditions of terrestrial gravity (Gorman 2009c; Gorman and Walsh forthcoming and Jeevendrampillai and Parkhurst 2020) and has little understanding as to how our traditional analytical frames can be conceived in microgravity, though Battaglia (2012a, b) has noted the importance of 'exo-surprise' in the unprecedented material conditions of microgravity and novel forms of post-humanist kinship (Battaglia 2017) and Valentine (2016, 2017) has observed how our ideas of nature and culture are disrupted alongside new forms of atmosphere and their attendant materialities, which challenge our terrestrially based assumptions within the social sciences. The extraterrestrial context of the ISS and its nexus provides rich evidence to interrogate these conventional terrestrially based understandings of daily life and the material world.

Acknowledgements

This chapter is informed and presents in parts materials from a recently awarded European Research Council Advanced Grant – ETHNO-ISS: An Ethnography of an Extra-terrestrial Society: The International Space Station. This project has received funding from the European Research Council (ERC) under the European Union's Horizon 2020 research and innovation programme (grant agreement No. 833135) that the author as Principal Investigator and its team of researchers have commenced work on. The author is deeply indebted to the extensive conversations and insights, provided by the members of the team, who have heavily informed this chapter: Jo Aiken, Timothy Carroll, David Jeevendrampillai, Aaron Parkhurst, and Aliça Zimmerlin as well as to the editors of this book and their very thoughtful and insightful suggestions.

Notes

1 See Gorman and Walsh's innovative archaeological approach through crowd sourcing and the systematic documentation of the material culture of the ISS through its extensive video and image archive, available at their blogsite: https://issarchaeology.org/blog-iss-archaeology.

2 www.ustream.tv/channel/live-iss-stream.

3 Being, being human, becoming beyond human

Timothy Carroll and Aaron Parkhurst

The London-based *New Scientist Magazine* has been publishing popular science and technology news since 1956. Each year it holds a large four-day conference in London, called 'New Scientist Live,' hosting talks and exhibitions from many of Europe's leading innovators and scientists, and attracting tens of thousands of visitors. The exhibition and speaker's space is divided into five main stage areas: *Cosmos*, *Earth*, *Humans*, *Technology*, and *Engineering*. While these categories have always overlapped to varying degrees, their distinctions are increasingly becoming blurred. In 2018, a talk on 'Boosting your brain with electricity and magnets' on the *Humans* stage was delivered simultaneously with a talk on 'Building bionic people' on the *Engineering* stage, as well as two talks titled 'The post-human future' and 'Our cyborg future' on the *Technology* stage. One of the implications of this overlap, and an intellectual challenge for scholars in these disciplines, is that the material form of the human body and external material forms of engineering and technology are increasingly imbricated.

This imbrication of the body and manufactured forms invites new bio-social approaches to investigating the role of materials within the sociality of the body. The body has, to varying degrees, always been manipulated and 'made.' The human relation to external technology, and the dynamism within the human-plus-external artefact (what we abbreviate as human-plus), is most important, not on the morphological level of the body's form, but on the technological level of the human's performative capacity. New human-plus designs, and morphological adaptations to the human body, create new forms of sociality. This is true across genres of artefacts, from fashion (Phillips 2005) to medical devices (e.g. Arteaga 2019) to body modifications in initiation practices (e.g. Ackley 2019). Such morphological adaptations to the body have clear health implications, and the design of the human-plus configurations rests upon socially defined notions of what constitutes 'therapy,' and is thus a means of restoring a 'natural' state versus what is 'enhancement,' and therefore is aimed at achieving something greater or new. The ideology of design intention is, therefore, an important aspect of how external objects are brought into orchestration with/in the body. As such, methodological approaches that investigate 'medical materialities' – that are looking at how 'the physical, the biological, the medical, the material, the social, [and] the symbolic are held together within a given context of care provision, well-being, and health intervention' (Carroll and Parkhurst 2019: 13) – are needed to push for an analytical model that allows for an understanding of the wide social implications and constitutions of human-plus hybridity.

The broad comparative method of anthropology – bound as it is in the ethnographic present – was expanded in the early theoretical framing of material culture studies to 'emphasize an approach that combined the study of both contemporary and past non-literate cultures from the point of view of the material record' (Rowlands 1983: 15). Peter Ucko, in his analysis of penis sheaths, frames this, saying the 'one thing which distinguishes the study of material culture from most other aspects of anthropological investigation [...] is that its data are manufactured objects, the result of some technological process' (1969: 29). In this view, the benefit of material culture within anthropology was that objects endured through time, allowing a synchronic analysis of cultural artefacts and their social phenomena. Objects could be revisited and used to revisit old questions with new analytical and scientific advancements. In this vein, the technology (in a broad sense) of one society at a given point in time may be used as a place of investigation and comparison with that of another society, no matter the time period or diachronic relation (Radcliffe-Brown or Morgan).

In the face of human hybridity, and the regular imbrication of the human body with manufactured material forms, the methodological investigation of material technology, following Ucko, should be extended into other domains of (post)human practices. While Ucko focused on the morphological differences of penis sheaths, we take his comparative approach to the body and its relation to external artefacts as a model to expand the material analysis from the form of the object into its transformative capacity of aspirational becoming, which informs social practice. This comparison shifts the focus from the morphology of the technology to the properties, affordances, and potentialities of the technology. The properties of technology – such as the opacity of a glass lens in a telescope – shape the possible kind of knowledge gained via, and social relations around, that object (Ihde 1998). Similarly, the affordances of an object – such as a chair to support a sitter – result from the relational complement between the object and its pair – as exhibited in a sitter's affordance to sit (Gibson 1979). These two levels of pre-social and social discursiveness give rise to any range of potential directions of social phenomena, some anticipated and designed, some that may 'torque' in excess (Pinney 2005), into unexpected directions. However, while the actual outcome of a material may be different from the design intention (see, for example the three 'relational functions' of technical objects as discussed by Coupaye, Chapter 4), the social anticipation of an object is nonetheless inscribed within the material form. It is on this level that we propose to compare a series of otherwise wholly unrelated case studies that, nevertheless, resonate with each other in terms of the aspirational anticipation of becoming, in some fashion, more than human.

We address the human desire to become more than human by examining two problematics, framed over three sets of case studies. The first two case studies drive at the problem of nature, religion, and the contemporary post-humanist desire to replace or become gods. The third set of case studies problematises the exceptionalism of the twenty-first-century drive towards post-humanism by comparing the use of new and emerging cybernetic technology with ritual magico-religious technological forms that have endured through the centuries.

The 'cyborg' as post-human

The imbrication of the human body and material forms of engineering and technology is an observation that has been given some attention in Science and Technology Studies (STS) and anthropology in recent years. (See, for example, Downey and Dumit 1997; Ott, Serline, and Mihm 2002; Suchman 2007.) Many drivers of these human-plus forms borrow the term 'post-human' as an explanatory identity, an ideal for the future, or simply as an inevitable consequence of the 'figural excess' (Pinney 2005) of exponential technology. Popular discourse on medical and 'enhancement' technology often creates and reinforces links between the post-human figure and that of the 'super-human.' For example, one exhibition at 'The New Scientist Live' demonstrates new prosthetic technology to replace human limbs, with an impressive range of motion and articulations. Others show emerging research in artificial brain material for cognitive 'enhancement.' The exhibitions are not sci-fi fantasy but rather are designed to be advertisements for international industry. They attract visitors with promotional images of classical comic book superheroes, with capes and costumes, arms stretched forward in mid-flight. The images are common visual metaphors for the technology being displayed.

As Lakoff and Johnson have famously theorised, common discourse is rife with metaphors that transform the way in which people think and engage with the world around them (1980). Emily Martin, in her work on fertility treatment, highlights how these metaphors impact upon the human body, especially as they are employed within the language of biomedicine (1998, 2006). Her work shows how the generative power at play between cultures of medicine and technology and the societies in which they are embedded flows both ways. In this regard, blurring the lines between the human body, technology – in a wider sense of a technical object (cf. Coupaye this volume) – and engineering also blurs the line between that which is designed to 'fix' the body (i.e., therapy), and that which is designed to 'further' the body (i.e., enhancement). The 'cyborg,' as both an ethnographic object and as an analytic, purposefully blurs these distinctions. Originally coined by Clynes and Kline (1960), the term was proposed simply to describe the type of cybernetic partnerships with the human body that would be necessary to live in environments harmful or impossible for normative bodily capacity. In recent years, however, it has become part and parcel of a desire for the 'post-human.' It is a way of thinking about technology and engineering, not as something separate from one's 'self,' but rather as a hybrid from which worlds are formed and performed. It is purposefully 'more-than-human.'

This emphasis on hybridity is not without its contradictions. In her 'cyborg manifesto,' Donna Haraway borrowed the concept of the cyborg to emphasise her rejection of binaries and boundaries ([1985]2000). There was, ultimately, little in her analysis of material technology. The cyborg, as presented in Haraway's work, is a device to illustrate a postmodern critique of nature/culture divides and, exemplifying a postmodern feminist perspective, it is a rejection of sex/gender in similar terms. She writes:

> The cyborg is a creature in a post-gender world; it has no truck with bisexuality, pre-Oedipal symbiosis, unalienated labour, or other seductions to organic wholeness through a final appropriation of all the powers of the parts into a higher unity.
>
> ([1985]2000: 51)

However, despite Haraway's emphasis on liberation and the rejection of traditional power structures, the concept of the cyborg plays upon hybridity to make claims of gestalt. This 'higher unity' is simultaneously abortive and generative, and many technologists reify and deploy the unification of binarisms Haraway attempted to overcome as evidence for post-humanism. Liberation through cyborgian identity, and the potential for technology to afford transcendence is, for many of Parkhurst's research partners, entwined. In this way, attempts to overcome nature/culture divides through a renunciation of, say, sex and gender, often ironically betray the postmodern mission, and instead reinforce the nature/culture divide through an emphasis on the hybridity of the postmodern body as a form of transcendence – something more than the 'natural' human.

The human's relationship with technology is often fraught with these contradictions. Some prominent British technologists who praise the cyborg for its capacity for liberation simultaneously envision and celebrate a future rife with inequality. They view the capacity of technology as permission for a mastery over others (Warwick 2003; Parkhurst 2012). They speak publicly, and often, of the *Übermensch*, Nietzsche's answer to those structures that he envisioned limiting humanity's potential, specifically religion. Kevin Warwick, a leading figure in British cybernetics, and author of *I, Cyborg* (2004), uses the term *Übermensch*, not in the pop-cultural sense of a superhero, but in its original Nietzschean sense. In Nietzsche's words, it will be the 'new, vast aristocracy based on the most severe self-discipline, in which the will of philosophical men of power and artist-tyrants will be stamped upon thousands of years [...] working as artists upon man himself!' (1927: 960). This self-discipline is envisioned as the will to work upon one's body with technology to break the ties that hold humanity back from its perceived potential. Many who work in this field in Silicon Valley, echo these ideals of discipline as an ultimate form of secularism. As Nietzsche writes:

> God is dead. God remains dead. And we have killed him. How shall we comfort ourselves, the murderers of all murderers? What was holiest and mightiest of all that the world has yet owned has bled to death under our knives: who will wipe this blood off us? What water is there for us to clean ourselves? What festivals of atonement, what sacred games shall we have to invent? Is not the greatness of this deed too great for us? Must we ourselves not become gods simply to appear worthy of it?
>
> (1974: 115)

The *Übermensch* is Nietzsche's own answer to this question. It is the individual who has the power to fill the void left from the metaphorical death of God through creativity, will, and sacrifice. Warwick, in famous experiments, had wired his nervous systems to that of his wife's, using implants and electrodes, opening up a new type of interpersonal connection through technological hybridity. He, and his students, are proud to have operated upon their own bodies, to feel these sensory alignments for themselves. It was for them post-human, but it was simultaneously art practice, extending their imagination into body transformation and visceral connection, with the goal in mind to take humanity forward into something new, and ontologically 'higher.' Embedded then, in this idea of the cyborg post-human is the artist-tyrant, the artist who uses the human body and mind as canvas to draw the future of the post-human person. The spiritual connotations are important, and we will return to them below.

The concept of the artist within cyborgian practice appears in other subtle ways. For example, Warwick and his team at the University of Reading use tattoo artists as technicians for their post-human experiments. For the scientists involved, this is a pragmatic decision. Tattoo artists are experienced in working carefully with the boundaries of the skin, navigating the biology just under the flesh with the presentation of art on its surface. Yet, there is analytical value in recognising that the skills used in making traditional forms of body art are the same as those used in constructing the 'post-human.' The study of tattoos within anthropology raises interesting questions for the analysis of the cyborg. Tattoos are another kind of imbrication between the material form of the body, technology, and art that has captured the imagination of social science. Ethnography on tattoo practice has produced a diverse range of theory. Tattoos have been conceived as forms of self-regulation (Atkinson 2004), practices in autobiography (Oksanen and Turtianen 2005), phenomenological engagement of pain and emotional grounding (Ferreira 2011), commodification of persons (Blanchard 1991), commodities in and of themselves (Kosut 2000), or indeed as devoid of real meaning at all, exemplifying the idea of a Levi-Straussian floating signifier

(Benson 2000; Sweetman 1999). Given the similarities between this type of body work and that of the cyborg, important questions can be raised within the subdisciplines of material culture studies and the anthropology of the body. Why are these analytics that have been applied to tattoos rarely (if ever?) applied to the cyborg? Why are tattoos not analysed as post-human practice?

It is crucial to note, however, that many of these technologists are also motivated by a desire to help others, through therapies that alleviate suffering, improve medicine, or adapt the human body to address conditions of oppression. Neural implant research, for example, is funded to address cognitive disorder, neurodegenerative diseases, or sensory limitations. Kevin Warwick often highlights to the public how neurostimulator implants might radically improve the symptoms of Parkinson's disease. The effects of using the technology are immediate. The individual's constant muscle tremors, a defining characteristic of Parkinson's disease, are instantaneously relaxed. However, users themselves see the human/tech hybrid as transformative of identity and the self beyond the simple therapeutic effects of the technology (Goddard 2015). Similarly, technology has been designed for neurocognitive enhancement to address memory loss. An artificial hippocampus can assist with long term memory potentiation, storage, and retrieval, promising hope for the families, carers, and sufferers of dementia and Alzheimer's. It is difficult, however, to confine a technology of this nature to therapeutic uses. There are countless contexts in which these technologies raise questions beyond addressing illness.

In a very different context, yet one that still pushes at the secondary impact of human-plus technological interventions, Lockheed Martin – the American global aerospace, defence, security and advanced technologies company – has produced lithium bionic tech in the form of its Human Universal Load Carrier, or HULC for short – drawing again on the superhero discourse that accompanies the cyborg in technical demonstrations. As an attachable exoskeleton, HULC allows persons to carry extremely heavy weights long distances without breaking the human body. It was designed for soldiers, but is also repurposed for paraplegics who, through the use of the device attached to the back and legs, are able to walk again.

In contexts of care, the therapeutic possibilities of these technologies put an ethical burden on the professionals and the wider community in terms of striving for the best possible intervention. As Goold and Maslen (2014) pointed out, medical ethics and law are now forced into a position in which they must assess the potential liability and culpability of medical professionals for patient health in light of cognitive enhancement technology that is now available. Technological innovations force new ethics that must balance potential benefits to medical practice against personal agency. When a person's life is at stake in the medical theatre, are surgeons obligated to engage with cognitive enhancement simply because it exists? Particularly in light of the link between cognition and legal definitions of human life, the responsibility to preserve cognition by all means possible is an acute obligation. An artificial hippocampus could very well greatly bolster the ability of an individual to retain events, faces, language, and the general experiences of waking life. Yet, for the conscious world of the individual, there is perhaps value in being able to forget (Erden 2013). The philosopher Paul Virilio (2007) has argued that in the invention of each new technology is also contained its failure; so too each new technology, and especially human-plus imbrications, brings with it new ethical burdens and responsibilities. This is all the more true in contexts where the imbrication holds the possibility for reinventing, or surpassing, the very concept of what it means to be human.

The cyborg and transcendence

We have briefly outlined above how, for the cyborg, ideas of salvation and transcendence are often confused and interwoven in regards to the technological hybrid. It is a complex Icarian landscape, and it is not surprising, then, to see the human/technology hybrid reconstituted through religion. There is, perhaps, some irony as well. As discussed above, in designing a cyborgian identity, Warwick (and others) preface their work by invoking the *Ubermensch*. It is the articulation of their post-human desire – bringing humanity forward after what they view as the post-enlightenment death of God. It is, then, interesting to see how the post-human is interpreted by some as the desire *of* their God, and post-humanism as feeling and being *with* their God. The Mormon Transhumanist Association, for example, is a group of like-minded individuals who believe that to be a disciple of the Gospel of Jesus Christ is to trust in, change toward, and fully immerse our bodies and minds in the role of Christ, to become compassionate creators as exemplified and invited by Jesus. A basic tenet of Mormon transhumanist practice is that 'post-human' technology is not only a secular prerogative but also the practice of transfiguration.

> For some, prophecy is not a living proposition, let alone religion or God. They wonder if we've not heard that God is dead, and they're right to wonder [...]. If God is merely a supernatural superlative, he very well may be dead, but positing such as God misses the function of God. God always has been and is at least a post-human projection, an extension and negation of human desire, imagined and expressed within the constraints of human thought, language and action. If we can raise our eyes from the altar of religious and anti-religious dogma, we'll see that the hand raised to finish the dying God is the sign of the oath to the resurrecting God. If we can keep our eyes raised, resisting the carnage below, we'll also see the hand is our own and it holds a blade that's aged and stained. That's when we have a choice, either to repeat the old sacrifices of our ancestors, or finally to make the new sacrifice that they always implied: we can put ourselves on the altar and learn to become Gods. Put differently, the negation

> of one post-human projection always implies another, misrecognized until humanity chooses to become post-humanity.
>
> (Cannon 2014)

God, for members of this group, is not 'dead' in a Nietzschean sense. Rather, God is practiced through technology. Specifically, leaders of the MTA cite the future ability for human cyborgs to be connected to one another, through thought, physical feeling and emotion.

> We [...] want to make a better world. We can do that through engineering and governance, but it's also not enough that we can make a better world. We want to feel it, sometimes powerfully, and more: we want to share our powerful feelings with others in ways that move us together. As engineering and governance are action on science and ethics, religion is action on esthetics.
>
> (Cannon 2014)

The practice of technology here is, in practical terms, similar to Warwick's famous experiments on the nervous system, but the way in which it is imagined is radically different. It holds no residence in evolutionary theory. 'Feeling' what God feels is reimagined as God's prerogative for humanity, and the historical imperative of religion made increasingly possible through science. Connectedness and communication between individuals are not just discursive for the Mormon transhumanists; these constitute an embodied intersubjectivity. The individual post-human being is imagined as the human collective; it is at once both one and multiple. The image of the cyborg is understood through the symbolism of the Mormon church, and specifically the honeybee and the beehive. In Mormon thought, the honeybee (called 'deseret') is conceived as a model of productivity, collective goals, and self-sufficient labour. For the MTA, the post-human figure is likewise understood in these terms. It is the model of living set out by the prophets of Mormon history, and advocated by Brigham Young and his followers as they settled in the American West. As anthropologist Jon Bialecki has pointed out, the MTA is the religious group most in tune with secular humanists, emphasising that God will not deliver people to post-humanism – it is something that God asks people to do for themselves (Bialecki 2017).

Increasingly there are other religious groups exploring constructs of spiritualism through technological embrace and cyborg hybridity. The Christian Transhumanist Association has developed a growing following since its formal 2014 inauguration in Tennessee. Similarly, Buddhist transhumanism is promoted in both Silicon Valley and in spiritual groups in India, portraying the Buddha as a cyborg, and arguing for spiritual awakening through technological advancements. All these movements apply scriptural interpretation as evidence for justifying cyborgian movements as well as ethical and moral authority.

While some religious groups frame the cyborg as a *return* to God, many secular transhumanists imagine the cyborg as a future-oriented goal to move humanity *away* from the old world of gods, binarisms, and power structures. This variety in narrative aspiration is made more complex by another growing group of individuals, who call themselves artists, citizen scientists, and cyborgs. Abandoning the trope of the superhero, they have little or no interest in gods. They envision transhumanism, instead, as a return to nature. Their experiments mirror or expand upon the work on sensory play in academic cybernetic laboratories, but their emphasis on 'connectivity' between themselves, others, and the environment is imagined as a type of restoration of the body – 'regaining' sensory perception lost or deselected in the course of human evolution. For those seeking to 'return' to nature, cyborgian engagement is a reaction to (against) urban life. However, as the sociologist and philosopher of technology Lewis Mumford has written,

> The city is a fact in nature, like a cave, a run of mackerel or an ant-heap. But it is also a conscious work of art, and it holds within its communal framework many simpler and more personal forms of art.
>
> (Mumford 1938: 5)

Borrowing Mumford's language, the cyborg is both a fact in nature, and a conscious work of art (Parkhurst 2016). In each case, the aspiration to surpass normative human capacity makes certain kinds of claims about the morphological constitution of the contemporary, twenty-first century body and the properties, affordances, and capacities the person might possess if this body is tactically joined with technological enhancements.

The twenty-first/first-century body

We are not suggesting that these religious groups versus secular groups are in direct contrast to each other; indeed, the relations between nature and deities is problematic, even within the European intellectual tradition out of which most of these individuals arise. Motivations differ and specific 'types' of post- or transhumanism can be mapped, to varying degrees of success and precision. However, this level of precision often belies the social fact that the individuals within the movement often borrow inspiration (either as positive influence or reactive counterbalance) from other groups using similar (and often fluid) terminology. All these groups (and many others, for which there is not space to discuss) borrow terms such as the cyborg, transhumanism and post-humanism in ambiguous and often interchangeable ways. Our aim in this chapter is not to abandon these terms, especially in their ethnographic usage. Our primary concern is the relationship between human and post-human practices, afforded by human-plus imbrications. We are not denying our subjects their post-humanism. We are suggesting, however, that the narrative of post-humanism is, on an analytical level, fundamentally and deeply human. In this regard, the twenty-first century body is synonymous with the first century body. This final section considers the parallels between the promissory technologies of the cybernetic future and those embedded within traditions that have clear lines of continuity going back centuries.

In what is reasonably a radically different context, the Eastern Orthodox Christian religious tradition has passed down from the first few centuries of the common era a set of textiles that have a critical and essential role in the ritual practice of the Eastern Christian faith. Something about fabric – with its ability to fold and flutter, show and shroud, shift and shimmer – allows it to act as a technology of transformation, creating sacred spaces in a variety of contexts (Carroll 2017). As a technology of transformation, textiles (like tattoos and more broadly clothing) are also routinely used cross-culturally to enhance personal capacities. This might be as simple as sportswear designed to make the body more aerodynamic, or undergarments designed to make it less so. This final section examines the enhancement in religious contexts, where the capacity achieved by the technological union of body to fabric is not simply being 'better,' but being 'other.'

In Orthodox Christianity, the priest clothes himself in a series of garments, each with a range of symbolic associations drawing from both ancient Judaism and the imperial custom of the Eastern Roman Empire (for more on this, see Chrysostomos 1981, West 2013, Carroll 2018). When a new priest is ordained, there are three required components: a bishop to pray for and anoint the candidate; a congregation to give their assent to the ordination; and the vestments of office. After a series of ritual prayers and anointing, the candidate is brought to face the congregation. The bishop then gives the newly ordained priest each item of vesture due his new office, and helps him dress, there in the view of the people, who publicly confirm that he is worthy of his new office. The newly made priest is then turned, back to the altar, and helps in the consecration of the Eucharist, thus fulfilling a crucial capacity of his new role.

In Orthodoxy, the priest continues to need the people and the vestments in order to fulfil his priestly office; a priest without a congregation cannot perform the services. Similarly, in order to perform the rite of confession, the priest who hears the confession (including thoughts, concerns, and admission of fault) of a penitent individual must have an *epitrachelion* to offer ablution (forgiveness). The epitrachelion is a long strip of fabric, shaped like an ox's yoke, that rests over the nape of the priest's neck, and hangs down his front, roughly to his knees. While a priest may listen to the penitent person without the epitrachelion, when it comes time to pray the prayers of forgiveness, the priest must don the epitrachelion, and – usually asking the penitent to kneel – he drapes the long stretch of fabric over the individual's head. With the priest wearing the epitrachelion, and the penitent sheltered underneath it, the prayers of forgiveness are said, and the penitent then leaves having been freed from any burdens they may have had.

The technological hybrid of priest-plus-vestment is able to do something for the penitent that an ordained priest without vestments cannot do. What is particularly important for our argument here is that this action – namely the forgiveness of sins – is understood to be a more than human capacity. Using the voice of the Old Testament prophet Isaiah, God claims the sole capacity to forgive sins (Isaiah 43:25) and, in the Gospel account of Christ healing a paralytic, the Jewish scribes again affirm that only God can forgive sins (Mark 2:7). This divine ability to forgive sins is then conferred upon the apostles (John 20:23), and through the laying on of hands (i.e., the service of ordination) that ability is passed on to subsequent generations of bishops and priests. The ability of the priest to forgive sins is, however, only a potential ability until he is vested in the epitrachelion. Without the technical apparatus of his office, he is unable to perform the divine act of forgiving sins. Conjoined to the ritual technology of Orthodox priesthood, or 'vested with the grace of the priesthood' (as he says of himself in the liturgy), he is able to perform actions only capable of, the supra-human.

It is also important to note that this kind of transformative capacity of ritual paraphernalia, which renders the human more than human, is not limited to the monotheistic or salvific religions. As Pedersen (2007) has shown in the case of Mongolian shamanism, the kind of engagement with the spirits made possible when clothed in the shamanic costume is so potentially other so as to be a real danger to the human. He argues 'that the shamanic costume affords the shaman with a multiple, extra-human body, which, by inducing a momentary transformation of his or her corporal gestalt, enables the shaman to attain otherwise unattainable points of view' (2007: 142). At the completion of the shamanic ritual, the shaman's assistant must rush to take off the costume (headgear, then robe, then boots), as 'it is considered extremely dangerous to wear the costume without continuing drumming'; once disrobed, 'the shaman slowly becomes herself' (152). During the event, rather than being strictly human, the shaman is a 'knot of knots' – a sort of composite multi-event moment; as Pedersen summarises, 'By donning the costume, then, a Darhad shaman is transformed into the *ultimate multinatural entity*, as this hyper-surface is believed to transport the shaman to an immanent space of multiplicities' (159, emphasis original).

Discussion

In the phenomenological tradition within material culture studies, Jean-Pierre Warner (2001) draws upon the insight of Paul Schilder (1923) to help answer the problem of the Kabyle man in his slipper – able to run down a hill quickly, despite the kind of footwear – which confuses Marcel Mauss (1973[1935]). In Warnier's solution, he posits that had Mauss known of Schilder when writing techniques of the body, he would have understood the capacity of the mind, and thereby motor functions, to be extended beyond the body into the attached objects (shoes, walking sticks, etc.) as well as rooms and other material environments (such as a basketball court). The relationship between the human and the external technology must be framed in terms of the affordances of the materials at play. This includes both the human body and the specific artefacts being attached to the body in the human-plus imbrication. While this is, as seen in Haraway's work, relevant to the hermeneutic tradition of intellectual scholasticism, it is also driven by the material constitution of the bodies and external things. This 'material hermeneutics' (Ihde 1998) opens up new points of investigation into how human-plus relationships work.

In terms of the hermeneutics of technology, there is also an important distinction to be made between the potentiality and the affordance of a technical object. When something is framed in the language of 'potentiality,' it is inscribed with intention and futurity (Taussig et al. 2013) – which are sticky, messy concepts and, as we have written elsewhere, give rise to failure, as the materiality of the artefact will develop in unexpected and unanticipated directions, aborting the process of objectification (Carroll et al. 2017). The affordance of the material (including, importantly, the human flesh) to make certain kinds of sociality possible, is constrained by the historically contingent ways a society knows the material around it (Gibson 1979), but also holds the possibility of shaping society in new ways. As Taussig et al. argue, potentiality and the promissory hope it offers should be both an analytic and an object of study. In a post-phenomenological tradition (Ihde 1993), the discourse around the promise and hype of technology – such as biomedicine, cyborg enhancement or sacerdotal paraphernalia – must be taken along the 'material hermeneutics' (Ihde 1998) of the given technology in how it shapes the knowledge that is knowable. It is in this aspect that the importance of affordances comes to the fore. In the first instance, these affordances are, as can be seen in Ucko's morphological analysis, related to the form and fit of the technology in relation to bodily apparatus and function. However, by looking at the affordances of things in a comparative framework at the level of transformational capacity of technology – be it cyborian transcendence, medical enhancement, or ritual enactment – a different vantage point opens, allowing for new insight into the aspirational drive of humans to become more than human.

As stated above, each body has its own affordances, and these inform and advance social movements independent of meaning that may be attributed to it. The cyborg, as we have outlined, is a complex term, operating in different and often contradictory ways within the aspirational and promissory narratives of scientistic, religious, and academic circles; it offers a range of promises to different individuals depending on the systems of social relations in which those individuals are embedded. Yet, an examination of the cyborg done in terms of the constituent materials and the intentions of transformation shows how the cyborg is also very human. Despite the narratives of 'post' (modern, human, etc.), the twenty-first century body is the same as that of the first. While this is complicated to some degree when we think of partnerships and hybridities, such that the affordances of the body are changed (amplified, extended, enhanced) by the affordances of the technology attached to it, the fundamental basis of material engagement is the same. At the level of potentiality, and what it is that these human-plus imbrications are understood to do, the pre- and postmodern human is similarly reaching for something beyond itself. And in this regard, being post-human is inherently human.

4 'Things ain't the same anymore'

Towards an anthropology of technical objects (or 'When Leroi-Gourhan and Simondon meet MCS')

Ludovic Coupaye

Introduction: the printer, the car and the doll

Just before leaving the office to go and pick up my children at the nursery, I decide to print one of the earlier drafts of this paper, to edit it on public transport. I deftly press the 'Ctrl' and 'P' keys on the keyboard and a new screen replaces the one where the document was visible. After a quick glance to check whether the correct device is selected, I click on the small icon showing a printer. I then pick up my wallet, which holds my staff card, walk out of my office and along the corridor to the place where the multifunction printer stands. Opening my wallet to avoid interference with the signal sent by the machine with my debit and travel cards, I place it over the small square attached to the side of the bulky machine. With a beep, the small screen on the top of machine lights up, showing images and text, including my staff ID code. I touch the part of the screen that reads 'pull print,' prompting a new beep, and a new screen appears, showing a line with the document's title. I touch the side icon that reads 'print all' and the machine comes alive: a fan starts somewhere, clicking sounds come from its inside and the tray on the right side rises automatically up into position. In a fairly rapid and rhythmic succession, the pages of the chapter are expelled onto the tray as the draft appears in printed form. Suddenly, the machine starts emitting a sort of wheezing noise and stops, beeping alarmingly. Startled by the signals, I wonder what is going on. The printer's screen now shows a red band marked with an 'X,' and an image of itself, with animated instructions about opening specific panels in order to remove a sheet of paper that seems to be jammed in its complex series of gears and rolls. Deciphering the instructions, I bend, locate the panel and open it. Having found the guilty sheet mysteriously trapped between rolls, I pull it out. I close the panel, the machine beeps, but now the screen indicates other panels to open and check. Cursing more or less silently (I can feel that time is fleeting), I comply, not finding any paper jam, and after several frustrating minutes, the printing resumes. Once it is done, I grab the small pile of paper and run back to my office. I am now running late, and I consciously blame it on the machine.

An analogous reaction was recounted by Alfred Gell, in his posthumous ground-breaking opus on the agency of art (1998). He described how, should his car break down, he would consider it as an intentional act of treason, perpetrated by the car itself (ibid.: 16–19). These particular situations, where humans impute to objects forms of capacity to initiate events in their surroundings, invited him to use the concept of agency, summarised by Daniel Miller as a 'theory of natural anthropomorphism, where our primary reference point is to people and their intentionality *behind the world of artefacts*,' based on 'inferred intentionality' (Miller 2005b: 13, my emphasis). Gell's theory of agency was extended to other types of artefacts, such as the relationships a little girl has with her doll, or paintings. While it appeared as an almost universalist theory of things, agency had two sides: one was a sort of retro-projection of human-like (or living-like) capacities onto an object encountered; another one was about a direct efficacy of patterns and shapes on the viewer's cognition, a theme that Susanne Küchler (Chapter 9) develops in relation to the idea of objectification. In this chapter, I focus instead on Gell's conception of agency as being the effect of a '*causal milieu* generated in the vicinity of an agent [assuming] a certain configuration,' interpreted as the index of 'the presence of another agent' (Gell 1998: 20, my emphasis).

By contrast, Bruno Latour's examination of artefacts offers a view from the opposite end (1991, 1998, 2000). Actor-Network Theory (ANT, hereafter) investigated how objects such as the lock on the door at a youth hostel or an overhead door-closer translate social (political, moral, ethical) intentions into a 'programme of actions' (Latour 1991, Akrich 1992) designed in the artefact (see also Schüll 2012), orienting human actors' behaviours. ANT's project was profoundly and primarily political, based on symmetrical examination of the participation of both humans and non-humans in social life, with an analytical focus on social mediations derived and solidified by technical arrangements, while simultaneously balancing simplistic social constructivism and material determinisms with an ontological pluralism.

Both approaches were part of a wider anthropological project that investigated artefacts from the angle of their social (or cultural) roles, lives, effects, and designs, as Sautchuk summarised (2019: 179–181). It is fair to say, however, that for both, the main horizon was an anthropic one: Gell sought to investigate social relations in the vicinity of art objects, while ANT sought to demonstrate how society was made durable (Latour 1991). In other words, relations were primarily understood from the angle of human sociality, located within or around the material and the technical, either as an effect, behind or before the artefact, or as a type of assemblage of human and non-human actors.

Yet, this sociological focus, by placing the 'social' as their main horizon, also implicitly left 'the technical' as a secondary counterpart (a 'secondary agent' in Gell's phrasing, or as 'co-actor,' as Latour suggested), a sort of material figure out of a social background (see also Ingold 2008 and particularly Sautchuk 2019). As a result, it remains difficult to ascertain whether, analytically, the agency the printer (or a robot, or space station) is different (or not) from the one from a doll, a door closer or a pen. Their material and technical particularities are dissolved within the type of social relations/

lives they emerged from, manifest, generate, translate, transduce, and so forth. By qualifying them as social beings – or at best sociotechnical ones – their agency, in particular of the political kind, appears as affected by their materiality insofar as their design (a term that equivocates a social process of making and its result) more or less successfully manages to synthesise (more or less democratically) *predefined intentions*, be they social or psychological. Our investigations of politics, ethics, or justice, thus almost necessarily ends up seeking how the 'social' is indeed made durable, but within, behind, or beyond a 'technical' ontological 'other.' It is as if, wary of the spectre of technical determinisms, we would prefer identifying the main agency of technical objects (be they tools, instruments or digital objects), as having a 'social' source.

In this chapter, I wish to investigate the possibility of recognising and describing the alterity of technical objects' agencies, in particular their specific 'technicity' as an ecological and historical modality of relations qualitatively *distinct* (but *not* separated) from human sociality itself. Pushing further Gell's and Latour's initial leads to specify the relations happening in the vicinity of emerging contemporary technical objects might require that, alongside the category of inferred agency, I add an agency that emanates thus from the 'technical' object itself, not solely in social terms but also in its own 'object' terms. This includes thinking of processes of objectifications, as Küchler invites us to do, as well as elucidating the specific vernacularity of objects' lives – their modes of existence, which would not presuppose the uniqueness of human sociality. This means proposing 'a technical life,' not in opposition but alongside a social one (see Drazin, Chapter 5), a specific 'technicity.' This, is particularly crucial if we want, as Haidy Geismar suggests (Chapter 6), to elucidate further the actual 'role of form or materiality in constituting sociality and meaning.'[1]

'Technicity' was the central point of entry in both Gilbert Simondon's[2] (2005, 2017[1958]) philosophical work on the mode of existence[3] of 'technical objects' (TO, hereafter) and André Leroi-Gourhan's work on technics[4] (1971[1943], 1973[1945], 1993[1964]). For both authors, technics were a human horizon *in themselves*, not only because of the role they played in human phylogenesis, or the continuity it could present with non-human species, but because of the increasingly concretised, objectified and hegemonic place technics occupied in the mid-twentieth century.

This chapter is my first attempt at investigating further the ways in which industrial complex artefacts, though resulting/instantiating pre-existing social relations, as the Social Construction of Technology approach demonstrated (e.g. Bijker et al. 1987; Bijker 2010), nevertheless contains specific modalities of mediations, through a reading of Leroi-Gourhan and Simondon. I wish to try enriching Gell's conception of agency and ANT's conception of programme, not from a sociocentric angle, but as being materially shaped by them. In other words, investigating what would a proper non-anthropocentric relationality look like. I start instead, thus, from the agency emerging as a specific field or effect as the materialisation of sets of topological and self-regulated causal relations, or resonances, internal to the TO, which include processes independent of the human agent engaging with it. By doing so, I hope to delineate the extent to which TOs exert their own (vernacular) agency onto a human actor and beyond. I thus choose to start by focussing on the particular *functioning* of industrial TOs, as an analytical entry point.

The functioning of technical objects vs. their function

One of Leroi-Gourhan's and Simondon's main premises was to challenge the apparent discreteness and fixity of TOs, a conception that has long fed both the analysis and museum presentations of artefacts (see Schacter, Chapter 14; Coupaye 2018). For both, an analysis of artefacts could not do away with the history of their evolution. Yet, far from any evolutionist theories, their view was not teleological, but was based instead on a Bergsonian understanding of temporality and potentialities, as virtual vectors of change and exchanges between entities (organisms) and their milieus, from which specific objects emerge as 'points of concurrence' or temporally situated iterations (Simondon 2017[1958]: 25–44, 124; Leroi-Gourhan 1971[1943]: 21–42), a mode of becoming that Simondon qualified as 'individuation.'

As a result, 'technical objects' can be defined as things coalescing as a locus of exchanges between humans *and* their milieus, as physical and/or sensory mediators. As such, this category encompasses the range of tools, instruments, and devices created and used by human beings to materially interact with themselves and/or their environments, from a knife, to a camera, a ritual carving, or an algorithm.

This position as mediators of human agency is not restricted to 'modern' objects and does not allow a complete separation of industrial, mechanical, and automated 'hi-tech' objects from 'low tech' or even ritual objects (Simondon 2005: 86–99).[5] Instead, Simondon, in particular, is concerned mostly with objects intentionally taken, made or designed to play a role in what Mauss would define as technical actions (1973[1935]: 75), that is, actions that are materially 'effective' (an efficacy vernacular to the actor) and 'traditional' (the product of historical dynamics of transmission and change). The term 'technical' here occupies an important analytical position, as 'whatever is woven together by the highly specific trajectory of "technical" moves becomes "material" as a consequence,' as Latour noticed (2014: 508). 'Technical' objects are thus supposed to extend human bodily and cognitive capacities in order to perform tasks or a set of tasks, or more generally to engage with the world.[6]

Forging the tools for an ethnographic investigation of contemporary TOs that both outline their anthropological import, as well as their particularity within the field of material culture, thus requires focussing on the materiality they encapsulate and importantly the vernacular – ethnographic – specificity they manifest in themselves. In order to discuss the particular type of agency they convey, not in the form of intentionality inferred from the point of view of a human actor, but from the one built in or delegated, I suggest investigating the actual *functioning* of the object itself.

This is profoundly different from the study of *function* and/or use (consumption), both relations being inherently social, as Drazin reminds us (Chapter 5). As such, the functions and roles of whole TO might not be sufficient to reveal their historical specificity, or to show what makes the difference between a doll and a car, or a printer and a pen. The 'delegate' (or in Gell's terms, the 'secondary agent') does not passively convey the 'programme,' but because of its existence within a milieu, it also behaves following its own rules and orients the ways in which actors engage with them (a point adumbrated both by Geismar and Knox, Chapters 6 and 8). Within social relations happening in the vicinity of a TO, some are indeed the result of human inference, but other relations also emanate from the encounter of human actions *with* the TO and its behaviours.

Completing Gell's proposal, thus requires examining this *functioning*, that is the modes of interactions (relations), between the different components of a TO (at the most basic level, a blade and its handle; at a most complex one, the long and complex network that links together a keyboard to a mechanical device that grabs paper and burns ink on it). Taking my own vignette, these different components of a TO indeed fulfil specific *internal functions,* as organised relations between themselves (rolls and gears) as well as with the whole object (the printer).

This is neither an easy nor a comfortable task for anthropologists, who are predisposed to assume the pre-eminence of the social or cultural meaning within a given field of enquiry. Indeed, the difficulty lies in the way investigating a functioning requires a minimum of specialised knowledge (such as mechanics, engineering, electronics or programming) because the complexity of the type of TOs we are dealing with in this chapter often goes beyond the one found in a knife, a pencil or a spear-thrower. The discomfort comes from the spectre of technical determinism raised by the analytical moment, suggested by Simondon (2017[1958]: 25–26), which suspends social and cultural domains, discourses and interpretative moves, in order to examine the 'technical.' This 'suspension' is however crucial to avoid implicitly validating the social/technical divide and, instead, perform an analytical move in a dialectical process in which both thus can be put *back in phase* (Simondon 2017[1958]: 173–176). This 'suspension' of the 'social' is thus a requirement to reveal whether the type of relationships happening in the vicinity of the objects can be seen as a particular milieu generated by what is happening *inside* the object in order to identify its actual contribution to sociality.

In Gregory Bateson's cybernetics terms, this corresponds to paying attention to the *code* instead of the *message* (1972: 103).[7] Indeed, by suspending hermeneutics or semiotic analyses, the aim is instead to outline what type of 'agency' emerges from the medium, objectified through its functioning and the type of 'causal milieu,' as Gell coined it, it generates in its vicinity. For Simondon, the mode of existence of a TO comes thus from the emergence (ontogenesis) of a specific combination of several functioning principles (concretisation), as deposits of both the history of technics and the synthetisation of several modalities and principles of actions (actualisation of technical principles or lineages; in the case of the printer: digitalisation, mechanics, optics, etc.). For anthropologists, it means, investigating this mode of existence *as well as* the results of the intersection of human intentions, material conditions and forms of vernacular logic.

Inside and outside the technical object: at the intersection of milieus

I have sent the command from my computer, made sure it got accepted by logging on to the printer, and the rest of the process follows 'causal actions' that transform the digital information I have sent into a series of electronic signals which, running physically through a sequence of devices, activate the complex set of rolls, grab the sheet, and control the movement of the laser, which organises the distribution of toner to form signs on the paper. Once I have activated my 'user profile' and pressed the 'print' icon, I have no control in the sequence of events that occur inside the machine. All is dealt with by the processors following a predefined and fixed programme regulating in a predetermined way the circulation of electricity animating the rolls and controlling the movements of the laser beam charging the OPC drum, in turn capturing the toner before depositing it onto the paper. The printer acts as a mediator between my intention, the whole network and the printed paper. But it also directs my own actions: I have to start the process, to leave my office and go to the actual machine, to use my staff card and, in the case of a paper jam, to spend time sorting it out.

Such technical objects, as both Leroi-Gourhan and Simondon suggested, are not only mediators of human intentions, but their functioning also generate an operating *milieu* around them, through what and how they do things, what they demand and require to work, as well as what they afford. The concept of 'milieu,' by analogy with biology, plays a crucial role in both Simondon's and Leroi-Gourhan's understandings of technics. It can be summarised as a global field in which an entity (living organism or technical object) is immersed and with which it interacts but also upon which its existence depends (Bateson 1979). As an environment, which can be internal or external to an organism or, in Leroi-Gourhan and Simondon's work, a TO, the concept of 'milieu' implies relations of mutual and recurrent causalities within and around the TOs. As such, implicitly, both authors invite us to develop a form of ecology of TOs as the study of relations between machinic 'organisms' and their surroundings (including other organisms).

Both break down this notion of milieu into interrelated ones, in order to specify the different effects of relations of technical innovations. Simondon and Leroi-Gourhan both identify an 'external milieu' (fundamentally spatial, geographic and natural), to which Leroi-Gourhan adds an 'internal milieu' (cultural), and Simondon a 'technical' one, which includes existing technics and TOs (Leroi-Gourhan 1973[1945]: 333*ff*; Simondon 2017[1958]: 54–59; see also Audouze 2002).

To these three interrelated milieus, we can also add what Simondon defines as two resulting and related effects. First, an *internal resonance* of the TO – emerging from how the different parts of the printer have come to be coherently and synergistically organised to function together in relation to the whole object.[8] In other words, complex TOs coalesce as several technical principles brought in coherent equilibrium, but also potentially generative of further developments. Second, crucial for Gell's notion of agency, the TO also generates an external resonance, which Simondon coins *associated milieu*. This associated milieu is a sort of the vectorial product of the internal resonance of the TOs and the world outside its physicality, resulting in the (material, social, psychological) enrolment of the object's vicinity – that is, the vicinity around the TO becomes thus enrolled and part of the existence and functioning of the TO.

Simondon identifies the relations between a TO and its various milieus, in terms of mediations of two analytical categories, 'energy' and 'information,'[9] adapted from his critical reading on cybernetics.[10] For living beings in an evolutionary perspective, the requirements for these two exchanges led to the phylogenesis of 'effectors' (such as claws, beaks, fins, legs, hands) and 'sensors' (sensory organs, including those dealing with vestibular and proprioception). In human beings, both have become increasingly mediated as processes of externalisations (Leroi-Gourhan 1993[1964]). On one side, effectors became externalised both as prosthetic extensions (that is, material), and mediums of action (e.g. the knife as externalisation of the teeth; Leroi-Gourhan 1993[1964]: 90), that Simondon qualifies as 'tools.' On the other side, technical externalisations of the senses (e.g. the telescope as an extension of the eyes: 2017[1952]: 130–131, 2005: 88) leads to the emergence of 'instruments' (e.g. sounding line, stethoscope) or 'sensors.' This idea of externalisation the physico-biological role of effectors and sensors allows us to rethink the question of agency through, on the one hand, the ways bodily capacities (which includes language and memory) are not only extended beyond the body but also crucially reframed into specific modalities of internal functioning and, on the other hand, the material effects of delegations. Indeed, Simondon indicates (2005: 89–92), these modalities of externalisation become materialised in three 'relational functions' of the living organism and its milieu: (1) prolongations of effectors or sensors (that is the extension of the reach achieved by a rope, a stick, or a telescope); (2) transformation of the action (direction, velocity, intensity, and/or even materiality: such as from kinetic to electric or chemical); (3) isolation (such as fur, gloves or tongs that can protect from low or high temperatures, dirt, shocks, wetness, and so forth.).

TOs emerge thus not only as combinations of the two categories of relations (effectors, as medium of actions; and senses, as medium of information exchanges), but also as various combinations of their three 'relational functions,' as the materialisation of specific ecologico-historical relations. Importantly, to be able to act *in an effective way* (in a Maussian sense) as mediators between human actors and their milieu, these three relational functions and their combinations must be structurally organised within the TO so that their respective intended and unintended effects (heat, noise, vibrations, etc.) are in synergy and in equilibrium, in a way analogous to the organs of living beings, creating an *internal resonance*.

It is this generated internal resonance, a set of topological self-regulated causal relations, materially constitutive of the TO, and the ways in which its structure relates to its functioning, which potentially constitute the main analytical point of entry into complex TOs. Indeed, in simpler tools and instruments, the main sources of both energy and information are external to the TO, often in a human actor who operates the object, giving it its force (energy) and direction (information) and interacting with its operational mode (a knife held in the hand, see Sigaut 2012). In the medieval windmill, while the main source of energy came from the wind, the miller's action was required and determinant in positioning the tail pole to orient the arms into the wind and maintaining the internal components, making the miller the main provider of information. The human actor was thus a functioning part of the TO itself.

As complex objects become more autonomous in both information and energy transfers, as they become associated with both separation of knowledge and skills and hierarchical organisations, the role of human actors from essential co-actor of its functioning is increasingly transformed into ensuring and controlling that the 'vicinity' (material and social) of the TO support their proper functioning (clearing the paper jam). In the case of my engagement with the printer, my role as actor becomes limited to the supply of basic commands ('Ctrl+P,' click, going to the printer, press, log in, etc.) and material (replacing paper), and following the machine's commands to clear the jam.

In a way that feeds technical deterministic interpretations, the agency of the machine directs my bodily actions, an agency that I experience as a form of political effect of alienation (Simondon 2017[1958]: 16, 133–134), and which Langdon Winner summarised as reverse adaptation (1985[1977]: 226–236). Indeed, combined with the emergence of industrial capitalism and its particular ecological-historical ethics, research on 'efficiency' (as the ratio of exchange of information and energy in quantified terms and in velocity) lead to the design of complex TOs, in which these relations become more detached from the actors, and delegated to the modalities of functioning of the objects (Simondon 2005; 2017[1958]: 79–81).

If I were to copy out what I have typed on my screen with a pen and piece of paper, it would definitely take more time than printing it. Instead, I have delegated the materialisation of the words on the screen to a chain of causal actions and information transfer, which – using the electrical energy grid and the digital network set by my institution – is transmitted to the printer. In turn, the printer's chain of internal effectors and sensors, verifies that it is actually me (or a human delegate with my staff card) who is standing in front of it, before automatically operating the transfer of the digital code into a printed paper. If I were to use a pen to copy out the chapter, I would be involved in a bodily relation with the pen, the ink and the paper in each element of the scripting of the second copy, but with the printer, I am excluded from the functioning of the machine.

Processes of delegations in modern TOs include and go beyond the translation of a (social/moral) programme as defined by ANT. From this ecologico-historical perspective, increases in complexity require also increasing in the internal structuration and self-regulation of effectors and sensors, in order to deal with the counter-effect of each of its components and functions, which inherently also generate an exclusion (alienation) of bodies from the functioning of the machine. Not only is '[t]he machine [...] a deposited human gesture that has become a stereotypy and the power to restart' (Simondon 2017[1958]: 151), but it also implies new forms of temporality when it comes to memory ('information') and power ('energy'): indeed, among the main criteria that govern their conception is the velocity of operation, the limit of which was, until recently, set by human agents, as the factory line scene in Charlie Chaplin's *Modern Times* (1936) sarcastically shows. To fuel such velocity, power (and, increasingly, information) cannot come from human agents anymore, but must be *harnessed* from other sources. This is where automatisation, auto-correlation takes the shape of 'autonomisation.'

The functioning of modern TOs comes thus from the creation (the origin of which is definitely social and historical) of an *internal structuration* made out of increasingly complex assemblages of functions of prolongations (delegation), transformations, and isolations, all in charge of operating the exchanges of information ('commands,' 'programmes,' 'knowledge') and energy (causal actions, 'power,' agency), independently from human experience and actual agency. In turn, the functioning of TOs generates around it an *associated milieu* that extends its reach by materialising in complex infrastructures (see Larkin 2013; Knox 2017), notably to 'the environment.' It also both excludes and orients human actions in specific ways,[11] which cannot only be accounted from the angle a simple inferred agency or a hermeneutic of machines as metaphors for society. Instead, we might also need to take into account the ways in which our inferences, our bodies and our interpretations of the world are not only shaped by TOs but also excluded from their functioning.

The politics of technical objects and the object of technical politics: removal and isolation

I use 'politics' to refer to relations framed by conceptions of governance through control (power) and justice (fairness), and to interrogate what does Simondon and Leroi-Gourhan's analytical move could bring to the old question of the politics of artefacts (Winner 1986) in a way which would extend Gell's and Latour's existing analyses of agency.

It might start with recognising that TOs appear *intrinsically* as the actualisation in material forms of ecologico-historical relations between living beings – in my case humans – and their milieus – as they have been since the origin of the human species. This relationality can make them political entities, as 'society made durable,' indeed, but also as 'culture' given material (efficacious) shape as George Kubler wrote (1962).[12] As efficacious and historical (in the Maussian sense) concretisations of technical relations, their status of co-actors invites us to question simple issues of governance.

Take the agency of the printer/photocopy machine at the end of my corridor. It is not only different from the one emanating from a pen, but it also materialises a technicity – a historically situated 'mode of being,' as Sautchuk summarises (2019: 182) – as a specific relation explicitly political/ethical/moral: the amount of energy and paper consumption, and of resources for the maintenance of individual printers led my institution, in 2013, to decide to stop supporting individual printers for moral and financial reasons.[13] This is where, indeed, the social reappears, not solely as the origin of the design, or a pre-causal inference originating from a distant human agent, but in a Maussian physical and bodily technical form. The delegation of printing to collective devices might contribute to saving energy and paper, but it simultaneously has a direct effect on my body and my mind as I have to physically walk there to print (and queue, should someone else be already there).

In this bodily encounter, it is not mechanical relations nor automatisation inside the printer that are fundamentally anti-democratic (there are indeed mechanical relations in hand writing between pen, paper, ink, hands and eyes for the words to appear and ideas to take written or drawn shape). TOs' three relational functions of prolongation (externalisation, delegation, extension), transformation (translation), and isolation (protection) start to appear, however more explicitly political, when their effects extend to other entities, particularly living beings, privileging predefined forms of governance, as ANT has demonstrated (Akrich 1992).

From the angle of agency, TOs (particularly those that perform automated decision-making processes, such as algorithms), AI and their concretisation in smart machines such as driverless cars or targeted adverts, can no longer be seen as simply neutrally mediating power relations between what Gell identified as agents and patients, or ANT's 'actors.' Building on these approaches, one can try to specify what happens when the particular functioning of TOs relies increasingly on autonomous and self-regulation processes, not only translating but in fact removing further away the delegation of the programme from its social origin. The recursivity of the 'sensors'/'effectors' relations, which include knowledge acquisition, as Hannah Knox shows (Chapter 8), decision-making and action, make these fully externalised, translated and operated by, and through, the functioning of the TOs, and leaves humans' interpretative and imaginative capacities and their performance excluded from the actual operation (Simondon 2017[1958]: 139–141; see also Ingold 2000).

I feel the isolation might be the key relational function here, as it indeed can lead to a profound alienation (and at times submission) of *both* the actor *and* the milieu from this materialised ecologico-historical relation. First, within such technicity both living beings and their environments thus risk being confined in their role as being enrolled in the *functioning* of the TOs (as supervisors in charge of their maintenance, as resources that need to be extracted to provide energy and/or information, as users who have to adapt to the functioning). This 'associated milieu' can thus generate a form of 'reverse adaptation,' where the role of humans and non-humans as actual partners and co-actors of their effects is not solely constrained by the type of technicity manifested in the object but also directed by it. Second, when the TO both excludes the patient from its functioning of decision-making and isolates further the primary agent from the interaction, it obfuscates and befuddles the possibility of tracing agency back to its source – the actual agent accountable for the action. This can indeed be experienced as a form of unilateral violence – a 'sorcery of technics' of a sort, to reuse Gell's imagery (1992). I take my point from Gell's example of Pol Pot's landmines (1998: 20–21), which do not really compare *anymore* with smart weapons. The decision to place a mine in a specific spot was ultimately made by a human, even when ordered so by Pol Pot through a long chain of command that isolated the soldier from the dictator. When an algorithm can (and might) make landmines decide whether they should deploy and where, and whether they should explode or not under a pre-defined walker, accountability will become complicated to attribute: is it to the mine itself (which seems to have decided by itself to explode), the coder of the algorithm (who programmed the conditions for making the decision), the whole software (who executed the decision), the arms company (who designed the device), the military chain of command (who bought them), or to the whole political regime that authorised their use? Perhaps we should look instead into the technical politics, the actual historically situated *choice* made by a collective of humans and non-humans to generate a technicity where an algorithm *isolates and excludes human choices and responsibilities* from the actual gruesome result.

It is important to remember that, in itself, technicity, as materialised modes of relation made intrinsic to TOs, does not necessarily create inequalities, even when its agency comes from the translation of an explicit moral programme delegated by humans. But we might have to scrutinise further how collective and networked associated milieus, stemming from the reproduction of 'technological frames' (Bijker 1989), based on automatised (self-regulated) hierarchy and control, actually privilege the emergence (ontogenesis) of objects whose functioning and effects stabilise/justify power relations (see Schüll 2012), both intended and unintended.

It seems to me at this stage that inequality emerges when the ontogenesis of the TOs realises a form of mediation that presents *at least* the three following characteristics together: (1) when it excludes human actors to a point that their main relation with them is of subordination to its functioning rather than as a collaborator; (2) when the direction of the agency is unilateral from the object onto the body and the mind of the human actor; and (3) when the 'primary agent' from whom originates the agency delegated is so removed that they are unidentifiable.

In addition, as Andrew Feenberg remarks,

> [t]echnology is power in modern societies, a greater power in many domains than the political system itself. The masters of technical systems, corporate and military leaders, physicians and engineers, have far more control over patterns of urban growth, the design of dwelling and transportation systems, the selection of innovations, our experience as employees, patients, and consumers, than all the electoral institutions of our society put together.
>
> (Feenberg 1999: 131)

Today, this might apply to how big-tech corporates rely on the technical logic embedded within the TOs that we engage with directly (bodily) at every moment of our lives, to unilaterally decide to impose upgrades on globally used software and mobile devices, extending the reach of their 'agency' at a planetary level. On the one hand, the development of connectedness (globalisation of Wifi, Internet of Things, AI, etc.) simultaneously extends (prolongs) human agency and requires new centres of information 'control' (data-mining, surveillance, automatic and unwanted upgrades), and energy supply (see Stiegler 2009). On the other hand the milieu generated by contemporary TOs and their interrelations is so vast and complex,[14] that people are inevitably isolated – alienated – from their technicity as any complete knowledge of their actual functioning is impossible, consolidating existing forms of structural social (cultural, racial, gender) exclusions.

The particular forms taken by contemporary TOs (and their extended relational functions of prolongation, transformation and isolation) cannot be separated from the historical roles of rationality, industrial capitalism, colonialism, and global neoliberalism and sex- or race-based relations. Scholars, such as Brian Pfaffenberger (1992) portraying situations of resistance, or as Trevor Pinch and Wiebe J. Bijker, revealing the social construction of technology (1989), have adumbrated avenues for the emergence of possible forms of techno-democracy or techno-resistances. However, because of their complexity and coherence, the design of new TOs requires specialised knowledge and resources that are already shaping innovations in both science and industry. As a result, the extension, connection, and cohesion of contemporary TOs into global infrastructures might constitute an extended associated milieu within which citizen participation might be occurring everywhere in a *contact zone* (Pratt 1991; Clifford 1999[1997]).

Conclusion: towards a 'technography' of objects?

In this preliminary reflection, I have attempted to think through how technical objects manifest historically and ecologically situated forms of agency that coexist with the ones described by Gell and Latour, and have outlined ways to bring these forms to light, using Simondon and Leroi-Gourhan's *technologies*. I suggest that objects can not only generate feelings of desire, frustration, or hate, and they do not solely translate, as Latour pointed out, political and moral programmes of action: they also create milieus, in ways that reproduce, reaffirm and consolidate privileged forms of social relations or technical politics, adjudicating which place is to be given to human subjects.

However, extending the notion of agency cannot simply lead to a post-human attribution of artefacts' responsibility and accountability for the forms of violent governance they generate in their vicinity – however modern, complex, and automatic they are. As Carlos Sautchuk summarises, after Simondon: the technical object 'not only performs a reversible mediation, but it also comprises a paradigm of the relationship between living beings and the environment and a model of the collective relation' (Sautchuk 2019: 181). It might mean investigating whether their technicity, as *situated* ecologico-historical relations materialised and stabilised within their functioning, privileges some modalities of actions that emphasise forms of isolation, alienation, exclusion, and disempowerment.

That is why I might end up still missing my old individual printer.

Acknowledgements

This paper owes much to discussions with the students enrolled in the course of 'Anthropological Perspectives on Techniques and Technology,' and with the members of the Research and Reading Group on Ontogenesis and Simondon. I am immensely grateful for the comments, suggestions and discussions with Timothy Carroll, Antonia Walford, Raffaele Buono, and Rosalie Allain, without which writing such an experimental paper would not have been possible.

Notes

1 This approach is not entirely dissimilar to W.J. Thomas Mitchell's work on images (2005). However, 'agency' has also be taken into critical examination and debates in archaeology, though a series of papers, either in favour or reacting against what has been called the Symmetrical Archaeology or 'New Materialism' (Witmore 2014; Lindstrøm 2015, 2017; Ribeiro 2016; Sørensen 2016, 2018). While I am also advocating a focus on objects in themselves, I feel my position being analytically different from Graham Harman's Object-Oriented Ontology (2018), which can be construed so post-human that it runs the risk of underplaying the role of vernacular theories of language and praxis. This is perhaps because of the fundamental Heideggerian quest of [universal] ontology and a lack of anthropology, in contrast to Leroi-Gourhan's and Simondon's positions, for whom emerging processes and potentialities are so capital that they approached artefacts as both diachronic and synchronic evolving complex

entities in ecological relations with their milieu. As anthropologists, we cannot reject the relations between emic imaginations of language (categories) and practices (See Geismar this volume for a similar caution vis-à-vis this form of New Materialism). Most importantly, we should not forget that our own modern categories, though dominating the contemporary world, are themselves vernacular. *At this stage*, I am more inclined to examine closer Karen Barad's 'agential realism' (2007), possibly in a cross-reading with what Simondon and Leroi-Gourhan are suggesting.

2 See Yuk Hui's 2013 paper on the intellectual relations between Simondon and Latour.

3 The concept of the 'mode of existence' finds its origin in the philosophy of Étienne Souriau (2015[1943]), whose major influence on Bruno Latour, prompted the latter to call his 'sequel' to *We Have Never Been Modern* (1993[1991]), *An Inquiry into the Modes of Existence* (2013). Fifteen years after Souriau, Simondon published his own work, *On the Mode of Existence of Technical Objects*. There are relations between the two philosophical approaches. Grounded in phenomenology, the concept also deals with question of ontology, but both Souriau and Simondon followed Henri Bergson's views on the primacy of processes over essences.

4 I am using here the term 'technics' instead of 'technology' because of its epistemological fuzziness (Simondon 2017[1958]: 9, n1; see also Schatzberg 2018; L. Marx 2010[1997]; Coupaye 2021), which confuses 'the theory or study of industry and of the mechanical arts' with actual devices, skills and practices engaged in technical activities.

5 Ancestors of contemporary complex mechanical things go back *at least* to the 2nd c. BCE, with the Antikythera, but also include siege engines, Renaissance fountains, theatre installations, as well as fourteenth-century clocks and pre-industrial water and wind mills (see Gille 1986[1978]; Brun 1985; Bedini 1964). Arguably, Gell's little girl's doll is also a mediator but of a different kind of relations than those of, say a knife or a printer, and more akin perhaps to parental care (I am grateful to Timothy Carroll for pointing this out).

6 In these respects, the embeddedness of technical objects within human lives is not enough to analytically define 'post-humans' or 'cyborg.' But the reverse might correspond to a particular vernacularity as I can read through Timothy Carroll and Aaron Parkhurst's Chapter 3.

7 Compare with Jean Baudrillard 1996[1968], whereby technics are analogous to a language [structure] and consumption to speech. To push further this idea would require re-investigating McLuhan's classic study (1995[1964]).

8 For the sake of space, I will not deal here with the complex concept of 'technical individuation.' See Simondon 2017[1958]: 25–26; 54; Barthélémy 2012: 213).

9 Information is seen here 'as the arrival of a singularity establishing a communication between levels of reality' (Simondon 1964: 130). But note also how 'energy' and 'information' are also categories vernacular to modernity.

10 But see Hui 2015 and Rodriguez and Blanco 2017.

11 It also both generates and requires a general technical milieu which frames the reproduction and appearance of new technical objects in specific directions and along specific paradigms, only rarely shifted by true inventions (see Leroi-Gourahn 1971[1943]: 376–394; Simondon 2005). Hence, the limit posed by Moore's Law on the miniaturisation of processors, might indeed require the shift to quantum processors, instead of the electronic ones.

12 Importantly, but outside of the scope of this chapter, they also contain within themselves potential for other emerging forms and relations (Simondon 2017[1958]: 25–81).

13 In the summer of 2013, UCL Information Services, as part of a contract with Xerox, implemented the print@UCL service across the university, both in student and staff areas, removing their support to individual printers that staff had in their office.

> This service uses 'pull printing' this ensures that users have the conscious choice to release their document at the screen of the print device instead of it just printing out automatically as soon as the user submits it from their workstation.
>
> (UCL ISD, email of the 03/04/19)

Alongside concerns for sustainability, the choice also included financial considerations (a contract with Xerox). By the time this chapter reached its final form, in 2019, old printers were actually replaced by new ones, whose software maintenance is done at a distance by Xerox, removing the ability of the IT person in our department to intervene on issues stemming from connections or settings, which have become automated and unalterable by any users on site.

14 I am not dealing in this paper with the ways in which technical objects can form ensembles or technical systems. However, extending the reflection would need to shift the analytical scale up to wider [socio]technical systems.

5 The object biography

Adam Drazin

Whatever happened to commoditisation? *The Social Life of Things* (Appadurai 1986b) opened up the anthropological study of material culture in ordinary, everyday life in an important way. It contained Appadurai's much-quoted exhortation to 'follow the things':

> We have to follow the things themselves, for their meanings are inscribed in their forms, their uses, their trajectories. It is only through the analysis of these trajectories that we can interpret the human transactions and calculations that enliven things. [...] it is the things-in-motion that illuminate their human and social context.
>
> (ibid.: 5)

Not a specific methodology, nor part of an interdisciplinary field of MCS, this perspective can be used in anthropology to inspire critical analyses of cultures of value. Appadurai's essay is accompanied by Igor Kopytoff's insightful piece, 'The Cultural Biography of Things,' which traced oscillations in the social lives of objects, between states of cultural specificity and generality. When things are possessed by certain people, they present particular, even particulate, identities and relationships; whereas in other phases they become more generally exchangeable, like emblematic types. The generic and the specific were also explored by Daniel Miller's *Material Culture and Mass Consumption* (1987) which, following Georg Simmel, characterised culture as a process of objectification, between the problematic, overwhelming mass of infinite things and the personal intimacy of possessions.

These works opened up new methodological possibilities for the study of culture. Anthropologists study culture, but if culture does not reside only in humans there are ever more ways to investigate it, beyond sharing moments with people during ethnography. Nonetheless, there have been very few attempts to study the social lives of specific objects, and few have considered what this can and cannot achieve. Most follow-the-things approaches have looked at general classes of thing or things as they are experienced by people, not as individual objects. Perhaps anthropologists feel more comfortable when things are generic and fall into fairly clear categories. This means we can avoid challenging certain conceptions about culture as a patterning of meaning, and the capacity of persons to be unique and specific against the background of a typical world. The exciting prospect of examining the lives of particular objects was also to some extent stopped prematurely when Appadurai (1990) published his analysis of modernity, emphasising the idea of global 'scapes,' or flows, and the disjunctures between them. During the ensuing debates, Marcus (1995) set out several kinds of follow-the-thing approach: the commodity chain (Wallerstein 1991), the political economy of a commodity (Mintz 1985), studies of globalised taste or style (Feld 1994), and Latour's actor-network mode of critical inquiry. None of these focus on individual objects' lives as a priority.

The methodological opportunities of following things was thus co-opted into grander projects aiming to theorise culture. Writing object biographies does not necessarily advance a particular theoretical agenda in the way that Marxist or Structuralist methods support a particular theory. Rather, like Abu-Lughod's 'ethnography of the particular' (1991), it presents a critique of culture as systemic. Although object biographies focus on material objects, they demonstrate the insignificance of things as much as their significance and can lead to a range of ways of framing the culture concept, as this chapter will later discuss.

We have many ways of following things, but very few published studies of specific, ordinary objects' biographies in anthropology, unlike in disciplines such as archaeology and museum studies. One exception to this is Stallybrass's (1998) extraordinary commentary on Karl Marx's ideas of commoditisation and alienation, through tracking the social life of his overcoat. Karl Marx pawned his coat when he needed cash and then recovered it from the pawn shop when he had money again and felt the need of the thing. A smart coat was necessary to enter the library where he wrote his most famous works. Yet, ironically, in those same works, Marx painted capitalism as a world of anonymous commodities and one-way commoditisation, not a world where objects are commoditised and re-personalised over and over again. Other exceptions worth mentioning are the anthropology of art, of craft, and of heritage, each of which often examine individual objects, as does European ethnology, which commonly explores folkloric heritage artefacts. Yet these fields tend to examine things of communal or transcendent value, and handmade things, rarely the banal mass-produced objects that typify commoditisation and the intimate quotidian lives of most people. Fragmentary object biographies also feature anecdotally in many ethnographic works, but such partial biographies are asides, rarely recognised as a deliberate method.

This chapter discusses the Object Biography, an intellectual exercise examining and describing the social life of one particular object. To write an object's biography is a critical methodology, a deliberate tool more than a paradigm. It is a method in that it is an action, something for a researcher to do intentionally to build research information. When I assert it is critical, this contradicts the commonly held presumption that critique must be an act of theorising. Methodologies can be critical as well. Object biographies do not simply lay out chains of facts for people to think with. They deal with truths more than facts, like human biographies, histories, or the clandestine genre entitled 'true crime.' An object biography is a deliberate methodological undertaking aimed to raise the critical perspective and alternative frameworks for understanding culture.

Each year anthropology students at UCL write object biographies as a part of a course on mass consumption, and I draw upon these here. I describe one particular example, a biography of a toothbrush. This example allows us to examine what kinds of issues an object biography is good for thinking about.

What is this method?

Writing an object biography involves choosing a specific thing and describing a range of different moments or periods in its life. One quickly realises that every object is very different, and the ways the stories emerge and are written are different. Fifty cans of Coca Cola provide at least fifty separate biographies, so one must be selected. The story before arriving in a shop may be similar, but one can becomes part of a packed lunch, another a party, another is thrown away, another gifted, another damaged. It is impossible to know about every period in an object's life – you were not there – and so these are life 'stories.' This is part of the exercise, since moments and phases in a thing's life are witnessed and experienced by different people. These communities, people, places, and purposes comprise separate social environments, each of which makes sense in its own terms. When an object's biography narrates life phases, such as when it is made, when it is in a shop, or when it is in a home, that journey cross-cuts separate human worlds, and so some periods are often imagined, assumed or suggested. It is always preferable to write based on observation but, at any point, other past and future moments and social domains may appear as inferences, such that the imagination of a social pluriverse across the many moments in the life of a thing features large in any narration.

As the chapter by Jeevendrampilai, Burton and Sanglante (Chapter 7) makes very clear, the lives of objects necessarily happen in parallel with the lives of humans (Tilley 2007). A biography describes uses, identities, experiences, and relationships that different phases of its life enable. The thing might be at times a tool, a gift, a product, a good: such words describe its social state in relation to the conceptions of people. Often there are phases and states that are unique. The biography considers what aspects seem important at these different times, and how it represents something that seems like a general type or specific thing. The term 'biography' then is important because we are not only interested in setting out a sequence of events, or how objects age and develop over time, but in the succession of their social states. This is similar to the way that the biography of a person will not place equal emphasis on every year of their life, but consider momentous events and key achievements or difficulties in life that are particularly significant. A consideration of value(s) is especially important, because of the importance of moments when a thing moves between being unique and being a representative type. At such moments, we witness the interchangeability of things, as carriages of meaning and as mutually exchangeable.

Being intrinsically temporal constructions, objects' lives also feature futures. We do not know exactly what lies in the future, but because objects have effects (agency, we might say, although agency is temporally uneven), they seem to obligate certain futures. In the current environmentally conscious era, for example, many objects oblige people to dispose of them in certain ways, shaping peoples' actions. The futures of possible pollutants lend them a palpable sense of danger. Similarly, many objects have the sense of only being exchangeable under certain circumstances. The future of an object in a shop is to be bought or otherwise stolen, constructing ideas of theft. The futures implied at different moments in an object's life are suggestive of value, of morality, and proper actions, and how people ascribe these to the object. These as-yet-unknown futures form a legitimate part of biographies and illustrate the force of their lives, based not upon facts but on apparent temporal truths.

An object biography is mostly descriptive. Description is not intellectually neutral; it is a purposeful act. In MCS, objects, things, and materials are understood as active in human knowledge processes and essential to culture. There are multiple paradigms for this: culture as objectification (Miller 1987); Simmel's (1971[1903]) notion of objective culture; Gell's (1998) and Küchler's (2013) conceptions of the distributed mind or, alternatively, Malafouris' (2013); Knappett's (2005) locating of meaning in the affordances provided by networks; material culture as text-like (Hodder 1989); or Henare et al.'s (2006) arguments about things as thoughts in themselves. Bourdieu (1977, 1984a) should take most credit for making the point that material objects are intellectually active. Following these authors, to describe material culture is also to model, offering possibilities to test ideas and interpretations. Every anthropologist should be a competent observer of minutiae, but we are trained to value the grand interpretive statement above the more challenging and potentially more critical descriptive mode.

In spite of the range of work mentioned above, cultural theories of commoditisation and its diversity remain inexplicably neglected (excepting Miller and Bourdieu; and see also Hart 1982). Ethnographic methodologies often maintain an implicit fiction that material objects have arrived in a place in order for people to build cultural ideas around them, and the idea of the commodity remains culturally passive.

An object biography is therefore very different from other methodologies but can overlap with other methods. It is not a design history, commodity chain, or a structural analysis, each of which discusses generic classes of things. It is not a straightforward ethnography, which emphasises contextual and contemporaneous understandings. So too do phenomenological approaches that examine the 'emergence' of separate things into human awareness. It is not a *chaîne opératoire*, which scrutinises how things come to be as they are. Nor is it an actor-network approach, which incorporates many different agents, actants and objects into its scope. An object biography, by contrast, is focussed on an object, rather than persons, examining a specific life of a specific object over time, and is more about exchange and transmission than it is about productive work.

To illustrate, let me share with you the story of Anna's toothbrush[1]. It is retold here in my own words and, as will be clear, it became the story of several brushes' lives. Anna did not discuss the making of the object, but began with her own first encounter with it; and she only implied its potential future life. She focussed in on the brief period when a toothbrush encounters its 'significant other,' a particular person. Toothbrushes are very culturally active and visible for a brief time, generally only months. They can exist for many decades afterwards, and travel extensively, but their time 'in the light' of human life is all too brief.

The life of a toothbrush

The toothbrush was bought cheaply in a discount supermarket in London, one of two in a packet. The supermarket was selling 'two-for-a-pound' toothbrushes, one coloured pink and one blue, in each packet. Buying two was actually cheaper than going to a chemist and buying one good quality brush, but Anna was just thinking she needed a toothbrush for a short time, because she herself felt transient while studying in London. She was not concerned about quality, or the branding, and gave no consideration at all to the fact that perhaps cheaper goods might imply worse labour conditions or other ethical issues. It was 'just a brush.'

At home, she chose the pink toothbrush. This was in spite of the implicit gendering of the toothbrushes, which would suggest that, like the toothbrushes, people live in pairs, similar in substance but appearing different. Usually Anna buys blue toothbrushes. That has been her general habit, rejecting the 'pink-for-a-girl' notion but, in this instance, she felt the pink one was 'more her.' Hence the pink one was installed in a mug beside the bathroom mirror. She lived in a rented student accommodation in which the toothbrush was one of the few objects that was 'hers.' Like in a hotel, she reflected, where the toothbrush is one of the things you bring with you; toothbrushes are an indication of 'where you are staying,' carrying with them the intimacy of 'home,' while not constituting home as such. The toothbrush indicates a certain kind of dwelling and inhabitation, and the presence of a toothbrush changes a place.

The new pink brush displaced her previous (purple) brush. In London, one person uses one brush. This unexpressed one-to-one principle only becomes manifest in the comparisons made between brushes. The purple brush, yesterday seeming very clean and functional, suddenly looked old, worn, and useless by contrast with the pink. Nonetheless, she did not throw it away, haunted by a notion that plastic toothbrushes should be disposed of responsibly, as if every toothbrush has a turtle's name stamped on it. She also had a sense that the purple brush was still 'hers,' intimate to her, and that perhaps it still might have a use, perhaps for household cleaning. Even when demoted to pseudo-rubbish, the purple brush still exercised a feeling of possibility and obligation over her actions. The difficulty of throwing it away is a sign of the times, and the public consciousness about plastics as 'polluting,' but there was as yet no clear knowledge of how to properly dispose of an object like this, made from several compounded plastics. In the 1990s, it would have been easy to simply put it in the bin. By the 2020s, it is possible that it will, again, be unproblematic, if systems for proper plastic disposal exist. Yet in 2017, this purple toothbrush resisted disposal, and became mere context for the pink brush.

The pink toothbrush travelled with Anna. It went to France, mediating the relationship between Anna and different places, producing flexible and short-term moments of dwelling. The presence of the brush in friends' bathrooms and hotel rooms established a temporary anchor as she visited, and it was one of the things that enabled her to make those spaces, for a short time, her own.

The blue toothbrush, the twin to her pink one, Anna gave to her boyfriend to use. While he had stayed over before, the presentation of his own toothbrush to keep in her apartment felt like a significant moment, a small step from mere intimacy towards a slightly longer-lasting kind of relationship. It is difficult to see what other kind of object might perform the kind of relationship achieved in this action. It is also difficult to specify in words exactly what kind of relationship this act of giving achieved. It is a relationship between people who give one another toothbrushes to keep in one another's homes. Sometimes objects, and their lives in exchange, can be much more subtle and specific than language, and communicate for themselves.

At a later date, a different friend of Anna's stayed over after a night out, a girlfriend. Faced with a choice of toothbrushes to offer her, Anna decided the best one was the pink one (Anna's own), rather than the blue one (her boyfriend's), or the unfortunate purple one. Her girlfriend was unaware of the history of these toothbrushes. It made no difference to her, but for Anna there was a lingering sense of intimacy in the rinsed and washed plastic bristles. Anna preferred that she herself use her boyfriend's brush, which had been at times infused with his saliva, and that her girlfriend used the brush which had been infused with Anna's. This also seemed like the appropriate or polite way to use the brushes, an arrangement that would suit everybody, and nobody would mind.

Reflections on the life of a toothbrush

This narrative illustrates some of the ways an object biography can be used effectively. Although it does not depict the entire life, it takes account of the long-term through implication, of manufacture and of its life as rubbish (beyond 'disposal'). Importantly, even in a short space of time, the narrative considers several moments of exchange or exchange-like actions. These are not all moments that correspond necessarily to grand structures of value (spheres, following Appadurai), such as the transition from shop to home. There are also moments of exchange that are particular and intimate, and some of which seem transient, while others could be permanent. Other moments involve value transformations and are suggestive of exchange-like actions, but are not actually exchanges between persons, such as the displacement of one toothbrush by another. In such instances, culture is emergent and unpredictable, and yet still apparently patterned with meaning.

The narrative focuses on the object, but it never falls into the trap of thinking that objects live their own lives independently of people, nor that people are fully in control of them. The toothbrush achieves social and cultural effects, states, relationships, and identities that other objects might not, and it acts 'back' upon people, so to speak. There is serious anthropological content here: at the heart of anthropology are human relationships, but it is mysterious how one can actually build a relationship with another person. Equally mysterious is how relationships produce certain kinds of persons. In this instance, one way in which human relationships are built becomes evident. After Anna gives the blue brush to her boyfriend, their relationship is different from before. It is a small but significant reflection upon the 'substance' of enduring social connection. I would contend that in this instance describing the life of the toothbrush is more effective than using words to describe the relationship directly, as if the relationship were a thing in itself rather than a compounded cultural phenomenon built from experiences and things in motion as well as words. As anthropologists, when we model culture and society it should preferably be done on purpose, rather than an unconscious by-product of our word-based ways of working. Consequently, the way in which one privileges the word 'boyfriend' or the description of the giving of a brush – these matter. An anthropological description can claim to be either an observation or an analysis; to make the same point, but to specify the story of the brush is to make a claim for understanding that simultaneously situates the social scientist as observer, aware of their own viewpoint, while to emphasise 'boyfriend' as a term makes a claim to transcend the immediacy of context.

Importantly, in these engagements with people it is clear through the narrative that the toothbrush itself can move on. While it sometimes has a tremendous sense of human importance, because every toothbrush does have one significant other, the constructed quality of this cultural edifice is also clear. The brush's owner for this brief period, Anna, certainly considers it important, but for many phases of its life she does not feature at all. It may have existed for months or years previously, perhaps in many countries, and it can exist for a century afterwards. While Anna's memories may endure, and the changes in her relationships and experience may be permanent, the brush itself can move on quite easily, unencumbered by memory, regret or attachment. The narrative approach then manages to conjure an 'object perspective' through which it achieves the effect of decentring its cultural perspective away from the subjective point of view of one person who owns it. This viewpoint advances the critical approaches to personhood and self, as the editors of this volume discuss in the Introduction.

This 'object view' is also the anthropologist's view, a narrative device premised on the bizarrely common anthropomorphic idea of objects having 'social lives.' A toothbrush is not alive, but it does have existence and is involved in events. This device enables a positional shift of the anthropologist, situating themselves in articulation with, but separate from, informants. Hence, the object view can be an effective device for some purposes. This 'object view' is often created before the author knows what their analytical or interpretive argument may be, but it opens up a space. Rather than the relating of a single sequence of material facts, or the dual accounts of informant and anthropologist, divided by human difference, the object biography seems to produce mutual truths within a fictional framework – persons' and object's.

The fiction of anthropomorphism, whereby author identifies more closely with object, is often comedic. Much of the story of this toothbrush's life seems ironic or whimsical, even at times when very serious issues are addressed. Within one object, we allude to the destruction of the global environment by plastic, and of enduring human partnership, but these are microcosmic and underwhelming. Object biographies may seem serendipitous, risky, non-normative, even ludicrous, as they conjure their unusual object-centred lifeworlds, and this enables us to question cultural understandings that we otherwise take for granted. As descriptions, object biographies often appear as a mixture of moments and, like all biographies, they may end in pointless death, rather than at the moment of greatest success. As stories, object biographies may have a critical purpose, but it should be fairly clear that this purpose is primarily that of the storyteller.

Importantly, a good object biography does not make solid claims beyond the scope of its observations. It comprises a sample of one. Yet because typicality and specificity are social states, most objects evoke, or are in dialogue with, cultural typicality at moments in their lives. Any interpretation is limited in its scope. This is a method for challenging or testing typicality more than asserting it. And knowing the basis for one's knowledge and its limitations is just as important an anthropological skill as a talent for grand interpretations.

There are many different approaches to the study of material culture. Object biographies offer a distinct method, notable for five characteristics.

(1) An exoteric method

Most MCS use what I characterise as 'esoteric' methodologies. That is to say that interpretation usually involves reading 'beyond' or 'into' material things. There are different ways of doing this: some theorists consider (and critique) a tendency to separate out systems of signification into material 'signifiers' and immaterial 'signified's (Keane 1995, 2006).

The ontological turn (Holbraad and Pedersen 2017) works to collapse such distances. Other approaches consider the indexicality of things for human actions, intentions, purposes, or uses, as well as representation and semiotics (see Carroll, Walford, and Walton, Chapter 1). What many methods do therefore is hinge upon the idea that the interpretation of things means accounting for them in terms of persons, human bodies and minds, which may or may not be present. These kinds of approaches are 'esoteric' in the sense that they treat material culture as a mysterious object that needs to be accounted for in other terms, and therefore strive to create a sense of depth.

Although object biographies can suggest depth, they really come into their own as a methodology capable of collapsing depth into a more straightforward sense of being 'just things.' This is because most objects' lives involve sequences of different social spheres and states. While at any particular moment, for certain people, there is often a sense of deepening profundity and significance, nonetheless equally often the object progresses to a new domain, leaving behind the previous values and meanings.

An object can become very personal, intimate, and special. In many object biographies, people talk of love, of memories of special events that attach to souvenirs, of the inalienability of objects and the unthinkability that the connection of a person to a thing will ever be broken. But the unthinkable happens routinely as objects move to new people, and their past lives are forgotten. This kind of moment is one where the profundity of culture is rendered from deep to shallow. At one moment the thing can be explained culturally or accounted for through its relation to people and, at the next, it may be simply another thing, ready to be reappropriated. Some things retain human connections, many do not. Context evaporates. Such moments are exoteric in the sense of not being mysterious and deep, but self-evident and shallow, but nonetheless just as culturally operative and worthy of anthropological investigation and understanding.

The exoteric effect of biographies may be most evident around personal possessions, but similar processes happen around brands and objects in shops. To take just one brief example, one UCL student related how a pair of special branded Japanese socks, whose manufacture was inspired by the traditional techniques and knowledge of Czech grandmothers, and which evoked a sense of home, came to be used as cleaning cloths.

What this amounts to is that if you want depth, you may not need object biographies, but if you want to challenge depth, consider their merits as a critical method.

(2) A method that challenges norms

A second, and related, use of the object biography is to question the sorts of categorical norms that interpretation often evokes. In our toothbrush biography, there are several recurrent ways in which a toothbrush's life evokes gender norms but need not correspond to those norms because what happens is not predestined. Queer theory (Graham 2014; Boellstorf 2007) and feminist thinking (Davis and Craven 2016; Strathern 1987b) have worked to undermine and challenge normative presumptions about culture and revealed its creative potential. The many-headed Foucauldian currents in cultural theory have been especially important in this. Object biographies often scrutinise and challenge such norms 'from the inside.' A study of a single toothbrush cannot hope to justify any claims about grand categories like heterosexuality, coupledom, or other identities. But what is described is the constitution of a particular relationship and its unfolding, its expectations and understandings, and the particular way its course is developed and explored. We witness here a constitution of a heterosexual femininity which, in other methods, would appear as a grand cultural category, but here is a single instance of an emergent identity in the micropolitics around the thing. The biography casts a relationship, usually described as normative, in a decidedly odd (but not queer) perspective. Presented with a world of objects that constitute cultural resources people in a sense 'grow into,' but also are involved in creating, roles, identities, and relationships.

An object biography at its weakest describes the social life and then accounts for it using conceptual categories such as 'gendering,' 'masculinity,' 'femininity,' or a host of other grand ideas. A better biography asks how these grand ideas are constructed at a certain moment, and hence questions how they are valid. In a life that moves between generality and specificity, it is often the stereotypical potency of such objects that itself comes to be the resource for new meanings.

(3) A method that crosses human-constituted boundaries

It can be exceptionally difficult for us to clearly perceive the particular cosmologies and cultural worlds that surround us, but establishing some kind of objective perspective on domains of cultural existence is one of the prime exercises of anthropology. Breaking open such worlds is especially difficult around banal, everyday objects. In the study of consumption, there is also a prevailing and dominant 'consumer' narrative. The destiny of things in this narrative is to be chosen, possessed, and consumed by people, and to form building blocks in the domestic shells that constitute their social existence.

Many material-culture methodologies serve essentially to support the idea that objects' most significant role amounts to participation in singular arrays and contexts that have persons at their centre. Object biographies, by contrast, very deliberately construct a framework in which these boundaries and contexts are pierced, rendered permeable and flexible. Hence, it becomes possible for the analyst to be able to gain some traction in stripping away and dissecting these webs of significance, to paraphrase Geertz and Weber. As Jeevendrampilai, Burton, and Sanglante (Chapter 7) make clear, in moments of exchange objects can evoke networks of relationships and problematically cross the boundaries of human worlds and, by doing so, play with preconceived notions of commoditised and non-commoditised relationships.

The clearest example of crossing boundaries for most objects is when they transition from making shops to making homes, from institutional property to personal possessions. Yet this is only one phase in an object's life. A biography that tells us about how something is bought, consumed, and ultimately becomes a part of one's identity, and then ends at that moment, has probably only told us what we might already have guessed. The narrative becomes more about an individual's psychology, and at the expense of social understandings. Alternatively, we can frame that consumption story by examining the institutional worlds of factories, of shops, and of waste processing. In many shop cultures, people present themselves as 'staff,' wearing uniforms, and spend their time arranging and replenishing goods in displays that make the objects look similar and anonymous. These finely tuned cultural environments prepare goods for the possibility of personalisation and maximise their capacity to move between contexts. Most objects pass through many hands in their lives, moving between commerce and homes, between homes, between generations, in and out of heritage centres, exhibitions, and performances, between countries and across landscapes. National borders move around them, and geography changes.

Travelling through social worlds, a sock becomes a cleaning cloth, a rose becomes a perfume gift, a cardboard electronics box becomes a bed for someone who is homeless, a tacky holiday souvenir becomes your grandchildren's family heirloom (see Frow 2001). In examining the transcendence of boundaries, the parallel social pluriverses are counterpoised and exposed. Some cultural models conceive of a world connected by cultural flows, networks, or relationships. A biographical approach, however, comes into its own in a world constituted by partitioned bubbles (Sloterdijk 2016) or microcosmologies.

(4) An intimate methodology

When it is a commodity, an object represents a generic type of object, and it also reflects persons as types, such as when a person is a 'consumer,' a 'middle-aged male householder,' or a 'driver.' When the same object is a personal possession, it becomes irreplaceable, something no other object really can duplicate, and which fits into a particular world of objects. Specificity is a dimension some methodologies explore well, including ethnography, but not others, such as semiotics.

In examining the specificity of material things, object biographies cast things in a very intimate light. They are often close to home, capturing moments when the treatment of material things is (often unexpectedly) charged with the force of obligation, the requirement to do the right thing. In the toothbrush biography, we saw that all three of the brushes, the pink, blue, and purple, each oblige certain behaviours. Using the wrong toothbrush or allowing others to use the wrong brush feels forbidden; taboo. This quality of encountering an unexpected sense of taboo is not uncommon in object biographies. One way of thinking about this is to consider it in the light of Mary Douglas's (2002[1966]) famous analysis of dirt, as 'matter out of place.' Almost by definition, an object biography appreciates the movement of objects through space and how, at different moments, the social boundaries and rules change around them. There are moments when an object has to find a new normative position within a new social order, and the interesting thing is that the rules and social orders seem to change periodically and of necessity. The potential closeness of matter in an object biography can give an intimate sensation of potentially unpleasant self-awareness.

Appadurai (1986a) reminds us of Simmel's observation that the desire for objects sometimes lies in the degree that they seem to resist our possession. This observation does not only apply to single subject–object relationships, but characterises the world of the metropolis (Simmel 1950). As people pass through intensely occupied urban environments, the object world is pervaded by senses of possession and resistance. A complicated ballet governs as people pass, touching walls, not touching cars, selecting objects in shops, gazing at displays from a distance, brushing against other pedestrians, talking, not talking. Through this humanly constituted world of sensation, material obligation, and resistance, objects traverse and violate in ways utterly unlike humans would. Many objects are intimate, coy, and resistant one moment, and unexpectedly close and personal the next, testing social taboos.

(5) A methodology for the politics of the Anthropocene

If there are reasons why object biographies are good for examination of the intimate, private, specific, everyday life, what Miller (1987) calls the 'humility' of objects, they are also a methodology to examine contemporary public politics (see Carroll, Walford and Walton, Chapter 1, discussing participatory politics and temporality). The Anthropocene, of course, refers to a period of global history in which the nature of the globe is constituted and defined by human action. As a scientific phenomenon, the Anthropocene describes how measurable changes in the world can be explained as being caused by human activity. As a cultural phenomenon, the Anthropocene describes the global recognition of this fact and, consequently, a re-orientation across human cultures of what is 'natural' and how to relate to it.

The object world in this framework comes to appear as a set of manifest resources and materials (Drazin 2015). The plastic of toothbrushes thrusts itself into peoples' attention, for example, recognisable and meaningful alongside the object's role as a useful tool for certain short-term purposes. At the same time, objects are subject to evaluation in pan-human terms, according to parts they all play, however small, in the history of the globe. Like biographical approaches, the cultural Anthropocene always presumes that there is a moment 'before' the current one, and that there have almost always been human hands working in the world before our own. More importantly, however, we need to have methodologies that ask about what happens to things 'afterwards'? After their explicit and conscious role in human lives, what happens culturally to objects?

The object biography framework makes conceptions of waste, disuse, obsolescence, repair, and renovation more accessible to anthropological inquiry. Evidently, often it is speculative or fictional, as we have discussed, but it is nonetheless an important methodology to consider.

Conclusion: Why not object biographies?

Every object has a story, and there are many I wished to share in this chapter but have not been able to. The paper clip, the hair band, the seemingly immortal cardboard box, the inconceivably kitsch resin dog, the unsellable handmade shawl, the printer paper, the disposable razor, which is not. None of these biographies remotely resemble that of Anna's toothbrush. Biographies are written in the first-person voice (without being autobiographies), and in the second and the third. They are poetic, dry, witty, tragic, flat, ordinary, and frequently overblown. Every object's story seems to compel, and deserve, attention, because the richness of human experience only really emerges through looking at the material world in its detail. According to whichever object is selected, it offers the means to explore and question personhood, the body, identity, charity, architecture, property, suffering, memory, or a hundred other things. One of the methodological contentions of the renaissance of material culture writing in the 1980s and 1990s was that the project of building broader, culturally comparative interpretations, and the project of observing with clarity the minutiae of everyday life, should be mutual. The big picture and the miniature, the novel and the vignette, are different exercises in scale but not in critical essence.

At times, grander theorisations can obscure as much as they reveal. This has been the case with studies of exchange. A potentially infinite range of social and cultural events, relationships, modes of building identity, experience, memory and purpose, have tended to be condensed into ideas of commodities, gifts, and discourses of the consumer, the market, and the state. The language of values is much richer and more flexible than that of exchange.

Object biographies are not a catch-all methodology. They will not produce a major theory. They are a unique methodological tool, with a specific critical edge, to be used with some awareness alongside other approaches. They challenge thinkers to observe, and to acknowledge the extent and basis of their own knowledge. While in theory it is possible to write parallel comparative biographies, in practice one finds that mutual touchstones between two narratives seem rare and serendipitous. I have yet to read a single object biography in which the life of a thing unfolds in a predictable fashion, and I am tempted to suggest that stereotypical objects or categories of objects simply do not exist for the purposes of studying culture. If nothing else, writing an object biography or two will help shake up your thinking and compel the realisation that 'things' are not as you thought they were.

Acknowledgements

I would like to thank my many students – including Anna – for their creativity and ingenuity in exploring the possibilities of object biographies as a methodology, and some of whose work is mentioned in passing.

Note

1 I am deeply appreciative of Anna's willingness for me to draw upon and elaborate her original object biography. Anna is a pseudonym.

6 A new instrumentalism?

Haidy Geismar

This chapter explores the relationship between digital anthropology and material culture studies, using the 'digital' as a useful tool to unravel the entanglements between theory, method, and the production of value around social research, both inside and outside of the university. The movement of digital anthropology from university-based research contexts into other, often corporate, research environments (where it is often transformed from digital anthropology into digital ethnography) is a good place to explore the real-world consequences of academic theories about the relations between things and concepts, structure and form, media and sociality. If Miller and Haapio-Kirk, ask in this volume how anthropology can *matter* and to whom, here I explore some of the analytic and interpretive tensions surrounding the ways in which digital anthropology and digital ethnography are entangled when they move from academia to become objects in the world.

The conflation of subject and object, theory and practice, form and content, which has been explored within the broad church of material culture studies (MCS), might be seen to find its apotheosis within the digital, which is simultaneously constituted as a context or subject for social research, as a method (such as digital ethnography or data visualisation), and as an object within social research (for instance big data or social media). I am caught here between celebrating the emergence of a specifically 'digital anthropology' (as articulated by Miller and Horst 2012) and the simultaneous need for critical engagement with the problems raised by the conflation of the digital as both object of study and analytic category (see Rode 2011). Rather than fixate on definitions of the digital here (see Geismar 2012, 2018), my focus in this chapter is to explore the fundamental tension between approaches that locate the digital as an object in the world (like denim jeans or a work of art), versus those that see the digital as a structuring concept (like media or materiality).

MCS is sometimes also described as a 'material turn' in reference to the range of different disciplinary and national approaches that are united by a shared interest in the material within social theory, and in the world. There are distinct threads within the turn. 'Thing Theory' emerged from literary criticism to explore narrative from the perspective of the material world (e.g. Brown 2001; Lamb 2011). 'Actor Network Theory' emerged within sociology and science and technology studies and has significantly worked to demolish ontological distinctions between subjects and objects by exploring the ways in which knowledge is produced within networked assemblages of diverse actants (e.g. Latour 1991). Anthropology's material turn is often recognised as a *return* to earlier preoccupations with 'material culture' that historically were formative for the discipline (Hicks and Beaudry 2010). Within anthropology, MCS therefore signifies both a renewed interest in the specific contexts and materiality of artefacts, but also recognises the made and material world as a fundamental component of human experience.

A further strand within the material turn is *new materialism*, 'a term ascribed to a range of contemporary perspectives in the arts, humanities and social sciences that have in common a theoretical and practical "turn to matter"' (Fox and Alldred 2018: 1). The work of the 'new' here signals a perceived departure from Marxist inflected political critique of 'old materialism,' with its primary focus on modes and means of production and the social structures these produce. If old materialisms argue that objects are lightning rods for the production of class, inequalities and human values, new materialisms seek to move beyond the determining factor of the human, arguing for an analytic approach that understands matter and materials as a priori, in worlds that may not be dominated by human processes of making and signification (see Ingold 2007).

Accompanying the new materialism is a parallel turn from an 'old instrumentalism' in which academic work was in the service of knowledge production in a variety of applied contexts (from the formation of cultural knowledges to assist with the process of colonial governance through to understanding of the other side in war) towards a 'new instrumentalism' of market-driven utility, which Miller and Haapio-Kirk also discuss in this volume.[1] We might link how some strands of new materialism reject the founding humanism of the social sciences and humanities, to the emergence of a new instrumentalism that rejects an 'old' notion of academic instrumentalism of critique – in favour of 'impact': value that can be measured in relation to a baseline of some kind of number (citations, hits, downloads, numbers of viewers/listeners, grant income, and so on; see Collini 2018). Digital media has been a lightning rod for the emergence of this new knowledge economy within the university, for instance in the emergence of digital humanities which, some of its critics declare, 'has played a leading role in the corporatist restructuring of the humanities[, …] pushing the discipline toward post-interpretative, non-suspicious, technocratic, conservative, managerial, lab-based practice' (Allington et al. 2016). Critics of Digital Humanities see this focus on technical problems and machine-based learning as a direct refusal of critical thinking, and also as fundamentally apolitical, chafing against critical enquiry by promoting a 'post-critical,' machine-generated positivism (but see Risam 2009 for an articulation of post-colonial digital humanities).

I argue here that these converging strands of new materialism and new instrumentalism find their apotheosis within many mainstream understandings of the digital, producing a kind of leviathan within MCS and, indeed, within anthropology more broadly. For instance, the shift towards seeing the material world as transcendent matter beyond the human is mirrored in prevailing popular discourses of the digital that interpret data as an objective view of the world, often obscuring the ways that algorithms and data sets internalise cultural values (see Crawford and Paglan 2019). In other instances, the emergence of software, or platform, studies naturalises a corporate infrastructure as the universe that structures activity both inside and outside platforms (see Boellstorff 2008). The inability, often due to lack of research access, to fully account for corporate infrastructures of digital platforms implicitly reorients the participatory register for research in ways similar to those documented by Marilyn Strathern in her account of how the choices around making family through new reproductive technologies naturalise consumer capitalism in determining analytic as well as political subjectivities around modern kinship (Strathern 1992). Within this dystopian digital landscape anthropology is often positioned as a form of critical engagement and a place to evidence alternative practices and experiences from within. However, it remains the case that the digital is increasingly naturalised as both a research tool and object of study, without enough attention to this broader contextual dynamic. In the rest of this chapter, I unpack some of the emergent tensions within digital anthropology that highlight ongoing concerns within MCS regarding the political implications of our theories of things. I then trace how these emerge, albeit often implicitly, within the nascent methodology of digital ethnography. Finally, focusing on a student project by way of example, I highlight the need for a more *applied* critical examination of the nature of the digital itself.

From material culture to digital anthropology

The Digital Anthropology programme was established in 2009 in the Department of Anthropology at University College London (UCL) by Daniel Miller as an explicit continuation of several theoretical perspectives he had already been developing for many years under the rubric of MCS. This new programme marked out a trajectory very different to other emergent fields of enquiry into the digital, which at that time were still largely preoccupied with questions of the virtual and the immaterial (see, e.g. Boellstorff 2008), and which rarely positioned themselves in the explicit context of material culture. In a manifesto of sorts, Miller and Horst (2012) positioned the digital as formally integral to the production of self and society in diverse cultural contexts. Viewing the expressive qualities of digital media within a Hegelian process of producing subjectivity or identity, they defined the digital as a form of mediation through practices of consumption and self-making, building on Miller's early work that had focused on the generative capacities and social resonance of mass-produced commodities (1987). Miller and Horst listed six core themes of digital anthropology, all of which emerge from their earlier ethnographic work on material culture. They articulated commitments to holism, cultural relativism, and to exploring how objects are able to activate dialectics of normativity within particular social worlds. Miller and Horst's vision of digital anthropology *qua* MCS was therefore not one particularly interested in the qualities of the digital as an object per se, but one focused on the ways in which digital media could be seen to refract broader cultural and social worlds and identities, and indeed help us to better understand them.[2]

Miller's earlier work on material culture moved away from the prevailing interpretive frames for objects of semiotics, symbolism, and signification, in which objects were illustrative of deeper concepts that structured social organisation. It also moved away from the tradition of Marxist approaches to the study of objects through processes of production, arguing for a focus on consumption and circulation, and for attention to the ongoing social life of things (cf. Appadurai 1986b). Over a number of studies, and in relation to a large array of different case studies (from denim jeans to Coca-Cola, saris to interior design), Miller and his co-authors have argued that objects and, most importantly, ubiquitous everyday things, played an active (rather than illustrative) role in processes of subjectification (e.g. Miller 1987; Banerjee and Miller 2003; Miller and Woodward 2012). In the same way, Miller and Horst's initial definition of digital technology broke away from several strands of contemporary media theory, responding to detractors of digital media, who present it as a radical break with past media traditions (e.g. Turkle 2012) to argue that digital technologies are fundamentally normative, mediating no more or less than any other cultural expression or communication. They argued, 'one of the major contributions of a digital anthropology would be the degree to which it finally explodes the illusions we retain of a non-mediated, non-cultural, predigital world' (2012: 12; see also Coupaye, Chapter 4).

Horst and Miller's definition of digital anthropology also responded to debates within MCS privileging what might be called a classic version of social anthropology over a broad clustering of perspectives that could be described either as post-humanist, or broadly technologically determinist in their orientation towards the material and digital world (see, for instance, Whitehead and Wesch 2012). For instance, in the multi-sited research project, *Why We Post*, also discussed in his contribution (with Haapio-Kirk) to this volume, Miller and his co-researchers have argued that Facebook (or QQ in China) could be understood as distinct within the particular local landscapes of, for example, Industrial China (Wang 2016) or south-east Turkey (Costa 2016), where culturally and socially defined norms were brought *into* social media. One of the primary findings of *Why We Post* was that 'it's the people who use social media who create it, not the developers of platforms.'[3]

This approach to digital anthropology is quite different to contemporaneous studies of the digital in other fields, which tend to upscale the form and structure of the digital from metaphor to structuring reality and often sublimate the social into a growing commitment to a more top-down technologically and materially led enquiry. Examples of this can be found in influential treatises on the digital, such as Bratton's formulation of *The Stack* (2015), part of the MIT Software Studies series, and is broadly mirrored within other work constituting the fields of media archaeology, platform theory, and format theory.[4] Miller and Horst's definition of the digital also is in tension with the broad approach taken within Actor Network Theory and Science and Technology Studies, which see the social and technological as co-constructed or produced within broader socio-technical networks or assemblages. Today, Digital Anthropology at UCL draws from all of these trajectories – with research focused on the anthropology of data, the materiality of the digital, the infrastructural quality of digital networks, and the politics of digital forms as evidential categories, as well as the communicative and expressive capacities of networked digital communication (see Walford, Coupaye, and Knox's contributions in Chapters 15, 4, and 8).[5] However, the peculiar location of digital anthropology within broader debates about material culture continue to both intrigue and confuse students, forging ongoing debates about the relationships between form and content, scale of analysis, and location of criticality within digital projects. These questions play out most explicitly in the emerging disjuncture between a critical digital anthropology, and the formulation and representation of research through digital ethnography.

From writing culture to digital ethnography

As the material turn broadly explores the role that objects play in the constitution of social worlds, sometimes even from the perspective of objects themselves, it is interesting how little critical thinking about ethnography within anthropology connects to the emergence of digital ethnography, and the elaboration of visual and other social or 'inventive methods' (Pink et al. 2016; Lury and Wakeford 2012). The set of reflexive explorations of anthropological representation precipitated by the publication of *Writing Culture* (Clifford and Marcus 1986) explored the nature of anthropology as a representational practice and challenged anthropology's claims to objectivity by situating and deconstructing the production of ethnographic texts, and later turning attention to visual practices of film and photography. For some reason, this critique is rarely discussed in relation to the form of digital ethnography, even as MCS highlights that representational technologies have the power to alter and affect that which they are trying to capture (see Marres et al. 2018: 24). With the exception of a few key examples – such as the theatrical, exhibitionary, and media lab work of Bruno Latour (e.g. Latour 2005; Venturini et al. 2017 or some of the work of the Harvard Sensory Ethnography Lab (e.g. Castaing-Taylor and Paravel's cross-over 2014 film *Leviathan*) – multi-media anthropology remains on the margins of the discipline and the digital remains largely unpacked in terms of its representational conventions – even as digital tools for anthropological research proliferate.

Clifford's genealogy for *Writing Culture* (Clifford 1988) drew on anthropology's links to twentieth-century modernist social theory, in which representational forms, such as writing and photography, were linked to other aesthetic avant-gardes in art and literature. Today, 'inventive methods' (Lury and Wakeford 2012) are also positioned at the heart of experimental and creative avant-gardes in the social sciences. Yet the emergence of digital ethnography increasingly draws not just on bespoke or artisinal technical projects (often funded for the short term by research grants), but by necessity on more commonplace digital forms: blogs, social media, databases, and algorithms, many of which have been developed primarily to produce value for corporate shareholders through the production of data related to consumer preference, ultimately to generate revenue through advertising. What then are the implications of these generic forms on the structure as well as on the content of our work as digital anthropologists? If, as Lupton (2015) observes, digital ethnography is simultaneously a form of research, analysis, and criticism, what effects does the black-boxed political economy of social media and its accompanying ingrained culture of obsolescence have on our work?

A case in point: learning from social media in the field

Digital anthropology is, for me at least, a site of struggle between totalising views of the social as they are constituted within anthropology, and as they are constituted through increasingly generic digital media. In the rest of this chapter, I use as a case study an experimental student project at UCL designed to explore the analytic limits of social media and to highlight how the methods and perspectives of material culture might enable a greater reflexivity within digital projects. By exploring the successes and failures of a student project, I want to briefly look at the ways in which understanding digital media as method not only provides an avenue for forms of social and participatory methodologies but, perhaps more importantly, provides a place for discussion of the relations between form and content, and between theory and ethnography. This is less a 'how-to' primer for utilising social media in anthropological research and more an exploration of how much work we need to do in order to understand the interpenetration of social media and digital methods within anthropology.

The project in question was built around the use of ubiquitous social media platforms. As will be seen, unlike other research projects focused on social media that present social media as an enduring part of people's social worlds (e.g. Miller et al. 2016), the ephemerality, contingency, and fallibility of these projects and their status as partial successes and partial representations, demonstrate some of the false dreams of digital and ethnographic holism as they are packaged up in alternative research environments. The project illustrates a dissonance in the ways in which it is often assumed that social networks map onto digital networks, and it unpacks the nature of digital participation and co-production as well as contributes to the un-black-boxing of technical expertise and skill around the use of everyday platforms that have become so ubiquitous that they are considered unskilled or banal by their users (see Geismar and Mohns 2011). Finally, observing the ways in which social media is transformed into method enables us to place digital practices and methods in specific cultural, and local, context.

Social media as community?

Our methodological exploration of social media was undertaken in collaboration with Dr. Rebecca Ross (University of the Arts: Central St Martins) and students from the 2013–2014 cohorts in the graduate degree programs of UCL Digital Anthropology and CSM Graphic Design and Communication. We proposed to link our students to two community associations in London in order to explore the capacity of social media to build community in particular neighbourhoods, with the intention that our students would 'hack' social media and push it to its limits. The aim of the project was to explore generic, popular social media platforms focusing on the free platforms that were the most popular at that time – YouTube, Twitter, Facebook, and Instagram (one group also worked with a fitness-specific social networking platform) – doing so in a way that challenged the conventional imaginaries of 'social networking,' using the trope of hacking (as a form of non-normative interventionist participatory methods of active engagement with form) to investigate our definitions of the social, of the network, of communication, sharing, neighbourhood, friendship, participation, authorship, ownership of platforms and content, and so forth. Our brief to the students made it very clear that there are multiple perspectives on social media, the stakes of which can vary across projects and fields of enquiry. We wanted them to be able to critically engage with social media and its conventions in the same way as they had been trained to critically unpack other kinds of media (e.g. documentary photography and film, ethnographic texts).

Our students teamed up into groups of four, and we partnered each group with two community associations in the London Borough of Camden. Working together, they determined the needs and expectations of the association and then brainstormed how to set up a social-media network for them to use over the summer of 2014. The aim was to co-produce with community members something that was at once productive to the community associations, creatively considered, and critically engaged with contemporary social media.

Community associations occupy a liminal place in the landscape of public culture in the UK. Over the past twenty years, community centres have largely shifted from being social or public services to independent, self-governing charities, as part of a broader scaling back of national support for local government and civil society. While community associations, as they are now called, are funded in part through the local authority, this is often with short-term or block grants, and they are expected to gain some kind of financial autonomy. This money is increasingly competitive and hard to access, and direct funding from government has been cut radically over the past ten years by the Conservative-led national government. Associations are expected to fundraise through social enterprise and by making connections to local businesses. It became clear throughout our project that the community associations we worked with saw our project as a way to participate within a broader agenda in the UK to provide social welfare and community services online. It was also clear they were very aware of the role of digital technologies in implementing the conditions of austerity in regards to public services, where online services were understood as both liberating and democratising, and as a way to transfer responsibility and activity away from state welfare services.

Here, I focus on what happened at the Castlehaven Community Association (CCA) (www.castlehaven.org.uk), an independent charity set on four acres in Camden Town just opposite the cluster of Camden Markets. The railway viaduct divides the neighbourhood, with gentrified railway cottages that are home to affluent professionals and the popular markets on one side and, on the other, a large cluster of social housing estates. Within these estates, despite the economic vibrancy of the market, unemployment is a significant issue. Association CEO at the time, Eleanor Botwright, observed that, in the face of little opportunity, there has been a growing sense of apathy from the younger members of the community.

CCA in 2014 had a garden, playground, outdoor sports ground and two main buildings, as well as managing a large tenant's association hall on the nearby public housing estate. The Haven housed an Internet radio station, recording studio, dance studio with piano, and well-equipped meeting room used during the day as a space for Age Activity (for older people), and in the evening and school holidays as a youth centre. Another building contained staff offices, a large meeting room used as a drop-in for under-fives on Tuesday, Wednesday, and Thursday mornings, and a meeting/activity space offering learning and social opportunities for Age Activity members. The CCA recognises that their spaces are intergenerational and has increasingly built on that in its programming. Many activities are developed in reference to mass media, for instance building on the popularity of reality television shows such as *Big Brother*, *X-Factor*, and *The Voice*, to develop youth activities.

We went into these projects with a series of issues to explore, clustered around the primary question, 'What does social media really want?,' drawing inspiration from W.J.T. Mitchell's question 'What do pictures "really" want?' (1996). We were interested to explore how much community engagement we could establish through social media. We also wanted to explore whether we could see social media as being part of a neighbourhood, and whether social media could in fact be a neighbourhood. The students' own activities showed us what we can and cannot do with social media, how much social media foreclosed certain kinds of participation, and helped us understand the limitations of the platforms as well as their affordances.

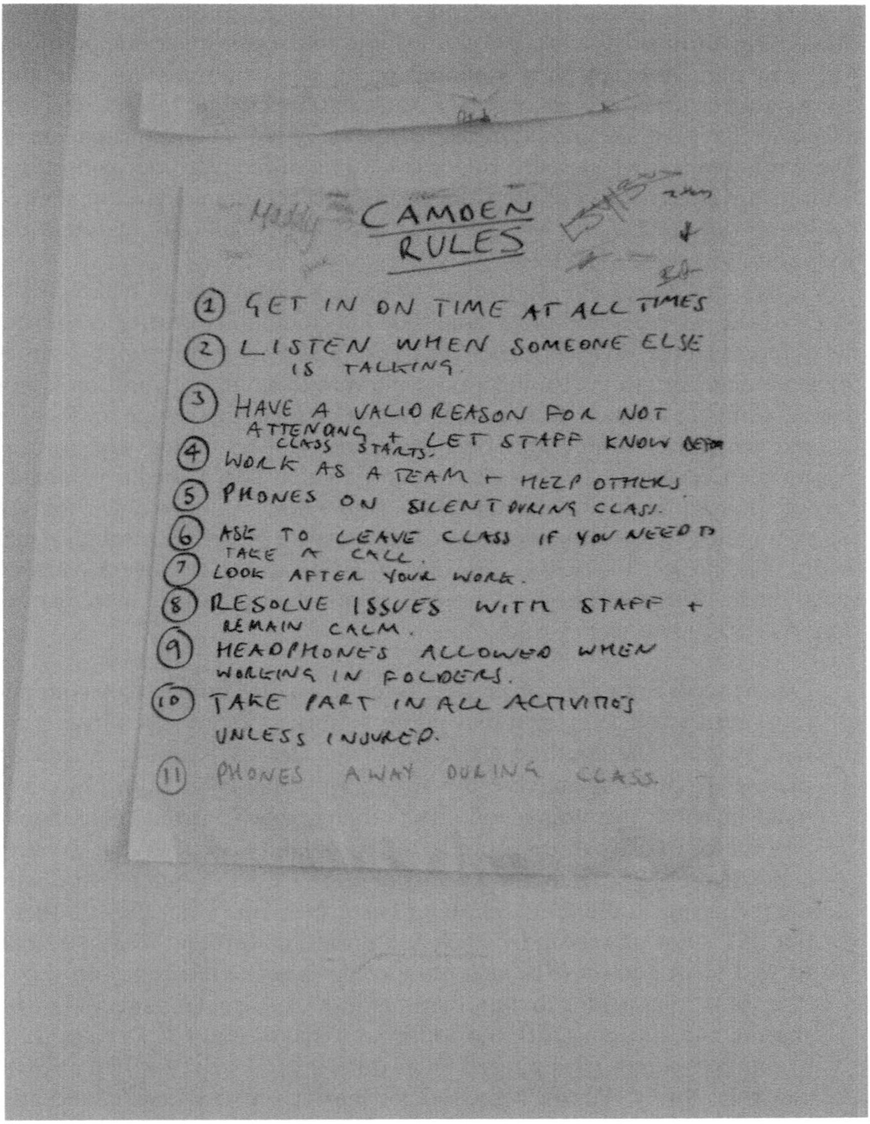

Figure 6.1 'Camden Rules' posted at CCA.
Source: Photo by author.

Camden Rules

One of the first things we noticed during our first visit in 2014, was a list of house rules in the second-floor club house room of the CCA, written by the

Youth Coordinator. The rules aimed to structure the engagement of teenagers in weekend and after-school activities, including the production of radio shows, Big Brother style sleepovers, DJ'ing, and open-mike competitions. Rule five stipulates that phones should be on silent during class, rule nine decrees that headphones can only be worn when working in 'folders,' and rule ten states that everyone should participate in everything unless incapacitated. There is an appended, eleventh rule, written in a different colour pen, which admonishes students to put their phones away. The organic development of the house rules reflects the struggles that the CCA is facing to maintain the attention of its networked teens.

We had challenged our students to leverage social media to make people aware of the activities and programmes of the community centre, to connect the different groups on each side of the track and open up more opportunities for lower-income groups to make people aware that the organisation is not part of Camden Council, but an independent charity responsible for its own income generation to develop social enterprise opportunities. The student project focused on developing the recognition that social media is already part of the community association. Rather than developing rules to limit the use of mobiles, the students developed strategies for inclusion through social media. The project focused on drawing attention to the creative aspects of social media and on social media as a form of attention rather than distraction. As the student team wrote:

> Given that one of CCA's objective[s] is to drive more youth to their community centre we've suggested they adopt a 'noisier' approach to Instagram and Twitter. The youth at CCA loved the idea of having and sharing pictures from CCA activities on Instagram. This takes both objectives hand in hand: increasing visibility and engaging youth at the community centre to drive more youth to the community centre. What we were concerned about was that the youth who use CCA's resources, especially the recording studio do tweet about being there and what they do there, but this is not picked up by CCA and promoted further. What we'd like to see CCA to do over the next few months is better integrate and showcase what is happening in the studios, offline, in a visual manner online by sharing and engaging with the youth on social media. CCA's rules seem to suppress sharing as opposed to encourage it, so we've drafted a set of new rules for CCA's which encourages sharing on social media and also asks them to tag CCA and use the specific #makenoise hashtag to promote CCA as much as possible for the youth as well as other members and stakeholders in the community.

The students redeveloped the classroom rules to make space for social media, encouraging the students to use social media as a form of engagement rather than as distraction, creating their own set of rules of engagement that incorporated rather than resisted social media. It became virtually impossible to critically engage with the form of social media itself, beyond the genres and forms that it afforded. Social media became both the start and end point of the project, the form and the subject, the primary definition of social engagement and the record of it.

One of the issues this project highlighted was the difficulty students had in developing critical toolkits around social media platforms, partly because of the blurring of boundaries that takes place when undertaking research on social media, while using social media as both a tool and as the form of a project. Despite our intentions to experiment with these ubiquitous platforms and to complicate the ways in which social media is mapped onto community, the default mode of operation for both students and community association leaders was to assume that social media is predominantly a form and method of visibility, especially useful in the competitive market-led environment that governs local government and community life in London today. The leaders of Castlehaven saw no disconnect between their desire to expand their social media presence and their desire to disconnect their teenage users. For them, social media was in the first instance a broadcast medium to afford new recognisable forms of visibility (and attendant forms of value that this visibility produces) for the Centre's activities to a wider constituency.

Our students, on the other hand, were being trained in a shared analytic between graphic design/communication and digital anthropology that explored how social media platforms might instantiate, and mirror, *different* kinds of sociality. However, it quickly became clear during the project that they also brought to the table a number of contextual or vernacular assumptions about social media. These assumptions raise a question regarding the neutrality of these platforms or whether their functionality has been designed with particular uses in mind, to the extent that they prefigure and pre-program our own capacities as users.

The anthropology of social media is still in its infancy, bringing together a number of different strands from within anthropology. One brings together classic interests in kinship, relationship building, and other forms of sociality to these new technologically mediated networked platforms (e.g. Boyd 2015; Baym 2010; Miller and Slater 2000). Another connects the longer history of media anthropology to unpack these platforms as mediating forms in and of themselves. This work emphasises the contexts of infrastructure and governance (Larkin 2013; Malaby 2009), the social norms that structure online interaction in platforms such as games and virtual worlds (e.g. Nardi 2015; Boellstorff 2008), and cell phones (e.g. Horst and Miller 2006), and the social conventions through which people inhabit these digital environments (Gershon 2010). Anthropologists disagree regarding the ways in which social media may be understood to either map onto social life or produce an inauthentic or alternative vision of the social (see Costa et al. 2016 versus Couldry and Van Dijk 2015). Most recently, as our awareness of how social media platforms function grows, there is a growing anthropology of algorithms and other digital forms that construct a socio-technical perspective on the ways in which social media mirrors or represents reality (e.g. Seaver 2019).

Within the teaching rooms of UCL and CSM we are relatively free to train our eyes away from the more dystopian aspects of social media: the corporatisation of social media and its role as a large-scale industry. Despite our intensive engagements with creative practices and grass-roots uses, for many people social media represents opportunities to convert strong or weak ties into valuable connections, to understand the self as partial data, and to participate in the processes of their own surveillance and commoditisation.[6] The anthropology of social media, then, seemed to have very little capacity to surprise us. Rather, like a hall of mirrors, one keeps seeing the same forms over and over again. Likes, retweets, and portfolios of friends and followers, like many kinds of cultural capital, can be understood by academics to fetishise and alienate the self in their conversion of personal data into financial capital. Even though many users within social media networks may not be directly engaging with critiques of the financialisation of the self, there is also a general awareness of many platforms as being vehicles for the production of value or using transactions to benefit themselves through the production of a networked form of visibility, that is, promoting their careers on LinkedIn, and creating a branded self (Gershon 2017). It is also true that, as well as importing the value structure of marketing and consumer capitalism into their logic, these instrumental socialities also refract social theories developed within the social sciences to produce what Kelty has termed a 'recursive public' (2008) in the specific terms of social media platforms.

Conclusion: social media imaginaries, failing methods, and the limits of objects

By paying close attention to the imaginaries of social media that emerged within our supposedly experimental project, we were able to understand more fully the assumptions that are developing within both communication design and digital anthropology about social media, as well as within the London Borough of Camden's community associations, who are struggling to keep afloat in the wake of massive budgetary cuts from local and national government. The project showed us how social media is simultaneously an object of analysis and a method through which people use their connections to other people to produce value by making these connections visible. By utilising these forms within our own methods for digital anthropology, are we uncritically constituting a new form of instrumentalism? And if so, what role does MCS play within this?

Marres et al. argue that 'social life is not something that simply exists out there, but [it] is *made*: the very existence of social life depends on specific practices of display, representation, accounting and enactment' (2018: 19). Their notion of 'inventing the social' recognises the creative work that underpins social research methods, and the performativity of research methods in which ‘to experiment [means], first and foremost, [...] to intervene in social life, not necessarily with an instrumental goal in mind but to highlight social formations’ (2018: 29). Here, I have focused on a particular project, in which students and community members tried to make sense of social media in order to explore the interrelationships between some of the insights of MCS and the new methodologies that are emerging within the domain of digital anthropology. The stalemate of method – in which digital ethnography has become a marketable skill, an index of the qualitative, used strategically to produce value within non-academic research contexts – depends on the complex materiality of the digital, which blurs the boundary between form and content, structure and practice, theory and method. I have also posited a divide between anthropological accounts of the social that emerge from within social media, and those that are understood to exist outside – or a priori to – social media. Marres and Gerlitz argue that social media is a place within which people experiment with the social (2018: 253). What we found in our classroom project is that the social was increasingly constrained by the format of social media. While this is a fundamental tension for the material turn – whose theoretical drivers rest on the question regarding the role of form or materiality in constituting sociality and meaning – the tension between these world views is also pivotal to digital ethnography and other digital methods (see Knox, Chapter 8).

Zuboff argues that what she terms ‘the age of surveillance capitalism’ marks the rise of what she describes as ‘instrumentarian power’ (2018: passim) whereby digital technologies are used to shape and mould people’s activities, extracting surplus value at every moment in which human activity can be converted into data – an economics and politics presaged on a kind of extreme behaviourism. Within this top-down, nihilistic view of technology, digital anthropology, and most particularly the contribution of ethnographic method, is often presented as a palliative – helping us to understand the experience of everyday life within these technical platforms, and sometimes arguing for greater local or grassroots agency than can often be imagined within perspectives that look down from afar. Given these stakes, I propose we follow Alan Liu’s call to ‘use the tools, paradigms, and concepts of digital technologies to help rethink the idea of instrumentality’ itself (2012). For Liu, this involves ‘thinking critically about metadata (and everything else related to digital technologies) in a way that scales into thinking critically about the power, finance, and other governance protocols of the world’ (ibid.). This requires us to *return* to anthropology as critique, not just description, and to unpack the epistemologies that are increasingly hardwired into our methods as they move into digital platforms. These approaches to digital media both extend and complicate the various turns epitomised by today’s MCS. Ultimately, it is helpful to remind us of the reactionary qualities of MCS, that it is the *return* to persistent questions about the entanglements of structure and practice, production and consumption, materiality and sociality that enable us to constitute a true anthropology of the digital as well as a digital anthropology.

Acknowledgements

Thanks to Rebecca Ross, Eleanor Botright, and our students (2013–2014) for the project that is at the heart of this chapter, and to Rebecca for critical engagement and contributing to the writing about community associations (all mistakes however are my entirely my own). Thanks also to Daniel Miller, Ludovic Coupaye, Antonia Walford, Shireen Walton, and Timothy Carroll for critical feedback on many drafts.

Notes

1 Anthropology, like other disciplines, has a long history of being 'put to use' or applied, from the role the discipline has played in supporting colonial government (Asad 1973), through to the embedding of anthropologists and their work in the military (Price 2011). A ubiquitous political economy of audit and value within teaching, research, and now increasingly engagement with the outside world (variously referred to in the UK as impact, knowledge transfer, knowledge exchange, and public engagement) increasingly structures the values of the modern British university. It is no coincidence that some of the most prominent theorists of MCS have also turned their attention to audit cultures that explore the ways in which new regimes of value produce the need for new knowledge and information economies and vice versa (see Miller 2005a; Strathern 2003). Miller's initial work on audit was part of a bigger project focused on 'virtualism' which linked a number of shifts in the production of value: towards audit rather than information, consumers (rather than commodities), and management as a particular form of labour borne out of the deliberate shift towards concepts rather than objects in late twentieth century political economies (see Carrier and Miller 1998).

2 In many ways, this approach mirrors contemporary debates within media studies and media anthropology in which media (or digital) is articulated as a 'total social fact' (taken from Mauss' articulation of 'the gift' as a total social fact, 2002[1925, 1976]). Miller works with a definition of the digital as 'all that can be ultimately reduced to binary code, but which produces a further proliferation of particularity and difference' (Miller and Horst 2012:3).

3 www.ucl.ac.uk/why-we-post/discoveries/5-people-who-use-social-media-who-create-it-not-the-developers-of-platforms, last accessed 31 May 2019.

4 Indicative references are in relation to media archaeology (Parikka 2012b); digital humanities (Lui 2012, Risam 2019); software studies (Manovich 2013); platform theory (Dourish 2016); format theory (Sterne 2012) and new materialism (Henning 2007).

5 More widely, see Knox and Nafus 2018; Walford 2015; Adriaans 2019; Miller and Sinanan 2013; Geismar 2018.

6 See for example this report in the Guardian newspaper, recounting how TikTok users are 'used to being tracked' and are 'unfazed' by corporate data collection: www.theguardian.com/technology/2020/aug/05/tiktok-gen-z-millennials-data-privacy-trump-china (last accessed 21, 08, 2020).

7 Objects of desire

Sexwork and its objects

David Jeevendrampillai, Julia Burton, and Eva Sanglante

Introduction

The 'Objects of Desire Collective' is made up of artists, anthropologists and sexworkers who take the lead in curating the collective's activities. The collective formed in 2015 around a shared concern for the form and content of public discourse and the emergent legal regulation of sexwork in both the UK and later in Germany. The collective felt that there was a distinct lack of objective understanding in these public discourses as to what sexwork actually is.

In the summers of 2016 and 2019, the collective curated shows in London and Berlin, respectively, comprised of a collection of objects given to or used by sexworkers in their work. The aim was to foreground narratives of sexwork as work through an explication of the forms of labour relations involved in sexwork via object biographies. The shows formed a critical interjection into such discourses and acted as a catalyst to public discussion and community building around the issues. This chapter works through how the exhibition was influenced by, and contributes to, material culture approaches to anthropology and ethnography. It relates particularly to the issues discussed in chapters by Drazin, in terms of object biographies, and Schacter, in terms of curation, as well as other chapters with regard to the conception of studying objects in anthropology.

The collective take as their starting point the critiques levelled at Durkheimian approaches to material culture (see Carroll et al. and Küchler, this volume), namely that objects need to be considered as more than substitutes for persons or as a repository for social relations to be collected and read as symbolic of various cultural groups. Rather, we recognise them as dynamically involved in the ongoing formation of social relations and power relations. Influenced by Mauss's discussion of the gift (Mauss 2002[1925]), we asked what is happening when a client gives a sexworker a gift? How the worker deals with the gift, uses, stores, interacts with it, is indicative of how they manage the relation not only to their client but to such things as labour, home, themselves and others. This work of managing and moving the objects in relation to its imbrication in a network of relations is part of the labour of sexwork.

The collective used object biographies as a starting point for the exhibitions. We interviewed around twenty workers in London and over forty in Berlin. With each interview, we asked the interviewee to bring an object to lend to the exhibitions. Interviews were usually semi-structured and lengthy, over an hour, often more. The narration of particular moments of an object's role in the mediation of a relation, as described by the interviewee, would accompany the objects in the exhibition, usually by a small text, video or a sound recording. In this sense our approach to the collection, display, and curation echoes that outlined by Schacter (this volume) in that we foreground the contexts in which the object was involved at specific moments that led to it being considered important for collection. Our roles, as ethnographers and curators go beyond the object into 'collaborating not simply collecting, enabling not simply exhibiting' (Schacter, this volume), and we approach curation as a 'highly charged opportunity to think and do otherwise' (Bouquet 2000: 228).

Schacter draws attention to the ways in which curation develops openness, prompts questions, and interrogates sites of knowledge production but, further, we purposefully emphasise the political aspects of our work with the target of asserting how sexwork is work through ethnographical detail drawn from object-orientated methods. Through specifically presenting objects alongside their narratives we focus on a specific moment in the life of the object, one in which it is involved in the labour of sexwork, to assert precisely how sexwork is work. We select a particular moment of its movement and mediation in social relations and, in taking this moment seriously, we consider each object as more than a signifier of social relations but as specifically active in particular moments of mediation. We are also influenced by approaches to objects that attend to their material qualities (see Witmore 2009) – such as smell or stickiness – and the forms of interaction they enable, or demand, in understanding their role in the labour of sexwork. To illustrate our approach, we describe our use of some objects from the shows, but first we outline the socio-political context in which the shows emerged and consider the role of anthropological collaboration in gathering politically targeted ethnographic detail.

The context of the shows

The exhibitions were curated in direct response to particularly fervent conversations around the regulation of sexwork across Europe. In the spring of 2016, the UK government reviewed prostitution laws while in Germany a new law regulating sexwork took effect on 1 July 2017.

In both the UK and Germany there was a range of public voices calling for different approaches to the regulation of sexwork. Some called for heavy regulation whereby sexworkers register on state databases and have to adhere to a range of very specific work conditions, such as sanctioned places of work and heavy surveillance (as in the Netherlands). Others called for implementation of the so-called 'Nordic Model' whereby buying sex is illegal, and laws and enforcement focus on the buyer – a model that assumes that sexwork is a form of gendered violence against women (see Bernstein 2007, 2010). The UK government review recommended more of a decriminalisation approach to soliciting and brothel keeping. This approach, while not perfect, was hailed as a 'radical moment' for sexworkers in the UK and was welcomed by many sexworker organisations (Eastham 2016). In Germany, the law, which has been promoted as 'The Prostitute's Protection Act,' was drafted without consulting sexworkers or their organizations. It enforces mandatory registration and identification of sexworkers and stipulates that they must carry a government-issued 'Prostitute's ID,' for which they must undergo a mental health examination in order to obtain. The law has come under fierce criticism from sexworkers, public health and legal organisations, and human-rights groups. They point to the collection of personal data, increased powers of surveillance, and the threat to privacy as issues that fundamentally impinge on the rights of sexworkers, thereby increasing vulnerability within some areas of sexwork. There was great uncertainty in the sexworker community about how these laws will be implemented. Our interviews in Germany were conducted in the year following the enactment of the law and we inquired about the law's effect on the daily lives of workers in Berlin. The exhibitions served as public interjections into discussion around sexwork in these contexts.

For all the public discussion on the ethics of sexwork, there was a noticeable lack of first-hand accounts of the everyday practice, particularly labour and workplace relations, of being a sexworker (Banyard 2016; Editorial 2016). The mainstream media frequently featured articles and opinions from people with strong views on what sexwork was, is, and should be, regardless of their expertise on the issue. For example, Hollywood 'A list' actors Meryl Streep, Kate Winslet, Emma Thompson and others actively took a position opposing decriminalisation of sexwork as advocated by international human rights advocacy group Amnesty International (Mandle 2015). In Germany, sexworkers anecdotally told us that many people had been registering their 'hooker name' as Alice Schwarzer as a form of ironic protest. Schwarzer is a prominent 'feminist writer' who views prostitution as violence against women and is an advocate of the Nordic model for Germany yet seemed to have little engagement with the sexworker communities, who were making clear their positions in support of decriminalisation and against the Nordic model (Khomami 2016; Freis-Bryce 2016).

Debates often polarised into positions that perceived sexworkers as either 'empowered,' or as 'victims' in need of saving (Burton 2016; Bindel 2016; Banyard 2016). Both positions take sexworkers as a homogenous group to be talked *about*, rather than a group of people involved in a wide-ranging and varied industry to talk *with*. As such, sexworkers had become objectified not only in terms the typically conceived objectified body or as a sexual object, but also as 'the sexworker' as a moralised object to be regulated, saved, or protected.

The collective formed around the need to refocus discussion on labour rights and conditions, which started with a campaign to ensure that sexwork is understood as work. This involved challenging and breaking down the objectification of sexworkers as a homogenous group to be managed. The collective's name, 'Objects of Desire,' is a purposeful play on the idea of what the object is in our shows. We draw attention not only to how objects such as jam, incense, and books are involved in social relations, but also to how the sexworker had become objectified in public discourse. One obvious issue when trying to explain in detail how sexwork is work is that not only do you have to overcome assumptions of what sexwork is but, further, many sexworkers would not wish to be talking heads or public figures in discussing the issues. Objects enabled a discussion on the empirical detail of the labour of sexwork and allowed the foregrounding of personal stories. In this sense, we also think we are speaking back to recent discussions in material culture.

Danial Miller states, 'the concepts of subject and object are failures to notice [the] process of objectification' (2005b: 37–38). While we consider the process of objectification, we also consider subjectification, that is, the process of reasserting the subjective qualities of a person or people in a discourse. This argument has been taken up by Severin Fowles who, in his call for closer attention to the processes of subjectification, has claimed that, within the 'material turn,' objects have become 'quasi-human subjects' (2016: 9) enabling anthropologists to obviate the colonial critique levied at their analysis of a subject-centred social world. Through substitution of persons for objects, Fowles argues, anthropologists were able to maintain a form of authority to write culture. However, whereas Fowles asserts that a focus on objects led to the obviation of the critique of power imbalances in practices of representation, we foreground objects in order to reassert the subject through the elicited object narratives.

A focus on objects allowed subjects to be made present via the stories of how persons and objects intra-act (following Barad 2003, 2007; see also Buchli, this volume) to mutually constitute each other through the negotiation of social relations. Our approach works to de-objectify those who have become bracketed as a homogeneous and unspecific group of 'sexworkers.' While Fowles might argue that the study of the object evades letting the subject speak, we focus on objects to allow stories to be told, thereby drawing attention to the ongoing processes of managing the ways in which objects and subjects are constituted, and the labour that this involves. It was a focus on the processes through which objects and subjects become objects and subjects that allowed us to interject into the representational economy surrounding sexwork and sexworkers. Attention to process is key, as Buchli (2002a: 19) writes,

> looking at what happens before and after the artefact is more significant than the artefact itself; that is, the terms of materiality rather than material culture itself and the differential ability of individuals to participate in these processes is more important.

But, whereas Buchli states that how people participate is more important, we feel that the materiality holds 'something in reserve' (Witmore 2009: 29). That is, materials have the capacity to provoke surprise and assert forms of relations through their stubborn material presence. To illustrate, we will work through examples from the shows.

The shows

The London show was based on interviews with predominantly independent and medium-to-high-earning workers. This demographic reflected the connections that Rori and Chelsea, who are active sexworkers and members of the collective, have access to in their ethnography as members of that community. The Berlin show had a much wider engagement with different types of sexwork made possible through funding from an Open Society Grant that enabled us to compensate for time, hire appropriate translators, and invest in workshops and a community building. We took ethnographic immersion seriously, and some members of the collective lived and worked as sexworkers in London and Berlin during the research period. The shows do not aim, nor claim to represent all sexworkers but rather aim to give a sense of the details of sexwork as work.

Figure 7.1 Eve's Mound.
Source: Photo by David Jeevendrampillai, © Objects of Desire.

Both shows were designed after gathering the ethnographic material. We would consider what themes emerged from the interviews and how we could best communicate a narrative of sexwork as work through curating the shows.

Upon entering the London show, the visitor would come across 'Eve's Mound,' a giant pile of gifts given to Eve by one client, who would show up at appointments with bags full of gifts that included tea, his ex-girlfriend's clothes, Tupperware, sandwich bags, and more. The sheer physicality of the objects, their bulky and enduring presence, means the social relations they carry have to be dealt with long after the client has gone. The objects needed to be received, recognised, stored, and later talked about or related to. The objects were more than a representational presence of the client, they demanded engagement and, through their very presence, a continuation of a labour relation.

Objects, or gifts, also make demands as outlined by the description of receiving a book where the worker was expected to read the gift in her own time:

> [Steven Pinker's *The Better Angels*] was given to me by a client who is a biologist and lover of Richard Dawkins. His views on the superiority of Western science were tough to swallow, [...] I would get drunk, accuse him of being trapped in a cage of post-enlightenment rationality, and say I didn't believe in evolution. I took it too far though, when I told him I didn't read this book he gave me because it's total SHIT! He got offended by that.

Books hold an imminent potential for something else, the communication of information; and knowledge of information can be used to test if the worker has engaged with the book. However, to read a book requires a large amount of time that the worker is not paid for. As such, books and asking opinions about the book constitute a way in which the client can transcend the boundary of the business relation and test the limits of the 'girlfriend experience,' posing a problem of authenticity for the worker. The book does more than represent client–worker relations, but works to force a production of them. These relations required careful management as books can challenge the space–time boundaries of what labour is and where it occurs.

Both exhibitions contained many domestic items, cleaning products, towels, bed sheets but also bric-a-brac, or 'klimbim' items, as they are known in Germany. Domestic items and their management give insights into the management of boundaries. Workers spoke of how, while they could usher a client out the door, their gifts often would remain. Lisa felt uncomfortable receiving an expensive scarf. She wondered what might be expected of her in return. She 'pushed back' against the feeling she should reciprocate, but felt that she failed to provide the authentic girlfriend experience. This posed a challenge to the transactional nature of the work, to the boundary of worker/

client. Lisa felt that gifts 'threaten,' through their ability to carry conditions of debt. The power of the gift (see Mauss 2002[1925]) had to be managed. When Lisa's boyfriend wore the scarf on his head as a joke, she explained that it had 'lost some of its mystique.' Just as the scarf was placed on the head, objects can be physically moved to manage their potency.

Sometimes the potent associations of a gift are brought forward through its assertive materiality as Elisa remarked regarding an incense burner. She was disturbed by the gift's domestic connotations. 'It's definitely a fantasy,' she said, 'this guy gives me this home object because he fantasizes about living with me.' She hated the thought of the rose scented incense that the client had selected moving invisibly through her personal space and into her body: 'It's a very intrusive thing; smell is so important.' Here, the object's materiality contributes to what Kathleen Stewart would call atmospheric attunements, in that it works to produce an atmosphere that evokes feelings. Atmospheric attunements are 'palpable and sensory yet imaginary and uncontained, material yet abstract. They have rhythms, valences, moods, sensations, tempos and lifespans' (2011: 445; see also Buchli, this volume). But who controls and manages this attunement is constantly negotiated.

The 'intrusion' of clients into the home and body was exemplified via the many luxury foods given as gifts. Rori explains she would regularly receive luxury preserves from a client:

> I've wondered why he always gives me preserves. I can see that he gets a lot of pleasure from giving me food and feeling that he is caring for me [...] but I've also noticed that he is very into bodily fluids and often says how nice it is that we 'taste' each other. I wonder if the jars of sticky jam are a symbolic substance, a kind of proxy for bodily fluids [...] if me storing them in my cupboard and ingesting them is a way for him to transcend the boundaries of our sessions, inserting himself somehow into my home and body.

While Rori enjoyed having reminders of the client in her domestic space, the very stickiness of the jam, the tackiness of the substance when in the mouth, gave rise to new and unwanted forms of association. Rori explained that for obvious safety reasons she maintains highly policed boundaries of bodily fluids and exchange at work. In this sense, the jam served as a substitute and morphed from being a pleasurable conserve to being an index of the body and its fluids (see Warnier 2007). The very materiality of the jam is important here as its stickiness mobilises the index and enables the transference of the relation (see Keane 2013). It was through both its social context and its material qualities that the jam's stickiness animated a particular indexical relation. Thereby the jam became an object to be dealt with in order to maintain boundaries of work and non-work.

Just as Rori expressed how jam traversed the bodily boundaries of sexwork, Chelsea explained how an endoscopic camera could create new dynamics of bodily interactions when the client is not physically present. The camera was donated to the London exhibition by Chelsea, who frequently worked in camera-based chat rooms. Clients would send digital 'tips' that would result in clothes being removed or actions performed. Her clients occasionally sent gifts through the post to feel as though they had 'done something special.' The camera was the 'strangest' gift she received while camming. A client had often asked for close-up shots of her vulva in the public chat room. Chelsea said this was not unusual in itself, given that 'everyone is happy for a bit of a close-up.' He sent the endoscope for her to make him a custom video of the inside of her vagina – 'not just close but right inside!' She reflected on how gifts like this function as a way for clients to indicate a fantasy that they might be too shy to verbalise. In cam-work the boundary between the viewer and viewed is mediated by the webcam. The camera offered new perspectives, new forms of penetration as it transcends boundaries and becomes a mobile mechanical consciousness (Deleuze 1983: 23). The camera was not just an inalienable object that carried the spiritual essence of the giver (see Mauss 2002[1925]), nor simply an object that instantiates a social relation (see Strathern 1990), rather it created a new form of labour practice and labour relations.

The exhibitions demonstrated how objects (books, clothes, food, technology) produced and were imbricated in, rather than represented, sexworker–client relations. These relations extended beyond the presence or even the intention of the client via the assertive materiality of the objects themselves. At times clients would use objects to evoke others who were not present. Maria says gifts reflect the sorts of stories that clients like to 'spin about themselves' when buying sex. She recalled one client used gifts of particular clothes and perfumes to revisit memories of his late wife. The gifts had to be performed, touch flesh, be embodied in order for the client to remember. This enabled a simultaneous subjectification of his wife via object engagement with the sexworker who, through a performative act, animated particular indexical qualities of the objects and transformed the sexworker into a body, animated and with a vitality of life, which enabled these indexes to be efficacious in evoking memories of smell, touch and personal relations with the dead. In all cases, the objects themselves did not represent a person, but were brought into efficacious indexicality through the performative acts of the worker. This animation was showcased in the 'surprising' banality of the majority of objects on display. As one reviewer of the London show noted:

> As an outsider, you can imagine the gifts that sexworkers are given by clients; Champagne, flowers, sex toys and lingerie all make the list. Marks & Spencer's jam, however, doesn't come to mind. Neither does a Twinings tea gift box, a Mizuno golf visor or a history book. But [the] 'Objects of Desire' exhibition proves these preconceptions wrong.
>
> (Lloyd 2016)

Both shows had a similar collection of objects with themes of domestic gifts, work tools (such as bags and coats), techniques of discreetly getting to work, and the interviews particularly discussed the process of switching between work and non-work personas. The Berlin show contained a dressing table adorned with annotated make up, perfumes and mirrors to demonstrate the point. All interviewees talked about how the stubbornness of objects often made it difficult for them to switch from working to not working, as outlined above, jams and incense can pervade private space. However, the Berlin show raised an interesting point that was absent in the London show. In Berlin the new law made it mandatory for sexworkers to carry ID cards. These force sexworkers to reveal private information and make them vulnerable to being identified by both clients and authorities, as one worker outlined.

> It will be weird to see my real name and my alias together on the Hurenpass. I do think that for those who worked in brothels before, there has always been a connection for the police between identity and work. I have heard stories where you're pulled over and the cop's looking at your papers, has called to check and then came back with a certain comment like, 'oh you also work in this brothel.' Maybe in front of your friends and family. This is because they collect data from raids, and brothels are always raided every once in a while, mostly for taxes. I don't know why, but I was already really intimidated by these tax people, the Steuerfahndung. It's so intrusive somehow. You're there in your thong, and you show your passport, which you don't ever show to anyone in this place. No one knows your name, and there you stand.

The ID cards not only identify sexworkers but enables an objectification and moralization of them as 'sexworker' at any moment. They can be identified via their real names when they are working. Here, the ID card as object is used as a regulatory object, an object of power.

Influenced by the idea of the contact zone (see Clifford 1997; Pratt 1991; Schacter, this volume) we used the shows to draw multiple communities together, from different sexworkers, service workers, authorities, the public and art and museum institutions. In the process of curation and display, we echo Schacter's perspective whereby curation is less a practice of presenting artefacts, but of creating spaces to generate, circulate, and co-produce knowledge. To provoke and assert questions of being human alongside your interlocutors, as Schacter notes, curation 'becomes a model in which the anthropologist engages in experimental, speculative, long-term processes wherein we can speak together with our interlocutors, mediating, not controlling, their own ways of seeing.' But further we can use our ethnographic techniques and our approach to objects in order to elicit stories and narratives of the complex process of mediating social relations so that they can be foregrounded to communicate things such as how sexwork is work, to a wider audience.

Objects, sexwork and material culture theory

The exhibitions approached objects, not as static representations, but as caught in particular contingent and contextual moments of making social relations. They are, as material forms, caught in a moment of objectification and subjectification. We lean on the object biography as a method of orientating the research, of disseminating information and of expressing how sexwork is work. However, we are less interested in what happens when the object 'moves to a new domain' (see Drazin, this volume) than we are in seeing how objects move into, and out of, those domains. What, we ask, are the motivations for throwing away jam, putting incense burners out or forcing someone to carry a card? What social relations does this produce, regulate, or manage? We are political in our ethnographic moment, that is, the moment where we have a need to understand (following Strathern 1996), but also there comes a moment that we need to make others understand in the politics and inherent labour of managing the social relations of sexwork and the objectification of the sexworker.

We recognise that our approach requires an ability to trace objects movements and the associated stories. Such contexts and narratives are not always available to the anthropologist (see Mercier, this volume) but is crucial in the focused political ethnographic moment practiced here.

As noted above, our analysis and attentiveness to the biography of an object was heavily influenced by the works of Arjun Appadurai and Igor Kopytoff – in particular, the edited collection, *The Social Life of Things*, where Appadurai argues that in looking for the value of objects, 'we have to follow the things themselves, for their meanings are inscribed in their forms, their uses, their trajectories' (1986b: 5). However, our approach places emphasis less on the meanings inscribed on the object, but on how the object is imbricated in a process of managing social relations. We focus on the processes through which particular subject positions, such as sexworker, client, and others are involved in a constant process of becoming un/stable in relation to the management, placing, moving, naming, and experiencing of material forms. In the moving, use and consideration of a material form, both object and subject become something in particular in relation to each other and in relation to a wider social realm.

Our approach to foregrounding objects via curation aims to interject narratives of labour, work and practice, into what Webb Keane (2003: 410) would call the 'representational economy' around sexwork, particularly where the 'sexworker' was an objectified figure. Keane emphasises the historical and social formation of different modes of signification asserting that underlying assumptions and historical norms determine how 'one distinguishes between subjects and objects, with implications for what will or will not count as a possible agent' (2003: 410). Keane argues that the relation between subject and object mean they are always imbricated in a dynamic economy of representation. Rather than drawing attention to the 'dynamic interconnections among different modes of signification' (2003: 410), OoD's emphasis was less on signification and rather on the ways in which the objects are involved in dynamic webs of relations – not only between people and people, or people and things, but also in the relation between the material world and concepts, ideas, and forms. That is, we examined how it is that objects are involved in the ongoing process of the experience and labour of sexwork and being a sexworker. With this, we push a form of political corrective to a material culture conversation that has been overly concerned with the notion of defining what an object *is* as opposed to considering why it is both considered an object in the first instance and what said object *does*.

Going beyond representation is not new to material culture theory, as the chapters of this volume attest, and the work of Alfred Gell (1998) has been hugely influential in shifting anthropological analysis towards a more performative and dynamic understanding of material/human relations (see Küchler 2002, 2005; Keane 2003, 2005b; Pinney 2002, 2005). Gell tackled Saussurean linguistics, which he felt was overly focused on representation, through his employment of Peircean linguistics, which allowed him to place greater emphasis on the performative, whereby meanings are felt as well as read. Gell's work allowed more space for the phenomenological and affective modes of explanation in the relation between people and materials. Through the notion of the index, Gell analyses why acts of iconoclasm, such as the slashing of Diego Velázquez's painting *Rokeby Venus* by suffragette and women's rights campaigner Mary Richardson, have such potency (Gell 1998: 64). The slashing, asserts Gell, was a form of 'volt sorcery,' whereby its slashing was an attack on the excessive agency that the painting held in the context of the imprisonment of campaigner Emmeline Pankhurst. In this sense objects, or art in this case, are more than a representation but are a 'constitutive act' (Gell 1998: 191), in that their potency goes beyond the symbolic to recognise the ways in which objects bring about an association of relations which can be either implicit, explicit or, as Küchler and Carroll (2021b) state, 'imminent.' Implicit here is a recognition of a wider field of power relations as well as the material qualities of the objects themselves.

To think through an example, the jam took assumed an excess when considered in relation to the client and their fetish for sticky fluids. The jam, without context is not particularly potent. One can be attentive to its material qualities, the stickiness, the sweetness, the glistening, but such qualities only become potent when contextualised within its social relations. Jam is indexical to the client, specifically to his desire to exchange sticky fluids. But further, jam is more than representational of the client; its very presence is a constitutive act. It brings forth indexical relations that are out of place. The act of eating jam creates, through the sensual experience of its materiality and the imminent relations it foregrounds, an excess of the client's desires and the associated labour. The jam becomes potent and is in need of policing in order to maintain stable boundaries. Transporting this relational nexus into a wider public discourse via the exhibition space affects what jam is and can be. As one exhibition goer remarked, 'I will never eat jam in the same way again.' Here, the anthropologist is not the passive describer of an object or a relation but curates relations alongside their interlocuters, in order to assert how sexwork is work and de-objectify the labourer.

Daniel Miller's (2005b) critique of Alfred Gell's *Art & Agency* (1998) is that, in recognising the operative act of art objects in human agency, Gell affords them a degree of second-order agency. Miller's humanist approach enhances the role of things by making them operative in the making of humans; he states, 'the things that people make, make people' (2005b: 38). Holbraad and Pederson (2017: 205) critique the ways in which Miller's 'humanist' approach leaves the ontological distinction between people and things unmodified. Following Viveiros de Castro's call to 'take people seriously' (see 2004; Holbraad and Pederson 2017: 291), they ask how could a thing which may be said to be a spirit, not be theorised as a *conduit* of spirit but *as* a spirit. Or rather, to take Holbraad's example, how could we think about powder in Ifá divination not as *representative* of power but *as* power? In essence, their argument for relationality suggests that, 'instead of treating all the things that your informants say of, do to and with things as modes of "representing" the things in question, treat them as modes of defining what those things are' (Holbraad and Pederson 2017: 213).

OoD presence objects in order to foreground their role in the mediation of a relation, what they *do*: their efficacy. This involves understanding them as symbolic and representational but also imminent, and as operative through their material qualities. Chris Witmore (2009) asserts that how a thing becomes a thing is dependent on a set of contextual relations. Labour is needed to either maintain objects and subjects as stable entities or break them down. Witmore differentiates this approach from that of Appadurai's and Kopytoff's, asserting that, in their conception of object biographies, objects can fall into the trap of 'vacuous actuality,' (2009: 26) where things happen to them. Witmore stresses that the ways in which objects are 'actively *happening*' (ibid) must be addressed with a focus on an object's local conditions and how it is made and remade in a nexus of relations, of shifting associations with other entities; however, this is not to say that objects are understandable through their relations alone, instead they hold something in reserve (2009). Objects often have surprising or stubborn qualities through their ability to smell or stick, and they demand to be dealt with. The objects presented in the exhibitions were not passive representations of aspects of sexwork, but active mediators of relations that likewise demand to be dealt with.

Conclusions

This chapter has focused on the specific moments of interacting with jam, books, scarfs, and cameras to illuminate the forms of social relations active in negotiating the subject positions and labour relations of sexwork. These everyday interactions are a product of the wider context of objectification of both the subject and subjectification of the object. In this dynamic, the materiality, social context, and process are important. An object-orientated methodology of object biographies was used to trace the details of the everyday practice of sexwork as work. A heavy regulation of boundaries, such as between one's work and one's personal life, the physical spaces of home and work and between client and friend was evident throughout all the interviews, and the objects selected often troubled the maintaining of such boundaries. The stubborn materiality of objects and the potency of their relations required that they be dealt with. Where does one put a gift? Do I really have to carry this card? What name do I use? How does one engage with it while maintaining highly regulated boundaries?

Both the exhibitions and this chapter have been influenced by, and form a response to, debates within anthropology regarding the ways in which materials are conceived. Further, it is a response to the role of the anthropologist as a curator and active agent in foregrounding the ways in which social relations are managed, challenged and regulated through objects. There has been a call to focus on the materials themselves, 'in their own terms,' and to 'take materials seriously' (Ingold 2007: 14). Such a materials-orientated approach opens up new ways of thinking about how the materials have efficacy in a given social situation – for example, through the stubbornness of the jam's sticky presence. On the other hand, others (see Witmore 2009, Küchler and Carroll 2021b) have argued that an over-focus on the material properties negate the fact that such affordances only become social when within specific contexts, so jam can feel sticky but only when conceived of in a specific context does that potential matter to the anthropologist. In this sense, an object – say a cloak within a museum collection – requires specific attention so that it does not stand for all cloaks, or merely a cultural variation on a cloak (see Geismar 2018). Rather what (or when) an object is, what it does, and how it evokes particular experiences, subject positions and relations, are dependent on the context. The anthropologist, through the choice of an analytical focal point, makes a political decision on the ways in which we foreground the configuration of object, subject, and persons via writing, talking through, or exhibiting their making to new audiences.

Acknowledgements

We thank the participants of the research, the wider 'Objects of Desire Collective,' The Schwules Museum Berlin, the Open Society Foundation Public Health Program; anonymous private donors, and those who contributed generous comments on this chapter, in particular Rebecca Williams and the editors of this volume.

8 Digital devices

Knowing material culture

Hannah Knox

Introduction

This chapter explores the anthropological significance of contemporary digital devices as forms of material culture. Digital devices, from smart phones to fitbits, sleep-monitoring apps to air-quality monitors, are an increasingly important aspect of people's way of relating: to their friends, family, bodies, and environments. Such devices offer a new focus for material culture studies (MCS), extending key questions about the role that objects play in social relations, how they are interpreted and given meaning, and how they mediate relations (Appadurai 1986b; Harvey et al. 2013; Miller 2005b, 2008). However, they also raise some challenging questions for MCS, particularly as regards the relationship between their status as objects and their place as generators of information, data, and knowledge. The ability of digital devices to not only be *subject* to human interpretation, but also to *produce* their own ways of seeing and knowing, demands an approach that goes beyond the usual forms of material culture analysis to consider how devices operate, not just as objects, nor even as agents, but as empirical or knowledge-producing entities.

Prompted by the challenge that this aspect of digital devices poses to anthropological understandings of human/object relations, this chapter explores how such devices might fruitfully be studied by drawing on anthropological and philosophical approaches that have considered, not the sociality of objects, but the materiality of knowledge. There is a resonance between this chapter's attention to knowledge and the critical focus paid to technical agency in Chapter 4 by Ludovic Coupaye. To explore these lineages, further detailed below, and to look at how existing understandings of knowledge and expertise might help us understand the informational qualities of contemporary digital devices, the chapter draws on ethnographic research that I have been conducting in the UK with people who use and engage smart-energy monitors to explore their relationship with energy, houses, and the global environment.

The chapter explores how digital devices like smart-energy monitors come to participate in social relations by drawing attention to their role as knowledge-producing entities. In order to operate successfully as knowledge-producing entities, digital devices must be able both to *sense* material relations to which they are oriented and to communicate this sensory information to other entities (including human beings) who are invited to respond to the formal outputs of these devices displayed as numbers, images, text or electronic signals. Understanding the part that digital devices as producers of knowledge play as participants in human social worlds demands that we pay attention to not only their materiality as static objects, but also to the way they work through sensors, communication protocols, and symbolic representations to forge a relationship between environments and people. Digital devices, I argue, are thus revealed to be not only *products* of knowledge, nor objects of *interpretation*, but also *active participants* in the formation of social imaginaries and material worlds.

Digital devices

On the grassy banks of the River Rune stand twenty-seven two-storey white terraced houses. Wide windows flanked by tomato plants, and balconies sporting pots of garden herbs give a bucolic aspect over the water, which tumbles and bubbles over glass-brown pebbles and swaying green water weeds. The houses are part of an English co-housing site, an eco-community established to provide an alternative, more communal, less resource-intensive way of living.

It is a wet Tuesday morning, and Tom is walking a group of visitors from an EU-funded smart community energy programme around the co-housing site, explaining how the place is powered. We start our tour at the clubhouse, which is shared with a local fishing club. A fine drizzle of rain gathers on eye lashes as we blink upwards to look at the three solar panels installed on the clubhouse roof and on the roofs of houses on the other side of the riverside path. Then, we turn en masse to follow Tom and, trudging along the path by the river, he tells us about the history of the place as a nineteenth-century dye factory.

At the end of the path, a grassy space opens out as we stop in front of a large shed. High, heavy double-doors hide the hydro-powered generator, though we can hear it whirring inside. A cut-out in the side of the building reveals a computer screen showing information about the energy that the generator is creating and a graph of its generation over the past week (Figure 8.1). Inside the stone building we climb up onto a platform above the generator, which is all servers and computers with wires and monitors, with photos of the grand opening day for the generator pinned to the wall next to newspaper clippings of local and illustrious visitors. The hydroelectric generator itself quietly hums away below us, encased in a leaden-green iron shroud, while we glance at the computers around counting kilowatt hours.

The monitors in the hydro-generation room are the most visible to us on the tour, but we discover that there are many other devices busy counting energy and materiality around the site. Each of the houses has an electricity meter ticking up a register of energy used. Four electric cars are charged at another point, attached to a separate meter that counts their electrical charge, data used to calculate costs of hiring them out. An industrial unit housing several small businesses that are organisationally linked to the housing cooperative also draws electricity from the solar panels and water generator as well as being heated by a biomass boiler. At the moment, the data from each of these sites of activity accrues on individual analogue meters, while the whole site sits 'behind' an energy company meter used to bill the co-housing group for the difference between the energy they produce and the energy they use as a collective. It is Tom's job to work out the bills for the households and the businesses on-site.

Currently, Tom has to copy the details from each meter into a spreadsheet and manually calculate everyone's annual energy bills. The visitors he is showing around today are part of a project that is exploring whether smart meters that use sensors and collate and display data using online monitors might be able to transform energy communities, and how they might be of use to this particular co-housing group.

Detecting digitally

The digitisation of monitoring and metering is a ubiquitous, if underappreciated, part of contemporary life. While there has been considerable attention paid to data analytics and surveillance technologies that track and trace people's activities online (Amoore 2018; Amoore and Piotukh 2015; Boyd and Crawford 2012; Zuboff 2018), the more mundane kinds of ubiquitous monitoring upon which digital devices rely for their functionality are rarely given much attention in anthropological research. Nonetheless, everywhere we now find digital devices we also find sensors and monitors. On mobile phones, motion sensors collect traces of information about the movement of the user that are translated into data on steps taken and sleep quality, and used to power haptic gaming experiences. Combined with apps that monitor the GPS signal emitted by handsets, movement data is used to detect a type of activity being undertaken by the phone's user (e.g. walking, running, cycling, travelling by train). Digital augmentations of a more extended range of mundane objects, from cars to fridges to clothing, are referred to by technology developers and business enthusiasts as ubiquitous computing (ubicomp) or, more recently, the Internet of Things.

What, then, are we to make of the sensory and epistemological capacities that characterise these devices? In much anthropological work on the digital, digital objects have to been seen as equivalent to, or extensions of, other kinds of material culture, different only in the fact that they are made of 0s and 1s, which is often seen as no difference at all (Miller et al. 2016; Miller and Horst 2012). The importance of the augmentation of objects through sensors and monitors that display and collate information has remained for the most part sociologically and anthropologically irrelevant, hidden by rich user interfaces through which people engage in relationships with one another at the level of symbolic meaning, with the materiality of the devices and the relations they engage in themselves fading into the background (See, for example, Boellstorff 2008).

Yes, while the focus on the social and symbolic importance of digital devices and platforms has constituted the main focus of anthropological interest in the digital to date, I contend that it is in fact the ordering of the sensorial, informational, epistemological relationality of digital devices that most significantly differentiates *digital* devices from other kinds of material culture (see Coupaye, Chapter 4, for a more extended discussion of the role of 'sensors' and 'effectors' in object relations). Indeed, it is these sensors and the monitors that display and collate digital information from such sensors, that turn objects from things into seemingly 'empirical' technologies (Marres 2015, 2017) capable of participating in social relations, not as mute things, but as knowledge-producing entities. I argue that, as scholars of material culture, once we notice these apparent knowledge-producing qualities of contemporary digital devices, it becomes incumbent on us to explore what the implications of object-knowing might be for the way in which we conceive of the relationship between objects and human social life.

Until now, the main options for looking at the role of objects in everyday life have been to either look at the social shaping of objects, that is, to attend to the ideas and concepts that go into and are carried by design (Latour 1991; Schwartz-Cowen 1985; Winner 1986) or, alternatively, to look at the way in which objects are used to carry social meanings in particular cultural settings (Appadurai 1986b; Bourdieu 1984a). The concept of 'Actor-Network Theory' put on the table the notion that objects also have agency (Latour 1999; Law 1999; Law and Hassard 1999), but although this extends agency to objects, this approach has tended to stop short of attributing these same objects the capacity to 'know.' Indeed, in a critique of overly humanistic accounts of how social effects come into being, actor-network theorists have located the value of their method more squarely on the 'ontology' side of an ontology/epistemology divide, preoccupied with how things or assemblages come to *be* (Latour 2005). But what if we were to extend the teachings of ANT even further to ask not only how do objects have agency, but how might objects also figure as forms of non-human *perception*? When objects take on capacities for measurement and description, I suggest that this opens up the possibility that objects are now able to create not only new kinds of relations but also new ways of what we might call 'knowing' (Küchler 2008; Thrift 2014) due to the way they materialise methods of social and physical analysis into their design (cf. Coupaye, Chapter 4). Just as ANT's attention to object agency has led to a refiguration of the idea of human agency, so I suggest that an attention to 'object-ive' perception holds the potential to opens up new ways of thinking about and locating the nature of knowledge.

Knowing with devices

Several months after the initial tour of the co-housing site, I return with Tom from the community smart-metering project to explore, with residents of the co-housing site, how smart-energy monitoring might help them understand and manage their energy use. Each of the homeowners is going to have a smart meter installed. This will take the form of an 'extension' that will be attached to their existing electricity meters. This SMX (Smart Meter eXtension) has an electrical pickup that detects the electrical current and voltage passing through the wires that lead from the houses into the non-smart meters. This signal will be converted into data and sent to a server, where it will then become available for the co-housing residents to look at on their computers or phones. Tom explains that, as well as SMXs being put on the individual household energy meters, these devices will also be installed on the meters that manage electricity generation via the solar panel and the water mill.

Tom shows everyone a mock-up display of the energy graphs they will see via the user interface. It is still in development, so a little glitchy, and people take a while to adjust to the graphs. However, as Tom begins to explain the meaning of the numbers displayed, people become more interested. Looking at the data, people start to raise questions and engage with the graphs. These include queries about such as the terminology used to describe what the meters are measuring (What is a baseload? What is causing it? Is there any way of using data to interrogate this?); the referentiality of the information displayed on the graphs (Is there a time delay in the feed or is it real-time data? How can the data differentiate energy from different sources?); and the relationship between the smart metering and the existing energy grid on-site (Is the washing room currently on the same meter as the lighting? Does anyone know how much energy is being generated from the solar panels versus the hydro plant? What is the relation between the amount generated by the co-operative and the amount drawn from the grid?). As people begin to engage, even with just these hypothetical read-outs from an imaginary meter, they start to read into these digital traces new possibilities for their relationships with each other, the environment in which they live, and the technical object itself.

While everyone in the room already had an electricity meter that clocked up ever-accruing numbers that registered the energy they used, this information was located on the meter itself in the form of a clicking analogue dial. To know how much energy one had used required physically going to read the numbers that were being displayed at any point in time. To make sense of this information required organising into a temporal array in a spreadsheet or database in order to allow rhythms and patterns to emerge. The work of turning these occasional numbers into streams of information that could be organised into stories that travel as knowledge was currently done by Tom, who read the meters at regular intervals and put the numbers into a spreadsheet in order to calculate their electricity bills.

Before the possibility of smart metering, then, the passive analogue meter was made active and participatory because Tom went around all the houses with his laptop and manually compiled the data into a spreadsheet, typing in each reading one by one. This was time-consuming work and intrusive – requiring that Tom knock on people's doors and enter their houses. Tom was particularly interested in what the smart-meter extensions could offer – initially because they promised to operate as a digital proxy for his labour, allowing him to 'delegate' (Latour 1994) the labour of organising data to a machine. Just as Tom had done before, the smart meter would turn these clicking numbers into a stream of data that could be displayed, collated, and arrayed in another form.

The smart meter here promised to take on Tom's agency in digital form. But, interestingly, it also promised to do more than Tom could do. For, while he could only transcribe the numbers that indexed total energy counted, the smart meters promised to provide a more constant stream of this indexical data and new ways of visualising and interrogating it. Moreover, this new set of data was to be not only visible to Tom for the purposes of billing, but would also be displayed to the homeowners and, Tom imagined, might potentially become a public resource that the co-housing residents could engage with as a collective. The sensory capacities of the meter, combined with techniques of information transfer, collation, and display, promised to turn the meter from a mechanical counter into an entity with the capacity to both reproduce and also extend existing ways of knowing energy.

In a 2010 working paper by Savage et al., the authors set out to explore what is new about digital devices (Savage, Ruppert, and Law 2010). Analysing a wide range of emerging digital platforms, objects, and datasets, Mike Savage, Evelyn Ruppert and John Law suggested nine characteristics, or 'theses,' that they saw as relevant to understanding the sociality of the digital. They suggested that digital technologies do not represent some kind of epochal shift in social relations, as articulated famously by grand sociological theorists from Jean-François Lyotard to Manuel Castells, but that the digital transformation of social life is primarily to be found in the digitisation of often already-existing empirical methods. Discussing digital processes from big data analytics to transactional data production, from visualisations to the mobile tracking of goods and objects, they argue that the digital is characterised by a heterogeneous and non-coherent reappearance of pre-existing methods of enumeration, inscription, visualisation, measurement, analysis, and representation in new configurations and formations.

This has resonance with Tom's experience of digital-metering technologies, which were not a complete break from prior empirical methods he had used, but a material transformation of these methods from non-digital to digital form. If, in the work of Bruno Latour, material objects have frequently been shown to operate as delegates for human action – the sleeping policeman for an actual policeman, or the door key fob for a written sign – here, the digital device appeared to operating as a delegate, not only for acting, but also for knowing. This raises the questions of how the social production of knowledge has been theorised and understood in existing literature, and how this might help us as we seek to consider in what sense digital devices might be said to be coming to operate as knowing objects.

In arguing for attention to the way digital devices build on previous methods of empirical practice, Savage et al. draw attention, in particular, to the importance of the key social theorists of knowledge, including Michel Foucault and his analysis of the techniques through which governmentality is performed (Foucault 1977, 2007, 2008; Foucault and Gordon 1980), those working in the tradition of the social studies of scientific knowledge (SSK), and laboratory science and its explorations of the emergence of objectivity and nature as a realm of knowledge (Latour and Woolgar 1979; Law 1986), and sociologists who have studied the emergence of disciplinary knowledge forms (such as Abbott Andrew 2001). We might add to this list the important work of Ian Hacking (1990) and his study of the emergence of statistical probability as a form of knowing that continues to shape data analytics; the work of Timothy Mitchell (2002) and his analysis of the material and social techniques by which 'the economy' became formatted as an object of governmental attention; and the interventions of anthropologists the likes of Anna Tsing (2005) and Marilyn Strathern (1991), who have shed new light on the specific histories and trajectories of knowledge forms, such as scientific 'universals,' interdisciplinary knowledge, and the network as an imaginary of social relations (See also Barry and Born 2015; Jasanoff 2002; Riles 2001).

These lineages help us locate the empirical qualities of emerging digital devices as part of a long history in the development of techniques and methods of knowing, such as counting, enumeration, calculation, and abstraction. This literature also draws our attention to the way these methods have widely been used as instruments of power – for example in attending to how methods of constructing and stabilising knowledge have been key to the way states, corporations, and economies have come to gain power and know citizens, customers, and users as social entities in the world. This literature highlights how the question of knowledge is always also a matter of who or what can be conceived as an entity with the capacity to make decisions, frame problems, and shape worlds, and with what effects. Turning this understanding back on digital devices, then, suggests that we might need to consider, not just how devices become delegates for human ways of knowing, but what the effects of this delegation is for the location and enactment of power.

Back at the co-housing site, the SMX was, as we have seen, promising to create new kinds of knowledge for the residents. As methods of empirical observation were moved from a practice of manual enumeration to the device, data's potential uses seemed to proliferate. The proliferation of data as a material to be worked on and ordered in turn created an opening for a new kind of social role for Tom and for others with whom I spoke who were interacting with these devices and the empirical methods they embodied. This new role for the person *using* the digital device emerged between the device and the person, as people came to learn from, and relate to, the device's representations, incorporating and coming to understand new connections and in turn forging new relations with the version of the world that the displayed and collated data denoted. People subsequently incorporated the methods of data description, collation, and analysis out of the digital device and into their own practices and modes of imagining – for example with talk of how to better 'balance' electricity supply and demand, or how to indicate or flag modulating grid intensity so as to be able to be better citizens in their energy use. The device thus became a tool through which practices of everyday control, management, and negotiation were being rethought and re-enacted.

Interestingly this reworking of practices of enumeration, collation, and analysis in the device, and their subsequent incorporation into the social imaginaries of those using the devices, seemed to create the possibility that the person who now had data-analytics tools at their fingertips might come to take on some of the qualities of the public figures previously associated with similar empirical methods – the manager, the bureaucrat or, in the case of energy, the grid controller. The relocation of methods onto the user of the device simultaneously served to relocate the expertise necessary to use those methods from offices and laboratories into other spaces and relations.

A key lesson that seems to emerge here about digital devices as empirical technologies, then, is that when they carry with them methods developed in other spheres, they bring with them not only the method but also a residue of the expertise and status that method confers. For anthropologists interested in studying digital devices, our first focus might therefore be to try to understand what social, relational, and political possibilities are opened (or closed down) when empirical methods move from the offices and infrastructures of corporations and governments into technical artefacts like the smart meter.

Materialisations of knowledge

If social studies of knowledge have shown that epistemic techniques have histories and politics, those interested in the way that methods come to have social lives have also demonstrated that such methods not only describe and order the world but also have often unintended world-making effects (Bowker, Star, and Press 1999; Merry and Conley 2011; Strathern 2000). We only have to consider the way statistics and indicators are 'gamed,' abstractions are subverted, or accounting creatively reinterpreted to realise that methods unfold the world as much as they describe and contain it.

This has taken on a more radical tenor in the work of scholars who have been studying the relationship between scientific methods and the creation of scientific knowledge, and who have become aware of the problem of analytically setting up an opposition between a stable world 'out there' *studied* by science and the representational practices that scientists use to bring that world to light. Karen Barad's (2007) philosophical engagements with the realisation by quantum physicists that the outcomes of their experiments are affected by the presence of the devices used to measure that outcome, led her to develop a social theory that tries to break down the opposition between objects and representations to focus instead on what she terms the 'intra-actions' of people, devices, and things. Similarly, Isabelle Stengers (2010), in her study of chemistry, has used the concept of 'cosmopolitics' to denote the interplay between material properties that, she argues, do not just feed representations but, through their formal qualities serve to 'force thought' in scientific settings. In both the work of Barad and Stengers, matter, method, and thought are complexly entangled and co-emergent (see also Coupaye, Chapter 4, for further discussion of how technical objects combine materiality and thought). Following from this, they argue for a new approach in the social sciences – one that no longer simply studies the social construction of scientific knowledge but, instead, repositions methods – devices, enumerations, calculations – as techniques that work to bring the world into being at the same time as they do the work of describing it.

The focus of this work on methods as world-forming processes as well as knowledge-creating practices, suggests a second lineage in our analysis of digital devices – that is work that has previously focused on the relationship between matter and mind. This traces a lineage of thought that incorporates Henri Bergson's materialist philosophy of memory (Bergson et al. [1911] 2004) and surfaces in Gregory Bateson's work on the ecology of mind (Bateson 1972) and in the work of continental philosophers Michel Serres, Gilles Deleuze, and Félix Guattari and their interest in the co-emergence of forms of being and forms of thought (Brown 2003; Deleuze and Guattari 1987). Within anthropology, attention to the interplay between matter and thought has recently appeared in work in environmental anthropology by the likes of Marisol de la Cadena, whose *Earth Beings* utilises Stenger's notion of cosmopolitics to attend to Andean ways of being with, and knowing, mountains (Cadena 2010, 2015), and Eduardo Kohn's *How Forests Think*, which draws on the work of Gregory Bateson and Terence Deacon to argue for a need to extend the capacity for thinking from humans to non-human forms of life (Kohn 2013). In the work of each of these scholars the processual, emergent, and transformative aspects of social life are highlighted through their attention to the co-relationality of non-human materials and humans, reconceived as a relation of meaning or intersubjective thought. By attending to how people and things exist in a sensorial set of sign relations, these scholars provide us with a further set of resources for understanding, not only how digital devices detect the world around them through empirical techniques of ordering, but how in doing so they bring about its very emergence. In the final section, I turn to the way smart meters, in the act of trying to represent electricity's relationality, also come to bring into being the very gridded relations that they aim to describe.

Productive devices

I returned to the co-housing site six months after the workshop to talk to Tom's wife, Maria, about how the smart-meter installations have gone. She has been the project contact and helped with the installation of the meters in the each of the houses. My intention was to talk in more detail to her and some of the residents about the data they have been using and how they have been relating to it, but as I try to arrange the interviews it turns out that, while the meters have been installed, they keep turning themselves off, and no data has been collated. Maria installed 21 meters and at first they were all up and running, but one by one they went down. We meet anyway to talk about the process of trying to get the meters to work, about what she still hopes for the smart-meter devices, and why she was interested in them in the first place. Here, loosened a little from the form of the data itself, Maria begins to tell me what it was that these particular devices promised for her, and how it was that she had hoped they would become participants in the life of the co-housing site.

Digital monitoring devices were brought in to the site as a way of monitoring electricity in order to support the ends of communal, semi off-grid living. When we were being given the tour of the site in the rain, Tom told the group about a time a few years earlier when the city near the co-housing site had lost power for three whole days due to a storm that had caused flooding that knocked out the city's substation. Pointing to the roof of the clubhouse, Tom showed us how high the water from the river was running at the time, but he also told us that they had their own generator on-site and were able to get it up and running, providing them with energy when the nearby city had none. To be on a local grid was a way of being resilient and 'energy-independent' from the socio-technical entanglements of infrastructure and the breakdowns that might result from social or ecological unrest.

Or at least this was the hope. In fact, the traces of electricity data coming through all the meters was raising new questions for the co-housing group about just what independence should look like. For, as electricity was being monitored and tracked by digital devices, its peculiar properties were, to quote Stengers, beginning to 'force thought' in previously unanticipated ways.

Take, for example, an observation that the site produced slightly more electricity than it used. This raised the question of why they would need to be billed at all for electricity from the national grid. To answer this question required a sensitivity to the material properties of electricity itself. Electricity is produced by the movement of electrons as they leave and join atoms. For the national grid to operate effectively, it has to create an almost perfect balance between the amount of electricity being produced and the amount of electricity being consumed at any one time to balance the whole system. If there is supply without demand, or demand without supply, then the grid breaks down, either through overheating or lack of flow. The co-housing site was connected to the national grid, but also had the potential to operate independently if sufficient electricity were being produced on-site. If there was a lack of local electricity, their connection to the national grid would balance the system out by simply drawing more electricity. Conversely, if they produced excess electricity this would be pushed up into the national grid, and they would be compensated for this excess electricity through payments via a feed-in tariff scheme.

Realising this, one of the concerns of the co-housing group was how to keep more of their electricity on-site when it was being generated, and how to minimise their need to use the electricity grid at times of low energy generation. As a proto-grid controller, Tom was being drawn into questions, not only of social organisation, but also of electrical possibilities that emerged out of, and constituted an energetic milieu for, imagining what social transformations could look like. On the one hand the aim of the housing group to be increasingly self-sufficient was one of the reasons they had been chosen as a test site for the smart-meter pilot project, for they were seen as a potential future model of a micro-energy community that would be able to help to balance out the imbalances produced by a renewables-based electricity grid through local grid management. But, as they had learnt more about electricity, they had come to realise that smart metering and smart grid technologies more generally might allow local communities like theirs to become more energy self-sufficient, not only by knowing energy better and reducing their own energy use, but by rethinking the very idea of what an energy grid could be.

The capacity of the smart meters to pull into view the proclivities of electricity's materiality not only taught the co-housing community about the already-existing organisation of eco-communities, but was also creating new ways of imagining electrical power as a medium of social transformation. First, there had been some discussion about whether the co-housing site down by the river might provide some electricity to the village at the top of the hill. The co-housing group wanted to be energy-independent, but they were also keen for others in the nearby area be able to participate in this independence. However, not being 'behind the meter' like the co-housing group, other households in the village were not able to draw on the electricity produced by the co-housing group and had to get their electricity direct from the national grid.

The possibilities and limits inherent to the form that is electricity, as evidenced through metering and, in particular, the move towards smart metering, materialised a condition of possibility for imagining present and alternative kinds of electric collectives. This is not as simple as saying that the materiality of electricity was a determinant of social practice. Nor was it as straightforward as saying that the representation of electricity in meters and grids was a 're-de-scription'[1] of an already existing reality that was simply being 'seen' differently. Rather, the lineage of work that has highlighted the interplay between materiality and mind attunes us to how the knowledge-producing capacities of digital devices like smart meters work with, bring into view, and open up possibilities for transforming both the materialities they describe and the socialities that these material configurations enable.

I return, finally, to the failure of the energy monitors mentioned in my final interviews at the co-housing site. One of the failures was that the monitors kept switching themselves off, meaning that they did not fulfil their promise of providing a stream of continuous data to Tom or the residents. No one had the time to keep going to turn them on, so they just there lay dormant, as objects, but ineffective as empirical devices. Talking to Tom about why this was the case, he explained to me that one problem with the project was that it was run as an electrical-engineering project and not a community energy project. Here, he pointed to an interesting dimension of what we have been exploring here: what happens when methods are expected to be transferred from laboratories to devices, from idealised delegates of human knowing to active participants in ecologies of knowledge. In this project, the system had been tested first in a laboratory. But this had been done without engagement or conversations with the potential users of the devices like those in the co-housing site. The smart meters worked perfectly as knowledge-producing devices in the lab, but they had never been connected to the user interface and it turned out that even when working, the data was not being sent to the server in a way that the interface could interpret.

The failure of the meters for the co-housing group was, however, more than just a technical glitch. Having become sensitised to the entanglements of electricity, community, and the local grid, the residents were very keen to pursue new kinds of energy generation, distribution, and storage as a way of extending their aims as a community-energy group. The failure of the meters produced not just a gap in their knowledge, but created a block on the possibility of remaking themselves as a community through the materiality of electricity, its potentials, and its demands. Here, the energy monitor as digital device with empirical possibilities had created an opening to a newly materialised imaginary of a social and collective future. The question now was how to realise this emergent energetic imaginary, and whether functioning digital devices could support them in this endeavour.

Conclusion

This chapter has explored how we might approach the study of digital devices as empirical technologies. What we have uncovered in this brief exploration of digital devices is not only material artefacts that have social meanings ascribed to them, but world-making devices that are generative of a materially informed mode of social and collective imagination.

To help us think about how to approach the epistemological qualities of objects, I have drawn on the work of scholars who have long concerned themselves with the question of how thinking and the imagination shape social worlds, and others who have considered how thinking might be reconsidered in ways that do not reproduce the divide between the realm of materiality and the realm of ideas. Rather than seeing objects as material and representations as matters of knowledge, this chapter has suggested that objects can also be knowledge-producing entities, meanwhile the representations they create can in turn be analysed as forms of material culture. As Coupaye also argues in his chapter, this demands taking into account, not only the function or use of digital devices, but also attending carefully to what he terms their 'actual functioning.'

Building on, but also pushing beyond the idea that objects have agency, I have suggested that acknowledging digital devices as knowledge-producing entities allows us to entertain the possibility that when technologies take on the world through sensors, and in doing so transform the indexical traces of material properties into signals, images, texts, and visualisations, these devices become capable of becoming active participants in the making and remaking of the processes by which the social world becomes knowable. Here, digital devices bring to the world not only object agency, but also the possibility of alternative formations of thought, further challenging any simple opposition between human beings and material artefacts, and raising profound questions about where we as anthropologists should turn our attention as we seek to understand the material formation of social imaginaries and their role in the reproduction of social life.

Note

1 Playing on Madeleine Akrich's (1992) notion of the description of technical objects.

9 Rethinking objectification and its consequences

From substitution to sequence

Susanne Küchler

Anthropology has always struggled with accommodating objects into ethnographic analysis in ways that go beyond mirroring what people do and say. Despite its efforts to bring the object into the foreground of social analysis, and despite the fact that the material culture perspective has opened an invaluable new perspective on lived-in worlds richly exploited across the social and historical sciences, material culture has arguably not made a paradigm-shifting difference, theoretically and methodologically speaking. This is, I argue in this chapter, because the study of objects has remained firmly wedded to the social anthropological assumption of how objects work and what they do in society, an assumption framed within the theory of a dialectic constitution of subject and object relations. Productive of analyses that show how objects make people as much as people make objects, the definition of what has become known as the theory of *objectification* allows for an understanding of how social relations are articulated and maintained in the everyday. The work that made a lasting impact on how anthropology conventionally understands objectification and the difference it makes to society was Marcel Mauss's (2002[1925]) theory of the gift. Gifts, Mauss argued, are capable of substituting or standing in for persons, and are thus able to have an effect the reach of which is extended beyond the physical boundedness of a person. This definition of objectification, attentive to the classification of objects capable of standing in for persons became the foundational pillar of the anthropology of the twentieth century, framing the way it approaches objects to this day.

In fact, the definition of objectification as substitution set out by Mauss in his account of the socialising capabilities of the gift, has proved so productive in anthropological analyses that attempts at formulating an alternative definition have remained largely in the underground of the discipline. This chapter will retrace attempts at challenging the standard definition of objectification and set out the theoretical ideas that inform an alternative. This alternative is attentive to sequence and to relations between objects understood in temporal, rather than classificatory, terms. It argues that objects are not only capable of serving interpretation on account of their classificatory relation to one another, but to enable incantation on account of their capability to serve as a model of sequences crucial to the understanding of the working of complex systems. The kind of complex operational systems that objects make tangible and visually accessible range from kinship to genealogy to resource and land use, all sharing a common trait of being part of distributive rather than extractive economies.

The search for an alternative definition of objectification met with resonance among anthropologists familiar with ecologies in which societies invest in prospective strategies and operatives that are predictable across time and space and that sustain political economies in which distribution reigns paramount. One such ecological niche is wider Oceania. To explain the relation between the idea of objectification, of a mathematical (recursive and predictable) nature of operational systems and the real world people inhabit, it is useful to recall the words of the geographer Tim Flannery (1994) whose *Future Eaters* set out the distinctiveness of the ideas that enabled the settlement of Australasia. Europeans, explains Flannery, came out of an environment conducive to the exercise and use of raw power, as the landscape was rich and extractable. By contrast, the flora, fauna, and human inhabitants of Australasia had to learn how to make a lot out of a little, to husband meagre resources by seeing how far they could be extended rather than how quickly they could be extracted. The distributive economies of wider Oceania thus have in common a concern with the mapping of sequences that underpin the complex operations of distribution, making life possible, and creating wealth by moving people and produce along rhizome-like networks (cf. Bird et al. 2019). Objects in this setting show off how predictable operations of distribution work and demonstrate the success of polities that claim ownership over resources.

The story of the alternative definition of objectification is one that involves the retelling of the history of theory in anthropology, following ethnographic research into wider Oceania. The notion that seemingly small societies create wealth and fame by extending themselves beyond the reach of persons via beautified objects such as canoes and decorative artefacts, is now commonplace in anthropology (Munn 1986; Bird et al. 2019). Ideas of (dividual) person and of (intellectual) property emerging from anthropology in Oceania (cf. Strathern 1988) quickly became now-classic counterpoints to an anthropology that had been framed by a very different set of ideas of personhood and property, emerging from the industrial context of mass production, where objects as substitutive of relations of labour (Forty 1986; Bourdieu 1984a). Alfred Gell's (1992) now classic paper – written for one of the early seminars on the anthropology of art held by a group of students in 1985 – profiled this double definition of objectification in his sharp critique of technology. Resonant of the work of Roy Wagner (1975, 1986) and the extended body of ethnographies emerging from Oceania, Gell's paper redirected theoretical thinking in anthropology to the question of what kind of work objects do and what this work does in society. In fact, the timing of the paper proved perfect as, with the end of Fordist mode production in the 1970s (Martin

1992), the question of how objects capture the sequences and modalities of distribution, now at the heart of economies of production and consumption, quickly moved from a regional preoccupation of anthropologists to concern the mainstream in anthropology. Thirty years on, the questions now asked about infrastructures and operational sequences shown off by objects in a demonstrable fashion have been brought together in the present volume.

We might wonder why, in the face of ethnographies complicating the standard definition of objectification, the assumption that it can be reduced to substitution managed to prevail. One such explanation is given by Michel Foucault (1970) in his now classic, *The Order of Things*, in which he traces the theoretical and methodological appraisal of objects to Enlightenment's epistemological concern with Man (cf. Maniglier 2013). The study of objects as stand-ins for concepts and practices attended to by persons has drawn attention to the diversity of forms given to materials of the same category, in effect allowing concepts to be deduced from a comparison of forms within categories of objects, which in turn enable an inductive study of the technical processes of production. Branded by Tim Ingold (2012) as the 'hylomorphic model,' whose long tentacles reach down to Aristotle, the indexical qualities of form, allowing for an understanding of relations between persons and persons and things (Gell 1998: 148), have been commensurate with a rich body of method now synonymous with the study of material culture in anthropology (Tilley et al. 2006).

We arguably have had, however, hidden in plain sight, another still largely untold notion of objectification. Its tentacles reach to the depths of the anthropology of the Enlightenment, with the writings of Johann Gottfried Herder (1778) on emotion and cultural form and Gottfried Leibniz on the Theatre of Nature and Art (Bredekamp 2008), and forward into the twentieth century and the work of the art historians Aby Warburg and Walter Benjamin. This other sense of objectification is concerned. not with what is seen in the object and with forms that make visible relations of production (*sehen*), but with showing off in a demonstrable fashion (*zeigen*) what cannot be independently referenced from the object (cf. Ginzburg 2001). Objects of this kind officiate as diagrams, maps, or models that bring quantities into relation in a qualitative manner. Such objects capture, in their form, relations the nature of which entices what Gregory Bateson has famously called the method of 'double description' (1979: 79), in which abduction (the finding of similar forms) is followed by induction (comparison leading attention to shift from similarities to differences) and the formulation of higher-order concepts or normative rules (that explain differences). This means that relations immanent within such objects are not easily re-constructible, while they are intuitively recognisable and intersubjectively understandable.

The closest ethnographic account we have of such an alternative framing of objectification can be found in the little-known historical ethnography entitled *Rockdale*, which recounts the life of a rural Virginia town and its people on the eve of industrial revolution in America. Written by the

American anthropologist Anthony Wallace (1978), known for his work on religion, *Rockdale* recounts the practices and attentions to distribution, honed over centuries, that initially informed academic concern with the workings of machines.

Wallace describes in this study how the writing and receiving of letters was attended to with fervour in the days leading up to industrialisation. Letters, bound and kept, were tribute, trophy, or testimony to biographies whose strategic or fated unfolding they mapped. The life of letters attended to the sequences of sending and receiving letters, accompanied by an accoutrement of pens, tables, and chairs, of middlemen, and of collection points and modes of transportation. The operational capacity of machines initially was met with similar modalities of attention, directed as it were to the hidden sequences and connections, captured in diagrams, the improvement of which was the work of a new class of people known as the machinists. Wallace shows that machinists explained to each other the hidden workings of machines, not with words, but with drawings and models that rendered the inner workings visible and tactile (Wallace 1978: 237–238). The model or drawing of the machine at a size substantially smaller than the machine is shown to have been vital to allow the product of the machinists' thinking to be communicated in order to ensure that companions had approximately the same visual experience as the person operating the machine himself. What Wallace describes here is the importance of sharing the understanding of *operational qualities* among those working with and repairing machines, an understanding that drew on skills of mapping temporal relations via objects, in a not too dissimilar manner such as letters in pre-industrial times. Models and diagrams show off the workings of the machine, its invisible operational system, with an attention to the detail of connections that alone enabled subsequent breakthroughs in 'seeing' ways of improving its workings.

Wallace thus recovered for anthropology the way in which objectification can be seen to attend to operational thinking via the modelling of sequence, a notion that is also explored in Chapter 4 by Ludovic Coupaye. As models of operational sequence objects officiate as epistemic objects (see Hannah Knox and Rafael Schacter, Chapters 8 and 14). Unique to Wallace is that he draws our attention to the kind of epistemic work the modelling of operational thinking such objects accomplish. The machines he describes are demanding of care and attention – intellectual as well as physical – to stave off obsolescence (cleaning and repairing). They are also never completed, as every diagram or model is just a snapshot of a system that itself is opened to improvement as a result of the externalisation of its inner workings.

The objects that will be used in this chapter to explore the alternative definition of objectification are, thus, in more than one way, never singular and unique as they either capture different moments within an operational sequence or else project one of many possible perspectives onto the system as it is understood at present. This means that objects attuned to this alternative definition of objectification as sequence are demanding in that they require an ethnographic approach that is both time sensitive and time intensive. Time sensitive in that the attention to time, mapped in a modular fashion in an object, demands that we understand an object to unfurl as a sequence of images, the relations between which may show up as patterns of non-random variations when we attend to similarity and difference between one object and another.

By attending to the patterns of constancy and variation, we can then proceed inductively to arrive at an understanding of the normativity of sequence and proceed on to consider what kinds of operations the attention to sequence might capture. And this, more than anything, demands time – time to understand what operations in fact are being modelled. In Oceania, as we will see in the remaining part of this chapter, operations tended to differ widely and range from complex systems such as gardening, usufructury rights, and genealogy. What is attended to in any particular case study may also change over time and this, in turn, may lead to seemingly new ways of modelling sequences harnessing seemingly new materials and new technology. The alternative definition of objectification, therefore, demands sustained attention by the anthropologist to unpack clues, using objects and the images the systemic nature of which they reference obliquely, about operational systems that themselves demand to be attended to with equal dedication on the part of the anthropologist.

An example that illustrates the theoretical purchase, but also the methodological challenges, of a definition of objectification as modelling sequences underpinning operational systems is Yap stone money, described by the American explorer William Henry Furness III during his two months' visit to the Micronesian island in 1903 (Furness 1910; Martin 2013: 2–5). Given the brevity of his stay, Furness must have been gifted with extraordinarily powers of analytical imagination and clarity of observation as he was able to not just understand the complexity of the social system he found. He also understood how a system of credit and credit clearing that allowed a hierarchical system of relations to underpin the political economy of Yap society was modelled. The Yap had, he concluded, against his own expectations an economy based on money in the form of large, solid, and thick stone wheels, called 'fei,' ranging in size from between two feet to twelve feet. Although 'fei' were used to secure transactions, the stone wheels, taken from quarries by canoe from islands outside Yap territory, were never, or rarely, moved. This is because rather than serving as quasi commodities in barter-like exchange, they served to make manifest the potential for future transactions of the household, not its past executions. The measure of a stone wheel thus did not denote a relation to a hypothetical set of commodities that could be purchased with it. They were not mere stone coins, albeit of an unusual size, but instead were instrumental to a temporally structured system of credit and clearing – 'a tangible and visible record of outstanding credit the seller enjoyed with the rest of Yap' (Martin 2013: 12).

Furness concluded that the objects thus presented were not substitutes for other objects or forms of value, including persons in whose yard the boulders sat, but were manifestations of the system of social prestation itself. Yap stone money, differentiated in terms of scale, proportion, and multiplication, thus emerged as an objectification of distinct, temporally extended, forms of available credit. The idea that objects can serve to model sequences of exchange extended over a period of time and across a vast region was influential to my own analysis of objects that have been produced in their many thousands on an island known as New Ireland at the northernmost extension of the Bismarck Archipelago, which connects island Melanesia with Micronesia. *Malanggan* is a corpus of object, songs, and dances as well as sequences of rituals commencing with the burial of a person and concluding many years later in the carving, moulding, or weaving of a so-called *malanggan* effigy (Küchler 2002). Unlike the stone wheels carved by the Yap, *malanggan* effigies are not made to last, but are left to rot in the forest (woven effigies are burnt and moulded ones destroyed) or are sold to passing travellers via the island's mission stations that officiate as middlemen. The interest the fret-like carvings espoused among connoisseurs of Oceanic art led to collections numbering into the tens of thousands, the formal study of which had allowed me to formulate the hypothesis prior to fieldwork on the island. This hypothesis stated that effigies themselves were composite assemblages of motivic elements, and that the assemblage particular to each effigy was not random, but itself made reference to relations that were internal to the operational system of which they were a part. It took an extended period of ethnography study and renewed work on collections to understand that *malanggan* was not so much a system of credit bearing and credit clearing as a system the operation of which underpinned a complex leasehold system covering differentiated rights to land and its use. The transposition of a body after death into body politic, the image of which was intersubjectively shared by those participating in the 'work for the dead,' allowed a future directed and inherently resilient system of usufructury relations to span the entire island in a rhizomatic fashion that, like the effigies that serve to model it, was perpetually under construction. It took yet more perseverance to understand how it is that effigies reference such modular time maps, enabling people to navigate through time as confidently as they navigate the island, fending off unwanted trespass onto land as much as selectively forgetting retrospective leasehold relations to pave the way for prospective ones. Rather surprisingly, the act of binding turned out to be as critical as understanding the poly-perspectival nature of the knot and its manifold transformations (Küchler 2002), the geometric nature of which continued to inform hypotheses I began to formulate about seemingly very different objects fabricated elsewhere in Oceania.

The objects of Yap and island Melanesia are classic examples of objectification that frustrate attempts at conventional analysis, which assumes relations between persons and persons and things to be understandable via relations between objects, as these relations are referring back to themselves and to their systemic properties. Although the term is problematic in the connotations it conjures up, it is useful to recall the inflections given by the anthropologist Roy Wagner (1986), who refers to the self-referential nature and immanent relationality of objects as 'autistic' to emphasise the generative competencies inherent in such objects whose hold on imagination he was interested in understanding (1986: 11). I will expand on other examples in this chapter, each example showing, in the sense of showing off, the workings of a different complex system. The different kinds of understandings that their modelling makes possible will show that the analysis of objects that do the modelling is far from straightforward, demanding that we be sensitive to the workings of complex systems, the nature of which we may not be equipped to understand. Even if we are able to get our head around complex ecological systems, economic systems, genealogical systems, usufructuary systems, and kinships systems – to name just those that briefly will be touched upon in this chapter – the way their operational logic is attended to in the object might stretch the remit of our training and imagination to the breaking point.

While attention to sequence rather than to classification in our yet-to-be-fully appraised theory of objectification is fraught with difficulty, this chapter will try to explain the purchase this theory is set to deliver. In fact, by showing how objects and the particular qualities, symptomatic of their formal arrangement, attend to quantifiable data moved about in complex operational systems, we allow objects to be comparable to data compiled in charts and diagrams. A direct comparison can thus be made between the argument espoused in this chapter and the work of Antonia Walton on the aesthetics of 'good' science data. By studying objectification as qualitative modelling of complex systems, we can arguably begin to close the gap methodologically and theoretically between object collections and the understanding we bring to big data with the aim to understand better what kind of modelling works and the difference it makes to society. Perhaps most excitingly, we can begin to reach out to coding and computational modelling with a renewed zeal to match the thrill it offered as a new methodology back in the early 1960s, when Claude Levi-Strauss made the greatest inroads into our understanding of kinship systems and the transformational patterns of myth using early computing.

The life of working models

Rockdale is without a doubt not well known in anthropology. In fact, *Rockdale* is quite possibly the kind of books one will not ordinarily come across unless explicitly referred to it. Such is the power of personal reference; however, it should not surprise that the author's thinking had the most profound and lasting impact on the thinking of anthropologists working in places far removed from the small town in Virginia that was the subject of Wallace's archival research. This is because it so happened that two of the most prominent anthropologists working with acute theoretical intent on ethnographic data collected in Melanesia were Wallace's colleagues and friends. Of these, the Melanesianist Roy Wagner advanced Wallace's thinking in two of his most formative theoretical projects. These are *Symbols that Stand for Themselves* (1986), a book that followed his earlier work on *The Invention of Culture* (1975), while Frederick Damon tested Wallace's theory in his long-term research to understand the way people in island Melanesia model the constraints of living in a complex ecological system, recently published as *Trees, Knots, and Outriggers: Environmental Knowledge in the Northeast Kula Ring* (2016). The work of these two anthropologists has, in turn, had profound influence on a generation of anthropologists working in Melanesia.

In the introduction to *Symbols that Stand for Themselves*, Wagner considers the process of modelling in science and social science to show off familiar relations and orderings, in ways extended across analogous domains, as new understandings emerging through research that leads to a restructuring of the model. He uses the example of the idea of the double helix that is 'seen' in the sense of showing off and is thus able to inform the structure of the DNA, the subsequent remodelling of which instils a confidence in the model which, in turn, leads to 'paradigm certainty.' Wagner proposes that this modelling procedure, supporting paradigm certainty or trust, underpins the invention of culture (1986: 10–12). In paraphrasing the nature of modelling, he calls up the figurative usage of symbols, which cannot provide a literal field of reference, while 'figuring sympathetically by becoming itself that which it expresses' (ibid.: 6). Self-referencing in a generative manner – in short, the symbol that stands for itself – is simultaneously what it is and what it is about. As model, a sign is at once propositional and a resolution, bearing the imprint both of generic form and self-closure.

There are definite crossovers between Roy Wagner's thinking on modelling and the work of the anthropologist Patrice Maniglier (2006), whose ideas in *The Enigmatic Life of the Sign* are taking forward the immanent relationality and self-referencing of the sign in ways that constitute the ultimate paradigm shift in anthropological thinking. My own thinking on the peculiar role played by algebraic systems in supporting the generative and transformative qualities of objects (ethnographically informed by my work in the Pacific), has found a surprising echo in Maniglier's work, to which I will return in the concluding section of this chapter. For now, however, I turn to the work of Frederick Damon, which sets out the rather complex idea of self-referencing objects and their work as models of a system the complexity of which means that the model itself is perpetually under construction as much as is our understanding of it (cf. Ingold 2010).

Damon's book, *Trees, Knots and Outriggers*, on the Massim region of Papua New Guinea has been long in the making, as he himself admits (2016: 5). His interest in the canoe as model of a complex ecological system started when he first began to question how it is that people manage to live in the array of small and variously articulated islands clustered in the waters of Southwest Papua. To understand what people in fact know about the ecology of the area, and how these ideas are shared intersubjectively, Damon had to turn himself into a specialist of ethnobotany and ecology, with astrology to boot, as his research proceeded. In the small island environment of the Massim, a sailing vessel known as the *kula* canoe is central to connecting islands close enough to make a voyage possible and yet far enough to make travel by canoe arduous. The complex system of trade and exchange between the islands, each with its own ecological constraints on gardening and the harvesting of maritime resources, is the subject of the classical study of the *kula* (as the regional exchange system is known), carried out by Bronisław Malinowski (1922) in the early part of the twentieth century. Malinowski extensively reported on the minutia of the trade between the islands, constrained by both seasonal variations in growing patterns and the islands' peculiar environmental conditions that made some inhabitants entirely dependent upon trade of foodstuffs from other islands in exchange for pots and other articles produced by the resource-poor islands. There have been many subsequent studies of the *kula* canoe and of the relation between trade and the exchange of distinct valuables and their clockwise and anticlockwise exchange (cf. Leach and Leach 1983; Munn 1986).

Damon was the first to realise that the people there are attentive to the different types of trees from which the canoe is constructed and to the locations on an island where these trees grow. The selection of trees appropriate for particular parts of the canoe forms the backbone of the book, which is an exciting detective story into an unsuspected relationship of the trees, land, and canoes. The realisation that the structure of the canoe is mapped onto the island, and that this map reflects variations in seasonal and general intensity of gardening that significantly varies across islands, prompted further questions. Damon began to pursue the question of how a canoe is in fact working as a model of the spatio-temporal relations of seasonal resource use that enables islanders in the Massim to predict which island will have what kind of food ready for harvest at specific times of the year. The result is a fiendishly complicated study that shows canoes being refitted with new parts to serve as better models for subsequent sequences of its journey. The distribution of trees on the island follows the structure of a model canoe, invisible yet shared by all inhabitants as an idea that informs the relational nature of all actions in ways described by Nancy Munn (1986) as characteristic of the value-creating nature of the habitus.

Damon reported that he had almost given up ever being able to explain what he had understood about the intertwining of the operational qualities of the ecological system and the structure of the *kula* canoe when he read one of my own papers on the knot in Pacific imagination (Küchler 2003). What in fact had led to a breakthrough in my own understanding of how *malanggan* effigies model the usufructuary system was the decision to take seriously the fact that the effigy carvings did not look anything like the dead, whose passing they marked, even though photographs of people are available, and sculptors certainly were able to create life-like appearances from wood. What in fact was carved was a knot, or rather the inside of a knot transposed onto different dimensions to create systemic variations of one and the same object, casting shadows of its own life backwards and forward in time. In short, the knot and actions of binding turned out to be shaping the order of objects and their sequential relation to one another, allowing *malanggan* to model the operation of the leasing of land by relying on logic alone. It took me another few more years of immersing myself in the intricacies of differential geometry to understand the model I was dealing with even better. The object of *malanggan*, encompassing all possible past and future images, is seen as synonymous with the idea of membership in the leasehold system as each and every object is in fact analogous to the inside of an object such a Rubik's Cube that could be manipulated to bring in view multiple perspectives at the same time (Küchler 2014).

One of the complexities this approach to objects bring with it is that the analysis of working model requires a perspective unique to anthropology, namely the perspective encompassing the biographical life span. As anthropologists are rarely able to extend their research across time to allow for an appraisal of how time is mapped and modularised, understanding how one object unfurls into sequences of images that allow for retrospective and prospective strategies gets one partially there (cf. Gell 1993, 1998). Yet there is a further complexity, and this is that we need to take seriously the possibility that one object can be the gateway to understanding what in fact is manifold. This last section of the chapter takes this forward to explore a related challenge, namely, that objects made in Oceania have in common the peculiar capacity to 'show off' the capacity to contemplate a multiple as one, or as Anneliese Riles has said '[to] hold multiple levels of action in view at once' (Riles 1998: 379). The ideas exposed here are not for the faint hearted, touching as they do on some tricky geometry, thus raising the question of what the training of anthropologists should really encompass.

Polymodality: geometry, transformation, and translation

Sometimes the most perplexing aspect of anthropological theory is how well known a theoretical insight is and yet how little is done with it, for decades. In an era of postwar anthropology in Britain in which model building was all the rage, two publications stood out. The first was Edmund Leach's (1954) pathbreaking book, *Political Systems of Highland Burma*, in which he set out the use of operational, representational, and explanatory models underpinning the transformational nature of Kachin social structure. The second was Levi-Strauss's (1969 [1962]) *Elementary Structures of Kinship*, which used computational analysis to unpack the logic informing complex kinship systems.

The influence of earlier texts on both these studies, which espoused the mathematical nature of working models of complex systems and the difference they make to society and culture is, however, not well known (Leach 1961). These texts are the posthumously published work (edited by none other than Levi-Strauss) of Abel Bernard Deacon (1934) on *Malekula*, an island of Vanuatu on the southern fringes of the Bismarck Archipelago mentioned earlier. Trained in Cambridge in natural sciences and studying for PhD in anthropology under Haddon, Deacon's findings led to a breakthrough in our understanding what the various objects attended to by Makelulans, from cats-cradle figures through to sand drawings and woven mats, were all about. Deacon's training in algebra and differential geometry enabled him to translate the myriad forms of patterns, woven from fibers and drawn into sand, spun into fibres and danced into the ground, into an algebraic system. Rather than being mere illustrations, pattern making in fact was shown up as a model the Malekulans think with, play with, and with which they seek to refine the operational sequences underpinning the extension of affinal relations.

For Levi-Strauss, Deacon's insights into the nature of the model and its relation to the operational qualities of the kinship system proved the decisive impetus for his later work on the *Elementary Structures of Kinship* (1969) that enabled anthropology to study kinship with a method and a rigor that was transformative to the discipline. For Deacon had in fact recognised the use in pattern making of an algebraic system of numbers and rules for their combination – the logic of which is known in mathematics as quaternion – that enables number series to be translated into geometric objects and back again. The peculiarity of the quaternion, the operation of which we know well from the Rubik's Cube, is that it enables one to envision changes to the spatial location of elements, in the formula and its geometric analogue, as reversible and as logically predictable. Sequences of 'moves' can be executed in the mind and the result be imaged from many perspectives at the same time, allowing the quaternion number system to be the code of choice for computing rotational objects digitally. The deployment of quaternion number systems proved revolutionary to Levi-Strauss who used it to unpack the transformational logic of myth and to understand the workings of complex kinship systems, the most complex of which he testified to exist in Malekula.

The legacy of these insights is acknowledged in the work of the anthropologist Knut Rio (2007) who has correctly established Malekulan sand drawings as an indigenous model and also highlighted the lack of a concern with classification and with the substitution of subjects in the making of objects. He does not, however, pick up on the mathematical and geometric ideas exposed by the kinship diagrams and the underpinning workings, in part because the drawings on which these ideas were based were by Deacon and not by the Malekulans. When we see that the drawings model the sequences of the system in much the same way as the objects upon which they are based model the idea of the operational system itself, we can see how fertile the conclusions to which Rio was drawn in fact are.

An example of an earlier study grappling with the type of logic underpinning an attention to objects that exposes their sequence in relation to one another is the work of Remo Guidieri and Francesco Pellizi (1981), who wrote a fabulous essay on the tree fern sculptures of Vanuatu. Their essay explores the relation between the figures and sequences of ritual that map out the stages men are required to go through to acquire ritual power. The figures do not serve to recall particular stages of the ritual, but rather they are there to permit the contemplation of ideas underpinning a system that equates the highest rank with the position of the living dead. Their essay, seminal and poignant as it is, has been hugely significant to my own thinking about how objects model processes of transformation underpinning cosmologically charged genealogical systems across island societies in the Pacific, leading to my recent work on the coverlets stitched in the Cook Islands of Eastern Polynesia (Küchler and Eimke 2009; Küchler 2017).

The *tivaivai* coverlets whose stitched compositions of floral images and arborous patterns trap those looking at the coverlets as representations, as if they are made for seeing rather than contemplation. In fact, that the coverlets hold relations immanent within the assemblage of motivic elements and these relations affirm the cosmology of the body politic as much as they validate a concept of personhood that equates one with the many. Composed of iteratively replicated and transitively arranged self-similar motivic elements, themselves the product of recursive number systems and their non-commutative computation, *tivaivai* hold the clues to how they work and what they do locked within their own construction. Made as shrouds for the dead and fanned out during life in exchanges that punctuate the life cycle, *tivaivai* eventually are to be wrapped around the bodies of the dead in tomb-like graves, their assemblage quite literally reconstituting the relations cultivated by the deceased person in an artificial cloth body.

Invisible, and yet made for contemplation, the paradoxical nature of the *tivaivai* is paradigmatic to social life in the Cook Islands. Nothing matters more on these islands than a person's genealogical position in a complex relational matrix that connects foreign with homegrown ancestors to adjudicate who has rights to land and where, and who has rights to annunciate the resulting patterns of relations. The genealogical relational matrix that literally shapes a person's life is contemplated and shown off to others in the form of a pattern that is being stitched into the coverlets, itself remembered as a sequence of numbers made visual as coloured patches. While the sequences that reference the workings of the genealogical system are attended to as immanent within and inseparable from these patterns, the operational sequence of stitching itself varies across the life cycle of a household.

When a young family first sets out to make and gift coverlets, these tend to be of an applique type (*ta-taura*), meaning that a composite motivic element is arranged iteratively in a circular fashion on the planar surface of a monocoloured coverlet, showing off relations between households related to one another as affines, bound together by marriage. Relations that are not thought to have a lasting impact on the household are recognised with cut-out coverlets (*manu*), by folding planar fabric and cutting a pattern into it so that it unfolds into an image with rotational symmetry in eight parts. It is only when her own children have left the household that a woman begins to stitch piecework coverlets (*taorei*) that recount the relations that connect the living with the ancestral generation. It is on a piecework coverlet that a young man must sit for his first ceremonial haircut while a genealogy is recited by his female relatives. It is the stitching of a piecework that a young girl learns who was adopted as baby into the household in ways that recount the arrival of the foreign ancestor. As the girl grows up she becomes the officiating head of a new household which, in genealogical terms, replicates all households that came before in a collapsing of time that is made visible and tangible in a demonstrative fashion in the piecework coverlet. Composed of iteratively replicated motivic elements, each in turn divided into a triangular shape, each part of a motivic element is stitched together by a different woman, following a precise sequence of coloured patches. The task is to stitch with the exact same tension as all the other women (in total 8 or 16 will be involved), working on one and the same coverlet so as to create a flat planar sheet that does not show unevenness. The 'oneness' of the tension and the exactitude of the replication of the pattern on the surface of the coverlet testify to a social body remade with every generation.

The principles of genealogy are attended in the Cook Islands as women stitch new coverlets to mark the extending of relations beyond the household to be eventually returned to the household. These sequences of unfolding and re-folding are also mirrored in the actions women perform upon the coverlets each time they are taken out of and returned to their storage trunks. We can also see the geometry of image-based polities at work when women create variations upon the pattern they have inherited from their adoptive ('feeding') mother in that the shape of a pattern is projected dimensionally, or in perspective, to show different views upon the same entity. A pattern is at once an algebraic entity and a geometric shape and, as such, immanent within the pattern, as inside and outside views are taken as an index of the subtle nuances of relations between persons and their households or simply to test the recipient's visual imagination and capacity to execute translations back to the geometry of the pattern. The rotational geometry of the pattern thus allows multiple views to be projected onto the surface of a coverlet, either successively or at the same time, almost as if the flat surface of the coverlet is itself a rotational three-dimensional object rather than the planar fabric it appears to be.

This aesthetic of the coverlet thus underscores the topological nature of its composition in that what is most important is not visible, but is reconstructed in the mind using imagination and the logic implicit in operations of genealogy. That it is the aesthetics of objects that is contemplated for its hidden reference to the workings of complex operations upon which life itself depends, is one of the often-missed conclusions of Alfred Gell's (1998) *Art and Agency* and inspired its writing. It is now time to return to the theme of this chapter, which concerned itself with recovering an alternative definition of objectification attending to sequence. The Cook Island coverlets also remind one of Wallace's writings about the machinists' attention to the complex operations of machines in their care. For it is the trust in the inner workings of a machine which was inseparably bound up with understanding and replicating the sequences of the operations critical to the work performed by each machine.

Objects made to contemplate the work of the machine were shown by Wallace to make a difference, not just to the machinist's attention to the machine, enabling improvements to be made to its operational capacities over and above the care devoted to maintenance. Wallace shows how these objects (diagrams and models) also allow the community of machinists to separate as an identifiable social group from those who operated the machines. The difference that objects, attended to for the sequences they map and allow to contemplate, make to culture and society is central to the anthropological endeavour.

Conclusion

The examples this chapter has brought together are mere snapshots and sketches that outline how objects could attract attention, not because they stand in or substitute for persons, but because they attend to the inner workings of complex systems in a way that magically 'becomes what it expresses' (Wagner 1986: 6). We saw how this 'becoming' can involve the fracturing and sequential unfolding of parts of a composite image, capturing the transmission of partitioned land (in the case of the *malanggan*). And we saw the multi-perspectival shaping of a pattern stitched into coverlets turn many into one and one into many in ways that offer up an understanding of, and trust in, the faithful transmission of genealogy as the cosmological core of the image-based polities (in the case of the *tivaivai*). The distinction we conventionally make between diagrams, physical models, and objects was argued in this chapter to be unhelpful when engaging with the question of how objects can map out and enable the contemplation of sequences underpinning the operation of systems that may not be available to observation. The chapter argued that the consequences of a shift in attention away from classification of objects to the sequences of which they are a part, and which they themselves 'become' a part of as they make manifest this attention to sequence, may offer up a productive engagement with questions at the core of anthropological theory. Perhaps the greatest benefit, however, is that the definition of objectification allows for, and prompts, a literal retooling of methodology to aid the analysis of data that, like the operations they attend to, are inherently complex. It is in the hope to have inspired the rethinking of methods that this chapter has been written.

10 Looking at things

Delphine Mercier

In September 2017, while giving tours of the UCL Ethnography Collection to first-year students, I was asked by one of them – as we stood in front of a selection of artefacts – if objects were important to understanding material cultures. As a reply, I asked him to tell me about the outfit he was wearing on that day, why he decided to buy it and why he picked it up that the morning. I was hoping it would help him to understand how the relationship between people and objects is intricate and meaningful. Things matter, indeed. They are the materialisation of culture and, sometimes, objects constitute the only traces left by people. Observing the network of interactions between people and things helps us understand both a bit better, as long as we take the time to really look at them. As curator of the UCL Anthropology Department's Ethnography Collection, I spend a lot of time surrounded by, and working on and with objects. I teach students how to engage with and learn from things. In this chapter, I draw on several years' experience working with objects in the collection, critically reflecting upon the use of methodologies like object analysis or assemblage and their efficacies in the teaching and research of material culture anthropology.

Looking at things implies being physically close to them, enough to be able to touch, use, or smell them, to see all the details of their surface and of their less-accessible parts. The importance of this direct contact for a better understanding of things has been confirmed by recent neuroscience (Quiroga et al. 2011: 17) and social-science studies, several of them addressing the issue of authenticity of learning in museum contexts (Leinhardt and Crowley 2002; Hampp and Schwan 2014; Eberbach and Crowley 2005; Bunce 2016). The definition of authenticity used in these studies meets the characteristics of 'aura' by Walter Benjamin; 'the here and now of the original [that] constitutes the concept of its authenticity' (Benjamin 2010 [1939]: 13), and the uniqueness and permanence of things (ibid.: 16). As noted by Dudley, based on her experience visiting galleries (2011: 2), criteria like the rarity or the age of the objects changes how we perceive them. Being aware of the authenticity of an object not only transforms the experience of engaging with it as something more valuable, but also, as Benjamin says, it puts the subject in direct contact with 'the quintessence of all that is transmissible' (Benjamin 2010[1939]: 14). As such, the aspect, weight, size, and smell of an object are all clues that can help us to understand the interactions between things, people, and the wider environment.

Engaging with objects is therefore a key tool in museums and collections, whether they are more traditional ones (Bunce 2016; Leinhardt and Crowley 2002) or, as with university collections, when they focus on students learning (Barnes and Lynch 2012; Vartiainen and Enkenberg 2013; Adams 2015; Kreps 2015; Krmpotich 2015). Helene Chatterjee, for example, has been doing pioneering research on the subject of objects and education, publishing multi-disciplinary research on the benefits of physical interaction with objects (Chatterjee et al. 2008) and working on objects and collections as powerful pedagogical tools for students (Chatterjee and Noble 2013; Chatterjee and Hannan 2015).

Considering that being in direct contact and engaging with original objects is a crucial way for people to understand them, the dearth of projects that focus on or start from the materiality of things is surprising. The vast majority of the sources referenced above highlight that the handling of objects is essential to understanding them. In her recent book, *Museum Object Lessons for the Digital Age* (Geismar 2018), Haidy Geismar details an object-based workshop, 'Properties and Social Imagination,' organised at UCL in 2012, which focuses on the materiality of the objects 'to explore the different ways of understanding and interpreting objects' (Geismar 2018: 28–62). Participants, including Geismar herself, focused on materiality, which in her case included a bark cloth, as key to understanding the objects. Physically engaging with the object is hereby not only a pedagogical tool used to illustrate ideas, but it is a method in and of itself, used to understand the object; to take 'the objects seriously in their own terms – a material-oriented perspective,' moving away 'the tendency of previous interpretive projects to incessantly convert objects into symbols' (Geismar 2018: 36–37). In this perspective, objects do not exist to illustrate ideas; they are the beginning of, and central to, the process of enquiry. In this vein, I elucidate the purpose and efficacy of an object analysis as an anthropological methodology.

Object analysis

An object analysis can be performed without having any particular knowledge of the object. The UCL Ethnography Collections is characterised – for various reasons relating to the history and provenance of the artefacts – by a lack of archives. This complicates not only the historiographic work one is able to do in and on the collection as a whole, but also with each object individually. Less than a quarter of objects in the collection have any documentation related to their provenance. This means that the information we can gather from the objects themselves is limited to the information provided by the labels, which constitutes the object's only first-hand information. As such, the object-analysis methodology becomes extremely important in the pursuit of understanding. This methodology, which is the focus of this chapter, is central in the teaching of anthropology at UCL. The experience of working in a collection environment with museum objects is generally new for the students. As Richard Handler describes it (Handler 1993), a collection that is a storage for objects is also a 'social arena.' Entering in the space of the collection, which has its own rules – for example, no food, no drink, specification related to objects handling – helps students to set their minds on getting ready for engaging with original objects.

As a methodology, one of the key questions of object analysis could be, paraphrasing Tilley (1994: 1): Why are particular objects made the way they are as opposed to others?[1] 'Phenomenology involves the understanding and description of things as they are experienced by a subject' (ibid.: 11–12) and, object analysis is based on a similar notion. Subjects enter into proximity with an object they know nothing about, with at times limited and/or wrong information. They therefore need to engage with the object using their senses, which helps them not only to find clues that they will interpret at a later stage, but also to develop an understanding of the experience of using the object. As Dudley says, sensorially exploring an object consists in 'want[ing] to touch it,' to understand 'how it would feel to stroke it, or how it would sound if I could tap the metal, or how heavy it would be if I could try to pick it up' (Dudley 2011: 1). According to Dudley, drawing on Tilley (2004) and Merleau-Ponty (1962), 'it is in the engagement between object and subject, in their very confluence, that sensory responses, emotions and ideas are generated' (Dudley 2011: 8).

As objects 'cement the ways people live together' (Lemonnier 2012: 13), the object analysis favours the development of a relationship between the subject and the object. Consequently, if the way of comprehending objects borrows from phenomenology methodology, the interpretation 'involves understanding the interconnections,' which are of a changing nature, between people, their observations and tools, the objects, the material cultures the objects belong to (Jones and Alberti 2016: 21–22). In turn, as people engage with objects as material things, their qualities become fundamental to the interpretation. At the same time, interconnections between objects, people, and their environment are also essential, as 'it is not that a material contains a finite list of inherent properties that are elicited by a knowing subject; rather, properties or qualities are the product of the interaction among subject, technology, and material' (Jones and Alberti 2016: 24 referring to Connell 2011). Thus, analysing the material, formal, structural, and functioning properties of objects can reveal localised (vernacular) relations of people with their environments, which André Leroi-Gourhan qualified as 'milieux' (Leroi-Gourhan 1971 [1943]; Leroi-Gourhan 1973[1945]; see also Coupaye, Chapter 4).

UCL Ethnography Collection[2] was created by Daryll Forde, the first head of UCL's Department of Anthropology, following its establishment in 1945.

The collection is made of seven thousand objects and photographs, mostly collected by Forde, who stayed head of the department until his retirement in 1969 (*Africa: Journal of the International African Institute* 1973: 281). Forde also managed to secure donations from other collections inside and outside UCL (Petrie Collection, Wellcome Collection, British Museum, Cambridge Museum), and a long-term loan from the Cuming Collection given back in 2007 (Were 2010: 300). The objects in the department's collection come from all over the world, with emphasis on West Africa and the Australian continent, following the orientation of the research at UCL Anthropology around the time of Forde. According to Professor Michael Rowlands,[3] the collection of objects was more spontaneous rather than scientifically organised, with little to no information recorded. After Forde's retirement, objects regularly joined the collection, but at a much slower pace, as less than a hundred objects have been given to the collection since the 1970s (Were 2010: 300).

Among the very few existing documents relating to the history of the collection are letters exchanged between Ashworth Underwood, director of the Wellcome Museum, and Forde between 1951 and 1954 – a time when this museum decided to re-orientate its refocus on health, the body and medicine, and to give away the rest of its gigantic collection. When he got in touch for the first time with Underwood in 1951, Forde described the collection as a teaching collection 'in connection with the teaching and research in the department.' Seventy years later, these words are still true. The collection gradually became, and still is, a valuable teaching tool, not only for various departments of UCL, but also for wider audiences including source communities and schools. Following a theoretical parenthesis (Buchli 2002a), a 'material turn' (Tilley et al. 2006) characterised by the 'rediscovery of the value of museum collections' (Adams 2015: 89) developed again from the end of the 1970s. It has been followed by a sensory turn, focusing on re-engaging with 'objects and their materiality on multisensory levels to overcome dependence on text and the purely visual' (Kreps 2015: 96–97), consequently putting 'the world of museum Anthropology and object-focused studies to a more central stage' (Adams 2015: 90). It is in this context that students learn how to engage with things using their senses and methodologies developed in the following case studies all kept in the UCL Ethnography Collection.

While the first case study exemplifies the object analysis methodology, the second and third ones illustrate the combination of object analysis and assemblage and how it allows to go a bit further in the understanding of objects.

Case study one: object K.0050

When starting an object analysis, first observations generally relate to the overall aspect of the object, such as the materials used. K.0050[4] (Figure 10.1 to 10.4)[5] is in a good conservational state and made of dense material. As such, it is possible not only to look at it, but also to touch, handle, and lift the object without damaging it. The object is made of one piece of wood, a dense

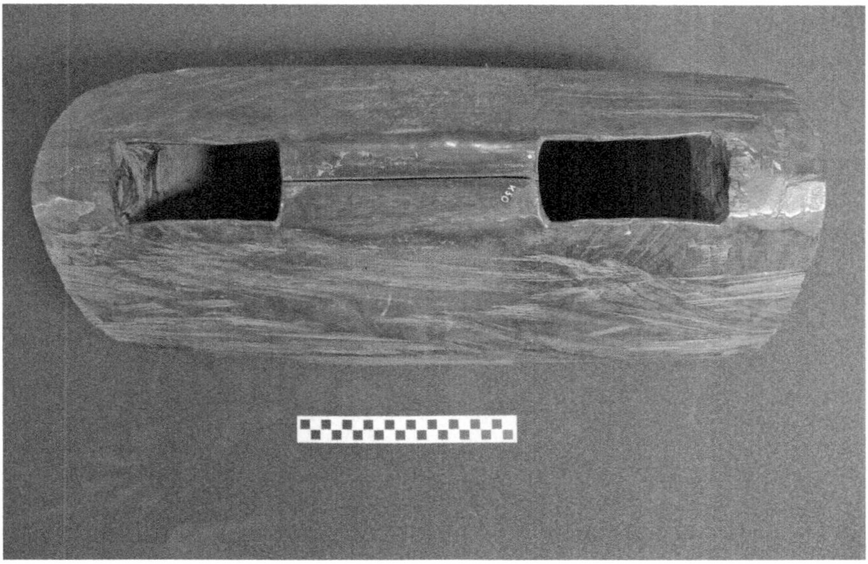

Figure 10.1 K.0050, top view.

Figure 10.2 K.0050, inside view.

Figure 10.3 K.0050, view from one of the small sides.

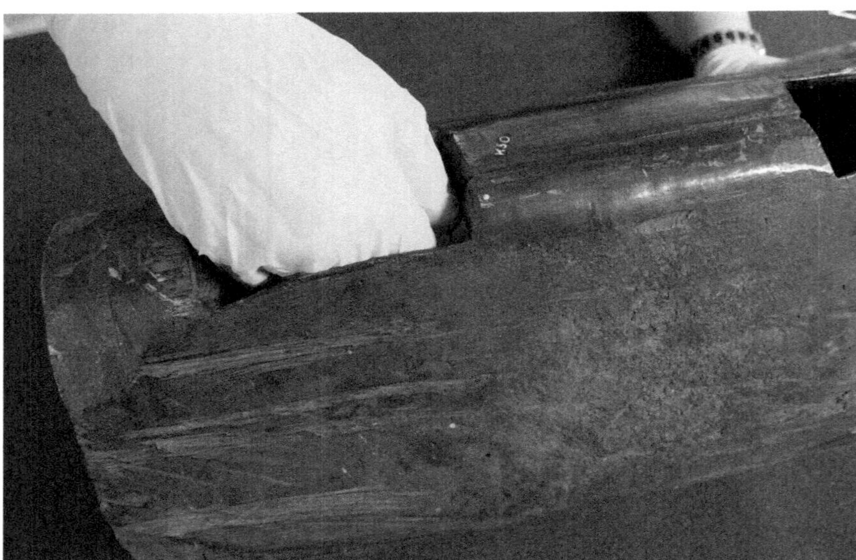

Figure 10.4 Attempt to introduce a hand inside of K.0050.
Source: Photos by Timothy Carroll ©UCL Ethnography Collection.

type, which one gathers from the weight of the object when attempting to lift it. The weight and shape of the object show that it has been made from the trunk of a large tree.

An object analysis methodology starts with sensory engagement in order to collect data that can then be analysed in a second phase. An object made of one solid large piece of tree trunk leads to the hypothesis that it was produced in an area where, indeed, there would have been trees, possibly in significant number. One could speculate that wherever K.0050 was produced, it was likely in a place where people could access or were surrounded by forests and trees of all ages. It might also indicate, or at least point, in the direction that wood might have been one of the primary local resources available, since people were able to use a sizeable part of a tree to craft one single object. This information also affords some speculation into the organisation of the material culture that produces K.0050, since the felling and transport of large trees is rarely handled by one person for safety and practical reasons. Hence, the existence of forests and the use of wood as a raw material also indicate the necessity of organised labour, or at least a need for communal work to be undertaken in order to obtain and transport the raw material to be used to create the object.

After focusing on the material itself, the next step of the object analysis focuses on how the object was made. The absence of bark on K.0050 means that it was removed before the trunk was carved. It was then carved into a pear shape (Figure 10.1). The surface of the object is covered with small incisions (Figure 10.1 to 10.3), which are two to three centimetres wide. This seems to be quite narrow for having been left on such a dense and thick material. These traces indicate that the tool used to carve the wood was made of a stronger material than the one used to make the object and had to be thin enough to leave precise and clear incisions. This could point to the use of a sharp blade of metal from the chisel type, and to the additional use of a hammer in having provided the necessary strength to leave these precise incisions. It is noticeable (Figure 10.3) that, even if these traces are mostly vertical, a lot of them have angles varying up to 90 degrees. This detail is extremely significant as it helps to mentally reproduce the gestures of the craftsperson who made this object, the movements of the body turning around the object to be able to carve it on all its sides, and the noises produced by this work.

The second part of the technical evaluation of the object would focus on the hollowing of the piece. At the top of it, there is a slit which is enlarged in a rectangular shape in the two extremities of the object. These openings (Figure 10.4) are narrow enough to make it complicated to pass a hand into them. Despite this, the inside of the trunk has been almost entirely hollowed (Figure 10.2). On one of the small sides of the object (Figure 10.3), there is a circular shape carved in the wood which is visible inside, too (Figure 10.2). This is certainly a sign that the craftsperson, after hollowing the object from the top, needed to dig a hole on the side of the object to manage to finish the work. They just had to put the piece of wood back at the end of the process. However, it was not only a work calling for strength: the craftsperson who made this object was also able to carve with delicacy, as shown by the very thin walls of the two top slits (Figure 10.3). The outside of the object is smoother than the inside because it has probably been roughly sanded to reduce asperities and to avoid splinters and other minor injuries. This technical evaluation of the object takes its cue from the *chaîne opératoire* (technical sequence), both from the point of view of archaeology (Walls 2015; de Beaune 2004; Schlanger 2004), as it starts from the object, and from the perspective of anthropology (Lemonnier 2012; Coupaye 2013, 2015), as it focuses on the uniqueness of each object. The direct engagement with the object and its uniqueness offers an experience also underlined by Drazin (Chapter 5). Object analysis, via the extrapolation and associative logic of abductive thinking (see Bateson 1972, 1979), is also an open door onto the intangible parts of life, like technical movements of the body (Leroi-Gourhan 1971 [1943]) – such as hollowing and smoothing, or carving in sometimes uncomfortable positions as well as the noise of the chisel, the time invested, the smell of the wood and the shavings piling up.

After having explored this object from a technical perspective, the next stage of the object analysis would turn to exploring its uses. Except for some scratches, traces of use are visible only at the top of the object (Figures 10.1, 10.2 and 10.4) – in the middle of the long and thin slit between the two rectangular openings. Here the wood is lighter and shinier than on the rest of the surface. Even felt through a glove, the texture is extremely smooth compared with the rest of the surface. By trying to touch it with various gestures, at some point the person performing the object analysis will gently hit the surface and realise that it produces sounds, which are two musical tones as the object is a double-slit drum.[6] The size of the instrument and its weight[7] – trying to lift it makes one realise that it is a very heavy object – suggests that this drum was likely meant to stay on the floor, with a musician sitting in front of it. The drum produces two different tones and can therefore allow for the creation of a complex rhythm. Engagements with this drum appear to have been diverse – playing it required human effort, particularly if played for a long time, while it also involves the acts of listening to it, and possibly also dancing to and around it. Other questions could also emerge, such as when was the drum played, during the day or the night? This aspect provides wider insight into the sound of the environment, in the context wherein a material object produces specific types of non-monotonal and rhythmical sound. Finally, as this object is a musical instrument, the craftsperson who made it had to be aware of the musical properties of the wood, knowing exactly how much to hollow it, or how thin the walls of the opening needed to be to produce the desired effects – perhaps the craftsperson and the musician had been the same person.

As shown in this example, an object analysis does not allow for a complete understanding of the entirety of the object. Without the adjoining label, information like the country of origin of the object for instance would stay unknown. It is also not possible to generalise all the interpretations made on one object as each object is unique, and even if it has not been handmade but machine made and mass-produced, the uses of objects differ, in discrete contexts. Therefore, in analysing an object, as in engaging with them in an exhibition as exemplified by Jeevendrampillai, Burton, and Sanglante (Chapter 7), it is necessary to understand the relationships between object, people, and environment, such as: the interactions existing between the makers of the object and the object itself, the object and the people who used it, the makers and the users, including the dynamics between them, as well as the situated context of the observer (the person performing the object analysis).

A thorough understanding of these networks in as holistic a way as possible is important in performing an object analysis on one isolated object. However, an object is rarely completely isolated and, in the case of a collection for instance, it might at least be related to the rest of the collection. By using the archaeological methodology of assemblage in a collection, it is possible to create connections between objects. This method is defined by Joyce and Pollard (2010: 292) as 'a collection of materials related through contextual proximity [making] possible to interpret the group of materials as evidence for specific events, processes, or practices in the past.' The assemblages, whose 'properties emerge from the interactions between parts' (Manuel DeLanda in Bille and Sørensen 2016: 16), also constitutes a first attempt of a stratigraphy of the collection. As Jones and Alberti argue (2016: 29), 'rather than interpreting the meaning of the artefacts they excavate through contextual analysis, archaeologists shape and compose the assemblages that they excavate; through this process of composition, interpretation and evaluation arises.' They conclude by stating that 'we have shifted from a conception of archaeological interpretation as a largely cognitive or cerebral endeavour to a recognition of the physical and material aspects of archaeological interpretation.' This dynamic relationship between the cognitive and the material leads to a better understanding of both. This is reminiscent of Bateson's theory of double-description (Bateson 1979: 132), whereby the observational description of an object and the abductive inferences about the object produce a deeper understanding of the object at hand based on the relationships of resemblance and difference between the object and those previously known to the observer (see also Küchler, Chapter 9 and Küchler and Carroll 2021b). Thus, in coming to 'look at' or 'be with' an object in an analysis, new understandings arise with 'reference to a particular pattern of interaction.'

Following on the dynamics of Bateson's double-description, case studies two and three use assemblage method – based on comparisons, interactions and inference – for a better understanding of objects and their context of making.

Case study two: the Inuit collection

At this point, I examine the significance of the environment in understanding objects through assemblage. In the following case study, assemblage

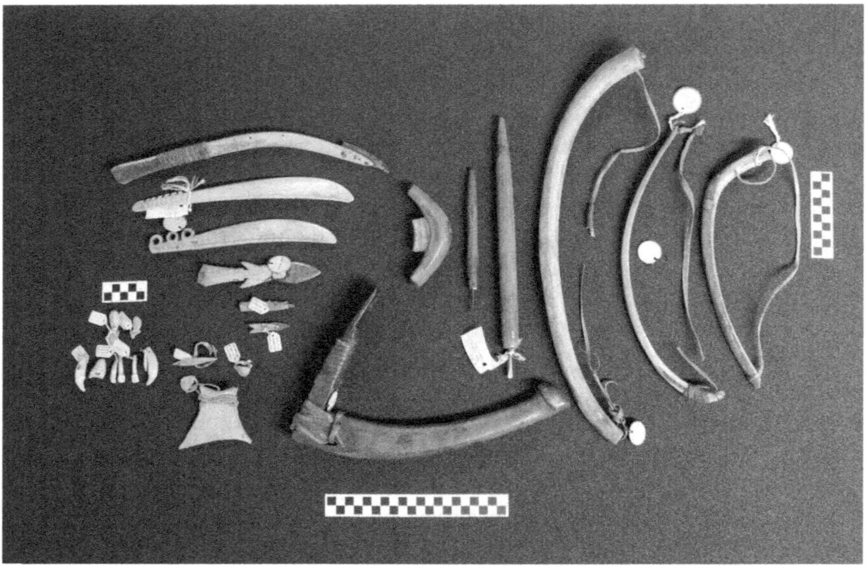

Figure 10.5 The whole Inuit collection kept in UCL Ethnography Collection.

Figure 10.6 Detailed view of the adze, showing the different layers of paint used to cover the wood every time it was reused.

Source: Photos by Timothy Carroll ©UCL Ethnography Collection.

methodology is used to understand the environment, in particular relation to the resources available and the interactions between resources and people.

The UCL Ethnography Collection keeps 24 objects from the Inuit/Inupiaq/Yupik/Eskimo[8] area.[9] By assembling and engaging with these items, we will try to understand how we can at least partially understand the environment in which they have been created and, consequently, the kind of techniques and knowledge needed locally to make things. The methodology of assemblage has been already used in the context of museums, notably to focus on issues like provenance and context (Flexner 2016). Here, empirical observation and sensory engagement will be used to highlight materials in relation to resources that were locally available at the time of the objects' production.

The Inuit collection is comprised of: one whittling knife; bow-drill tools consisting of two drills, three bows and a mouthpiece; a collection of five perforated animal teeth; a non-perforate animal tooth; two pieces on the *tingmiujang* game (Martjin 1964: 557); two *yaaruin* or story knives; a spear-head; a knife; a harpoon head; sewing equipment consisting of one buckle, two thimbles, a needle and a hand protection; an adze.

Table 10.1 Materials used to make the objects of the Inuit collection

Materials	*Number of objects in which the material is used on a total of 24 objects*	*Raw materials*
Animal bone	7	Animal
Animal skin	8	Animal
Musk Ox tooth	3	Animal
Wolf tooth	2	Animal
Animal tooth	1	Animal
Walrus ivory	5	Animal
Reindeer sinews	1	Animal
Reindeer skin	1	Animal
Wood	5	Vegetal
Fibres	4	Vegetal
Iron	6	Mineral
Native iron (?)	1	Mineral
Meteoric iron (?)	1	Mineral
Stone	1	Mineral

Table 10.2 Raw materials present in the Inuit collection

Raw material	*Occurrences*	*Different types of material*
Animal	28	8
Vegetal	9	2
Mineral	8	4

The two tables above detail the composition of the Inuit collection objects. It shows a list of simple observations I undertook in the collection, as seen in Figure 10.6, from which one is able to draw some conclusions:

- Animal materials are the most frequently used materials among these objects;
- All the animal materials come from mammals;
- The parts of animals used are diverse;
- The minerals used are diverse;
- Vegetal resources are minimal.

These considerations on materials are interesting primarily because they give us an insight into the environment in which these objects were made. As ethnonyms, ‘Inuit,’ ‘Inupiaq,’ ‘Yupik’or ‘Eskimo’ refer to several related people groups living across a very large area marked by long and cold winters and short and cool summers (subarctic climate). Within the assembled objects, and as shown in Table 10.1, a significant number of mammals and a variety of body parts have been used in the manufacturing of the objects.[10] Whether partially farmed like reindeer or wild like walrus, these mammals have been killed for at least two reasons in the context of our case study: to be a source of energy for their meat and fat; and/or to be a source of raw material, namely their bones, skin, sinews, and teeth. That these animals have been used in a variety of ways indicates that they are likely one of the most readily available sources of material locally. This would, for instance, explain the use of a mammal rib to make the bow-drill: a branch would have been much easier to use, just needing to be cut – instead of killing and butchering an animal before one could cut the rib and clean it. The quantity of animal parts used, in the context of the Inuit collection, strongly suggests that animals were among the main materials available locally in this context.

At the other end of the spectrum of available raw materials, there is metal. Only the inscription on the label allows us to identify the nature of the meteoric iron (Buchwald and Mosdal 1985) without any further analysis. Concerning the bullet, its size and shape enable its identification. At the top of the object, there is an inscription following the circular shape of the bullet, *W.R.A.CO.* on the upper part, and *.44 W.C.F.* on the lower part. Winchester Repeating Arms Company, Caliber 44 WCF were produced from 1873 to 2011, and Winchester used the ‘W.R.A.CO’ head-stamp until 1928 (Adkins 2011). The origins of the metal parts are diverse, which indicates that people tried to find metal wherever they could. Additionally, apart from the adze, the bits of metal represent only a small percentage of each object and are used only where they cannot be replaced by another material. A piece of stone replaces, for instance, what could have been made in metal in the mouthpiece of the bow-drill. This could be a sign of the metal’s limited local availability, and, consequently, this could signify the ability of Inuit people to find, use, and technically work with metals of different origins, thereby obtaining the best from their environment.

Materials of vegetal origin such as fibres and wood are commonly used by Inuit people, but in fewer varieties. Considering the specificity of the weather in the Inuit area, fibres could only be collected during the summer, which shows the importance of the warmer season regarding elements such as food diversity and the gathering of raw materials to survive the rest of the year. Wood is not a commonly known resource in Inuit areas. It is therefore interesting to notice how it has been used for objects or parts of objects, whereby the physical qualities of wood were needed to enable lighter, but solid, tools. When the qualities of wood were not needed, other more common materials would have been used. In the case of the bow-drills, the elasticity of wood might have facilitated their use and conservation as would have fibres, instead of using ribs and skins. The two later did not possess all the qualities of the vegetal materials but enough to fulfil the requirements of the object. They were therefore favoured, as they were easier to obtain.

With the noticeable exception of the adze, wooden elements do not feature much among the objects of the collection, which is another indication of the parsimonious use of this material. A sizeable amount of wood was used to make the handle of the adze, which is the result of very skilled work: the clear and precise shape, including its being slightly curved, as well as the finishing, all reveals the skills of the craftsperson who made it. Considering the quality of this work, it would have been simple for the craftsperson to carve the whole handle from one piece of wood, which technically also would have made it more solid and efficient. Instead, the craftsperson has used two different pieces of wood, both rather light. This reflects practical considerations, such as being lightweight to carry, while it also made the object less solid overall. The craftsman has tied these pieces together, using what appears to be leather cordage made of some kind of animal skin. A possible reason why the craftsman opted for this solution might have been the absence of a piece of wood that was sufficiently large, long, and thin enough to work with. By giving close attention to the piece of wood attached to the metal blade, it is possible to note traces of paints, in two layers, one light and one dark. These traces are visible below the skin cordage that holds the adze together. There is no paint in the most accessible part of this piece of wood, probably because it was sanded. The visible traces of paint signify that this piece of wood was used previously in another way before being transformed into an adze. In its previous uses, it has been painted at least twice, and recycled at least once – if not more often. Considering the lack of trees in the Inuit area, the most common sources of wood were indeed driftwood and second-hand wood (Alix 2012).

In this case study, assemblage has been used to reach a better understanding of the environment in which people were living and, consequently, the materials available to them and the techniques they needed to master in order to work these materials. Our last case study shows how the methods explained in this chapter can provide a framework for developing speculation that may serve as a hypothesis to test in future lines of investigation.

Case study three: the Fijian collection

The Fijian collection of UCL Anthropology consists of 18 tapas, a tapas beater, ceramics, and weapons. The main raw material used to make these objects is wood. The amount of wood needed to make all of these different

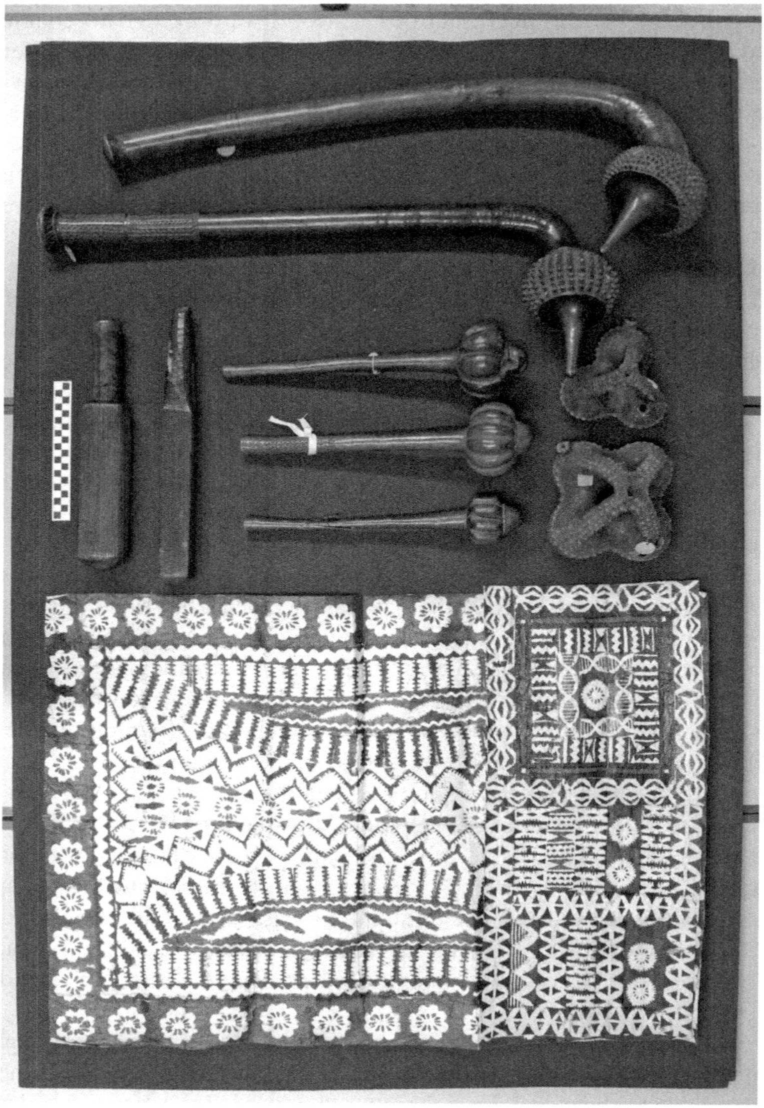

Figure 10.7 The major part of the Fijian collection kept in the UCL Ethnography Collection.

Source: Photo by Timothy Carroll ©UCL Ethnography Collection.

objects indicates, as per our first case study K.0050, an abundance of trees and, consequently, forests. Looking specifically at these objects, it is important to try and ascertain which pieces of tree have been used for which item. All the tapas are made of bark, and stencils were used to make the visible patterns. The stencils were just tools, which were not meant to be displayed but only to be used. These needed to be easy to carve with and use, and cheap to produce, in a material supple enough to be used several times without breaking and without taking the pigments in. Leaves would certainly fulfil these criteria. As for the pigments, they were traditionally 'obtained from plants, trees and earth pigments' (Flowers et al. 2019) and applied, as traces visible on tapas show, either with fingers or with some fibres, grass or thin pieces of wood when more precise design was needed. Tapas are obtained by beating bark and, when it is related to their uses, they are the equivalent to any other textiles in material culture practising weaving. Therefore, they are mostly used to ornate and protect the body.

The rest of the objects in this assemblage, apart from the ceramics, are each made of one piece of solid and heavy wood. Clubs are interesting as they follow the natural shape of roots or branches, their heads being carved in the nodes formed at junctions of elements. This is the reason why the straight parts of clubs are not entirely straight as they just follow the natural shape of branches. Therefore, these clubs are incomparably more solid than if they were made of two different parts put together. These clubs are weapons; the small ones are projectiles, and the large ones for contact at close range. Considering their weight, it was probably much easier for men rather than for women to hold and use them.

In this case study, trees are almost the only source of raw materials used. As trees were used extensively, we can assume that they were locally abundant. Therefore, people developed various techniques to utilise different parts of the materials to hand. At a wider level, it could be interesting to speculate that this might mean that trees could occupy a more symbolic place amongst these people. It is difficult to say, particularly from the object analysis alone, but if we follow this idea and try to picture the local society in a wider context, it might allow for new ways to drive ethnographic investigation. For example, does the source location of the various types of wood correlate with the social role of the people using these items? The tapas, made from the bark of the trees, wraps the body's 'trunk'; the branches, rendered into weapons, are used by warriors to extend their own 'branches' (i.e., arms) at length. While the object analysis and assemblage do not allow the analyst to go further than abductive suppositions, these methodologies of engagement make a strong case for understanding the relations among materials, techniques, objects, peoples, and society and point to its organisation at a wider, symbolic level.

Conclusion

Looking at things in a museum or archaeological context via a phenomenological approach starts from a place of engaging with things. As explained in this chapter, looking at things is used here in its mainstream definition. When we say we are looking at a piece of cake at the bakery before choosing it, we are using a range of senses in addition to visual. Similarly, with the object analysis, we are engaging with the thing using our bodies. What we commonly mean by looking in this context is thus actually wider than just seeing and, in the case of the object analysis, it is clear that the engagement of almost all the senses involves being with things rather than only looking at them. Objects are highly experiential, but being with things is not simply a subjective process – there are a series of deductions and analyses one can draw from the object itself, as shown. Paraphrasing Don Ihde, who argues that science is embodied through its instruments, we can say that technique, tools, knowledge, and knowhow are embodied in the traces they leave on objects (Ihde 2008: 7). For Ihde, the idea of interaction is key, as it is for object analysis, or for assemblage, whereby he insists on 'Humans and technologies' being 'interrelational and mutually co-constitutive' (Ihde 2010: 135). Therefore, unlike other pedagogical methodologies developed in relation to engaging with collections, being with things using object analysis and assemblage as examined in this chapter, is not only significant due to the investigative-analytical experience itself, but also because it is an effective way for non-specialists to understand objects and people in relation with all kinds of objects. Being with a thing, in other words, continues to be a powerful way of understanding material cultures.

Acknowledgements

I would like to thank Timothy Carroll and Shireen Walton for their attentive and kindly re-reading, which helped me to enlarge the critical frame of this study.

Notes

1 Tilley 1994: 1: 'The key question addressed is deceptively simple: Why were particular locations chosen for habitation and the erection of monuments as opposed to others?'
2 UCL Ethnography Collection belongs to UCL Anthropology.
3 Interviewed by Delphine Mercier in UCL Ethnography Collection on 18 November 2019.
4 Objects are more than numbers, but I voluntarily use here the dry appellation at the beginning of the object analysis, as I wish to emphasise the path of the relationship between the observer and the object, which becomes more intimate at the end of the object analysis.
5 The label of the object indicates that K.0050 has been collected in Nigeria among the Efik people. The UCL Collection does not have a record of how it was collected nor when it entered the collection.
6 This object is actually not a drum, but an idiophone, as its sound is created by the vibration of the wood itself, not by the vibration of another added material like a membrane.

7 UCL Ethnography Collection is unfortunately not equipped with scales with a range large enough for this type of object, so it is not possible to provide a precise number for the weight of K.0050.
8 On the use of *Inuit* and *Eskimo*, see Mailhot 1978 and Kaplan n.d.
9 UCL Ethnography Collection 24 Inuit objects: A.0030, A.0031, A.0032, A.0108, A.0109, C.0008, C.0036, E.0007, J.0108, M.0018, M.0019, R.0008, R.0009, R.0010, R.0071, R.0072.
10 On the articulation of vital and technical processes, see Coupaye, Pitrou, and Provost 2016.

11 Making things matter

Daniel Miller and Laura Haapio-Kirk

From 'meaning' to 'mattering'

One of the most concerning trends in global politics today is the way in which fear of the 'other' is operationalised for political gain. From the anti-immigration rhetoric of the pro-Brexit campaign, to garnering support for Trump's wall, the weapon of choice is often cultural intolerance. In many countries around the world stagnating economies give rise to increasing socio-economic and educational inequalities, which only fuels tensions. These conditions represent a historical moment that challenges the optimism of the Enlightenment (Pinker 2018), one of the main instruments of which has always been education itself. It is clear that in such circumstances a commitment to an engaged public anthropology matters. Furthermore, it is now possible for an anthropological education, and the humanistic principles it instils, to reach more people in this digitally connected age than ever before. It seems therefore a simple matter of responsibility that we directly engage with the question of how work in material culture and digital anthropology can actually matter beyond the academy. If anthropology is reduced to a discipline that justifies itself only as a 'critical anthropology,' but which merely critiques what is going on in the world for the benefit of other anthropologists, it is of little surprise if anthropology is largely ignored in the public sphere. Yet the current institutional forms of evaluation that (especially early career) anthropologists are subject to, are forces that tend to valorise this self-serving critique.

This chapter argues that we can address these issues directly through a focus upon what matters and to whom, and it demonstrates through case studies ways we can make our work matter. We document several stages in the trajectory of material culture studies and digital anthropology towards this goal of making things matter. A sub-theme will be the tension between things that matter to academia and things that matter to the populations we study. We start with a shift in anthropology from an emphasis on what things mean, to how they matter to our research participants. The next part examines how our research itself might matter to those same populations. The final part examines the pressures placed upon UK universities by the UK government's 'impact' agenda, and it proposes an alternative evaluation framework that also goes beyond education to include welfare as understood by the people we study. Finally, we consider how an alternative to impact evaluation can be an integral part of research design through the use of ethnography.

Modern material culture studies developed during the late 1970s and 1980s, inspired by the appropriation of new semiotic and structuralist approaches for the study of objects. Levi-Strauss's (1982) *The Way of the Masks* employed masks as analogous to myth in revealing structural transformations across societies. Books by Douglas and Isherwood (1978) and Sahlins (1976) addressed objects as categories, while Appadurai (1986a) examined the social life of objects, and Bourdieu (1977) addressed the materiality of habitus. Both structuralism and structural Marxism, however, emphasised the significance of objects for academics, while tending to ignore their significance for ordinary people. There was little place here for intentionality or human agency.

At UCL Anthropology during this time, material culture studies focused on modes of production (Friedman and Rowlands (eds) 1977); for the context see Godelier 1972; Hindess and Hirst 1975). Miller's book, *Material Culture and Mass Consumption* (1987), sought to re-think consumption from being conceived as the end point of production and distribution to being an active process of appropriation. In the introduction to the edited volume, *Material Cultures* (1998), titled, 'Why Some Things Matter,' Miller proposed a shift from an emphasis in semiotic and structuralist writing upon what things mean, to why things matter. The value of this term 'matter' was that it referred both to materiality and also to understanding why objects were important to people. Chapters in this volume examined material often considered trivial, such as paper in the office (Pellegram 1998) or shopping catalogues (Clarke 1998), and showed how they could be crucial in determining action and influencing people. The intention of the volume, and indeed of the people involved in material culture studies at UCL at the time, was to forge a distinct area of study focussed on materials themselves, rather than reducing objects to their linguistic significance or a reflection of social relations.

In recent years, the additional challenge has been how to use the methods and ideas developed in the study of tangible materials for understanding often intangible digital culture. The MSc in Digital Anthropology was established at UCL Anthropology under the umbrella of material culture studies to do precisely that. Not only can we draw on methods – including ethnography, cultural comparison, object analysis, and attention to materiality – but a material culture approach to the digital also means that we can find footing in an established body of theory. For example, advances in the study of offline visual material can be applied to understanding visual content posted online. In addition, we consider the materiality of technology such as computers or smartphones and examine their position in the context of people's offline lives. Materiality is important when thinking about the novel ways in which humans and machines are becoming increasingly intimately connected, whether through bodily augmentation (Parkhurst 2012) or, conversely, in the ways that data is constructed through material practices (Walford 2017). In rapidly ageing societies such as Japan, where a decreasing workforce and an increasing demand for care has driven the development of care technologies, a focus on materiality becomes especially important for understanding the affective capacities of technologies such as Pepper the robot, designed to elicit a sympathetic response (White 2018). Yet it is not only in the design of robots that humanoid characters are employed to communicate care. Haapio-Kirk's fieldwork in Kyoto and rural Kōchi (2018–2019) shows how LINE, Japan's most popular messaging app, which combines text with cute character images ('stickers'), has become an important medium for practicing care at a distance. Daily contact through visual messaging allows for affective contact while reducing the burden of care.

From 'us' to 'them'

Two decades after its publication, the limitations of the arguments in *Material Cultures* about making things matter have become clear. A shift from understanding what things mean to why they matter to the people we study was insufficient. The next stage would be to move beyond why things matter, to why our research itself matters beyond the remit of academic anthropology. For our research to matter in this broad sense, it needs to first be accessible to the general public. This implies a commitment to an intelligible style of writing and also a shift in emphasis in what anthropologists choose to write about. Much contemporary anthropological writing has re-orientated itself to a concern with relatively esoteric academic interests. Anthropology has passed through a sequence of theoretical interests, from post-modernism, post-structuralism, actor-network theory, and now to ontology and the Anthropocene, each in turn providing a carapace of obfuscating terminology that can be exploited in the manner described by Bourdieu (1984b), as each generation of anthropological students look for professional employment by establishing their reputation for theoretical agility and intellectual prowess.

A fixation with theory as an end in itself, rather than as a route to understanding the substantive, returns us to a situation where anthropologists develop their self-interest at the expense of those responsibilities outlined at the start of this chapter. Furthermore, most anthropological teaching is directed to small audiences of university students, who themselves tend to over-represent metropolitan elites. The situation suggests an urgent need for counter mechanisms developed for the express purpose of turning academic insights into education for a much wider public. Only then can we say that our work matters to people other than ourselves. This task is made all the more important by continued threats to the funding of disciplines that the general public may feel do not matter very much and therefore perhaps should not be funded. Anthropology is particularly precarious in the UK as it has not been taught at school level after the abolition of the Anthropology A-level in 2018. Making our work interesting and accessible beyond academia is actually vital to keeping academic anthropology in existence.

Two case-studies will be used here to illustrate the kind of processes material culture studies might adopt in order to facilitate a return of our research back to the original task of education and helping people to gain a more empathetic appreciation and understanding of other people around the world, which is perhaps the fundamental legitimation for anthropological practice. The first case-study is the *Why We Post* project, which was led by Miller and based at the Department of Anthropology UCL, concluding in 2017. The ambition to connect with a wider public is made much easier if we start with a topic of current concern and interest to that public. The topic of this research was the uses and consequences of social media – a topic that has relevance for a broad spectrum of audiences: parents worried about their children going online (Livingstone and Olafsson 2018), governments worried about security issues (Morozov 2012), or migrants trying to stay in touch with relatives (Dekker, Engbersen, Klaver, Vonk 2018). If we genuinely believe that the depth and commitment of ethnography leads to a more profound understanding of such transformations in the way that we communicate, then we need to convey these findings as quickly as possible to as many people as possible.

The *Why We Post* project comprised a team of nine anthropologists who simultaneously engaged in 15-month ethnographies in field sites that ranged from Brazil and Chile to India and China. Haapio-Kirk was employed in the final two years to develop the project's spectrum of public dissemination. From its inception, accessible and effective dissemination was one goal of the project, learning directly from the populations studied. For example, the project discovered that whether people were learning the skills of hairdressing in low-income Brazilian settlements (Spyer 2017), or learning about car mechanics or celebrity gossip in China and India (McDonald 2016; Venkatraman 2017; Wang 2016) short videos were crucial, especially for people with low levels of literacy. In an associated project, Sheba Mohammid studied the impact of these short videos on informal learning in Trinidad. *Why We Post* therefore made the commitment to produce ten short videos from each field site.[1]

The primary product of ethnography is the monograph; no other format speaks to our ethnographic method of holistic contextualisation (Miller et al. 2016: 28–29). Each field site resulted in a monograph with identical chapter headings to facilitate comparison. Five strategies were designed to facilitate wide dissemination. First, in emulation of popular historical genres, these monographs focus on conveying our ethnographic experience and findings with an emphasis upon storytelling. Academic discussion is largely relegated to footnotes. Second, the monographs were written in jargon-free accessible language. Third, we made direct use of innovations in Open Access publishing. Academic work is already paid for through grants derived from taxation. There are no grounds for authors to be paid twice through royalties. All our volumes were therefore downloadable for free from UCL Press. They were also available in an enhanced online format, with hyperlinks to videos and other content. Fourth, it was important that the audience was not limited to English speakers, but that the books should be available in all the regions where these studies had taken place. So, apart from Turkish, where there are currently certain political constraints, all the volumes are being translated in the local fieldsite languages, as has the comparative volume, *How the World Changed Social Media* (Miller et al. 2016). The final strategy was to embed the monograph in a holistic schema of research dissemination that also included our active social media, our films, a website[2] and our online course (MOOC).

The combination of these five strategies has proved highly effective. At the time of writing this chapter (July 2020) downloads for this series have passed one million. In 2018 books from the *Why We Post* series represented the first, second and fourth most popular open access book downloads, across all disciplines, on JSTOR, the primary online scholarly library containing 12 million sources from 75 disciplines[3]. Finally, and of particular concern to anthropology, the downloads are global, including many countries where sales of a printed ethnographic monograph were likely to have been very limited. For example, the first year's figures included 14,800 for the Philippines, and 5,068 for Ethiopia. The traditional ethnographic monograph usually sold in hundreds rather than thousands and it would be unusual to obtain any sales at all in many developing countries.

As part of this holistic research dissemination strategy we also created a free online course that could reach beyond the few dozen students we would normally address in a lecture theatre. *Why We Post* became the first MOOC created at University College London for the FutureLearn platform, which is part of The Open University. The MOOC movement represents one way in which technology has been harnessed for opening up education. MOOCS have received criticism for typically relying on videos and quizzes, and for not providing learners with a sense of community or enabling them to co-construct their knowledge (McAuley et al. 2010), but FutureLearn works more like a social media platform with considerable interactivity between students. A Russian translation of the course has been used for teaching over six thousand students within Russia[4] and a further translation into Thai has just been completed. At this point, more than twenty-two thousand students have been active participants on the *Why We Post MOOC*. This figure demonstrates the potential for MOOCs as a form of research dissemination that can bring anthropology to audiences who might not otherwise have considered the subject. Yet without institutional support, early career anthropologists face acute pressures to conform to job market demands which favour traditional lecturing over online teaching (Haapio-Kirk 2017). Finally, we also have a project led by Tom McDonald and Laura Poutney to incorporate the project findings within the school curriculum, initially in Hong Kong and within the UK as part of the A-level in Sociology by OCR,[5] where it fits well within an already existing unit called 'Globalisation and the Digital World.' For this purpose, we held workshops with teachers and provided template classroom worksheets.[6]

Our second case study is intended to represent the other end of the spectrum from this largely digital global dissemination, examining instead an immersive and experiential example of research dissemination. In 2010 Inge

Daniels published a book called *The Japanese House* based on many years' ethnography in Japan. Subsequently, in 2011, she was invited by the Geffrye Museum in London to create an exhibition, 'At Home in Japan.' She then returned to ethnography in order to assess the impact of her exhibition. The ethnography included observation of visitors as they interacted with the exhibition and each other, and post-visit interviews to assess their appraisal of the experience, conducted by Daniels and two research assistants, including Haapio-Kirk. The ethnography continued even after the exhibition ended, tracing the fate of the displayed objects, which were subsequently distributed through a raffle. This ethnography forms the foundation for another book, *What are Exhibitions For* (Daniels 2019). We will discuss how the impact of the exhibition was assessed in the next section.

The exhibition itself was intended to embody the principles that have been discussed here, in that the emphasis was on the possibilities of active consumption by the visitor rather than the production of the exhibition. Instead of treating material culture as art objects to be preserved behind glass cases, the exhibition was constructed as an ordinary home interior with the objects in drawers or on display as they would be in a typical Japanese apartment. The visitor could take them out, play with them, and if they removed them, these were cheap objects that could easily be replaced. Visitors were also asked to remove their shoes before stepping on tatami mats and behave in ways that would be expected of visitors to actual homes in Japan. The emphasis was on accessibility and intelligibility, based around stories of individual visitors. The educative possibilities included the provision of different depths of experience, since those wanting to have a deeper understanding could then read Daniel's ethnography. The style lends itself to popular education and also allows visitors more scope for exercising their own imaginations, which Daniels argues was one of the main consequences of this approach to exhibiting (Ibid. 201–203).

In this exercise, each element of exhibiting was reconsidered in terms of the potential for a more immersive experience. Instead of using photographs as framed artworks, in this exhibition they were full-sized and placed in a manner that extended the physical reconstruction of the house, or paired with displayed objects to help convey their original context (Ibid. 63–88). In turn, Daniels recognises that visitor photography has become an ever more important way that people interact with an exhibition, and her study examined where and why they take these photos and what happens to them after they are taken (Ibid. 171–193). So here the tradition of material culture studies has been employed directly to convey research findings through immersing the public in the material practices of other peoples of the world.

Beyond impact

All academics in the UK are today having to reimagine their work and its consequences in relation to a new agenda set by government. This is because a significant proportion of the funding of university departments now depends upon a demonstration that they have satisfied criteria that come under the label of 'impact.' What constitutes impact is decided by a series of rules and regulations. For example, the success of *Why We Post* in research dissemination cannot serve as direct impact, since the government's primary criteria is not educational, but evidence that other, non-academic bodies, have changed the way they behave as a result of this research. An example of impact would be that a commercial venture has put this knowledge into practice, such as marketing a product based on an anthropologically orientated semiotic analysis of that field of material culture. Many other countries have developed their own equivalent to the UK's impact agenda so that such concerns are becoming more important to academics world-wide.

The trajectory of this chapter has led to a point that is analogous, but also different, from the government's impact strategy. Having shown how we can study why things matter to people, and then how we can bring our findings back to people as education, we also face the issue of how we can make our work matter to people by contributing directly to the welfare of those populations. The problem is that currently we are doing so largely in response to the government's impact agenda. Instead, we need to formulate our own criteria for how things might matter for the wider world. We want to suggest two major departures from the government's impact agenda. The first is that anthropology should prioritise the welfare of the populations we study, and the second is that anthropology should contribute to ensuring that the populations themselves have a role in determining what we mean by 'welfare.' The example just given of helping a company improve its marketing would not satisfy either of those criteria.

What does it mean to suggest that populations should be included in this initial evaluation of what we mean by welfare? This proposal is derived from long-standing interventions by the economist Amartya Sen and philosopher Martha Nussbaum. Briefly, in opposition to the top-down determination of welfare embodied in the way economists or the United Nations ranked national welfare, Sen developed what became known as the capability approach: 'Our freedom to promote objectives we value' (Sen 1992: xi). But, as a paper on contraceptive choices by Mexican women (Beutelspacherm et al. 2003), published in Nussbaum's *Feminist Economics* (2003), showed, given the pressures people live under, it is very hard to understand what people would choose if given that capacity. Just as Sen showed that choice cannot be reduced to economic demand, anthropologists would argue that welfare values may not be reduced to choice – especially when we are talking about new digital technologies we have barely had time to assess. One legacy of anthropological work in structuralism and structural-Marxism is that we recognise that people make choices influenced by forces that they themselves may be quite unaware of. So merely asking people what they regard as their welfare would be insufficient. We would argue that it is actually ethnography, with its method of holistic contextualisation, that is best placed to assess the welfare goals of a given population. Furthermore, ethnography is rare in that it can accommodate contradictory ideals and practices, which is often the reality of welfare itself.

Soon after the completion of the *Why We Post* project, in 2017 Miller launched a new five-year project called *ASSA – The Anthropology of Smartphones and Smart Ageing*,[7] also funded by the European Research Council. At first glance the idea of welfare intervention sounds like the traditional scope of applied anthropology, which often meant working with medical or other organisations to help fulfil welfare aims with respect to a given population. But this grant proposal was unusual. While it made a commitment to the general domain of smartphones and health, it did not specify what aspect of health or smartphone usage we would hope to develop interventions around. The principled reasoning behind this omission was that if before conducting any ethnography we had already decided how we would intervene in developing a population's welfare, we would have been imposing our own criteria and concerns upon that population. If, however, we want to follow the strictures of Sen and Nussbaum, then we need first to ensure that the population is a partner in establishing the appropriate welfare aims. For these reasons, discussion of possible interventions began only after we had completed ten months of ethnography. This allowed us to make many observations – for example, teaching smartphone usage to older people in order to better understand how they struggle with the new technologies. Smart ageing for us is understood as 'smart from below,' the often ingenious appropriation of technology in everyday practice (Pype 2018), observable only through ethnography. The model came from Miller's previous study of the use of new media by hospice patients in the UK, which led to a series of suggestions as to how hospices night employ new media, all of which followed, rather than led, the research (Miller 2017).

We began writing this chapter precisely one year after the start of our ethnographies for the ASSA project: Miller in Ireland and Haapio-Kirk in Japan. This period has allowed for a natural evolution from the top-down perspective of writing a grant proposal to the bottom-up perspective based on the team's respective ethnographies. The original grant proposal mainly discussed potential engagement in terms of a huge industry that is churning out smartphone health apps. These generally follow from the interests and ideals of professional IT developers and medics, as well as commercial forces that recognise this is potentially a hugely profitable field. All three represent the top-down initiative normally associated with the word 'smart.' The top-down agenda also applies to academic work in this area, including that of anthropology. The dominant writing and interest in this field of smartphone health has so far been aimed at a particular set of practices that are associated with terms such as 'self-tracking' and 'data visualisation' (Dow Schull and Ruckenstein 2017). There are good reasons why academics find this a fascinating development. But in our ethnographies we see very little evidence that these constitute a significant aspect of mHealth (mobile health apps) in terms of actual consequences on populations. Some people use fitbits and step counters, but they are not widely used in most of our field sites, apart from Japan. One problem with current academic work is that it has tended to seek out and study the most extreme examples, such as people who belong to the 'Quantified Self' movement (e.g. Ajana Ed 2018; Lupton 2016). The problem is that such work takes the topics of most interest to academics and finds the requisite population.

By contrast, the ASSA project started with a deliberate attempt to forgo any academic agenda at all and, instead, allow the experience of ethnography to create a collaboration between the academic and those being studied to try to determine interventions based on aligning ourselves with what matters most to the people we work with as it emerges from that ethnography. As a result, over the course of the year the focus has largely moved away from dedicated smartphone health apps of interest to medical, technological and academic top-down agendas to focus instead on the usage of ubiquitous platforms such as Google and WhatsApp. Marilia Duque, working in Brazil, has developed a manual for health workers on the potential of WhatsApp for health, based on the observations she has made of both medical practitioners and her informants. We are also exploring the effects of Googling for health information. Charlotte Hawkins is working with an initiative led by Ugandan doctors rather than foreign NGOs in the assessment of possible mental health interventions in Kampala. Laura Haapio-Kirk is using her ethnography to unite patients, doctors, and state health officials in an effort to improve the experience of rural health care in Japan via the mobile phone. This sequence also ensures that the original ethnography retains its integrity as a project of enquiry, only after which we try to develop a strategy to make sure that the research matters.

ASSA's welfare goals are shared with many others, being akin to what Smith (2013) terms an 'interventionist design anthropology.' For example, the recent work of Jerome Lewis (2014), also based at UCL Anthropology, has extended the remit of digital anthropology at UCL in a similar direction. His team have worked with indigenous forest dwellers in West and Central Africa to develop a mobile application that enables them to report poaching and illegal logging, based on GPS and mobile photography. He classes his work under the umbrella of 'extreme citizen science,' which calls for as much involvement of the populations we study as possible in determining and securing their welfare. Research is conceived as a constant collaboration with participants in deciding what data should be collected, how it should be collected, and in determining its consequences. The chapter in this volume by Geismer approaches the issue of how to make Digital Anthropology matter through an engagement with the local community. There is, however, one final stage in this discussion of making things matter, in the context of the 'impact' agenda. To have our interventions accepted and valorised means we also have to consider how they will be evaluated. If we are unhappy with an increasing reliance on quantitative measures of evaluation, and we really want to demonstrate why anthropology matters, we should integrate methods of anthropology into impact assessment. The ASSA project is approaching potential mHealth interventions after a year's fieldwork, but it also includes provision for continuing to collaborate with the populations we work amongst for a further year after fieldwork, including an ethnographic assessment of those same interventions.

Fassin's edited collection (2017) presents the opportunities, challenges and limitations when ethnography 'goes public.' Yet, it remains the case that anthropologists have largely been absent from public debate, perhaps with the exception of Norway (Eriksen 2006, 2008). Despite anthropology's combination of deep and grounded observation with global comparative analysis, other disciplinary perspectives, notably from psychology and economics, often dominate public discourse (Ingold 2016). This may be a result of anthropology's withdrawal from public engagement during the twentieth century and its increasing institutionalisation (Eriksen 2016) . Encouragingly, today there is evidence of a growing movement by committed groups of anthropologists all over the world towards this endeavour, such as EASA's Applied Anthropology Network, which organises the annual conference, Why the World Needs Anthropologists. The authors in Fassin's collection demonstrate how ethnography has given them nuanced insight into often turbulent and precarious situations, where their intervention can have the potential to amplify under-represented voices. Yet there is an extra step that could demonstrate even further the value of anthropology: in the evaluation of public engagement itself. In the manner of Daniel's ethnography of her exhibition, the wider incorporation of ethnography of dissemination into our research design would allow us to learn directly from the public themselves how our efforts toward engagement are received, and ultimately how to do this better. So, in contrast to the government's Impact Agenda, anthropologists should consider the role of ethnography, both in determining what matters to a population, but also in the evaluation of our contribution to their welfare. Daniels's 2019 book demonstrates how an ethnographic evaluation of anthropological dissemination allows us to better understand how we might further innovate anthropological education by paying attention to this recursive nature of what matters.

Conclusion

This aim of this chapter was to demonstrate how a trajectory towards making things matter is not simply aspirational but is actually something we can hope to realise and that could drive material culture studies and digital anthropology in the future. The trajectory began in the 1970s, when after a long period in which material culture had been seen as peripheral to anthropology it started to emerge as of interest within structuralist and structural-Marxist perspectives. These perspectives established why objects might matter to academics but ignored the perspectives of the populations we study. Gradually, material culture studies at UCL developed an agenda that re-focused on why things mattered also to people other than academics. The next stage in this trajectory was to pay equal attention to the problem of how to make our research matter, which meant formulating strategies that could bring our research findings to the widest possible global audience and convince them that our work is significant and worth funding. For example, by making our work freely available, composed in accessible language and in an array of formats to suit different potential audiences. We have then moved on to consider how we can create an alternative to agendas such as the UK Impact policy. The ASSA interventions are intended to directly benefit the welfare of populations and also include that population's own understanding of what welfare means. Finally, we advocate for the use of ethnography also in the assessment of these contributions. This trajectory is not one of displacement; each of these goals is additional to the others, creating a more comprehensive model for the task of making things matter.

A sub-theme throughout this chapter has been the tension between whether things matter to academics or whether they matter to the population studied by academics. The problem we face was most fully theorised by the philosopher Hegel (2008) in his *Philosophy of Right.* All human institutions tend to migrate from the original legitimation for those institutions, usually in projects of reason, into something that is more self-serving. Law is supposed to adjudicate justice but can become a way lawyers amass a fortune. Shifting away from what matters to academics, or what is in their self-interest, does not mean reducing our aims to 'giving voice' to a population. Those earlier approaches from structuralism and structural-Marxism made a strong case for academics to excavate forces of power that populations may be entirely unaware of. Material culture ethnographies have never been mainly about asking people's opinions: they have always focused more on long-term observation of what people actually do, their practices, and extrapolating from our observations values that may not emerge in language. Our point is that the problem posed by Hegel in the *Philosophy of Right* will always be present. However well-intentioned, academic work will always tend to circulate back to the self-interest of academics and the agendas that promote our own well-being, rather than those of the populations we study. Much of this is inevitable if we want to obtain jobs and be successful in our careers. But this means we also need to engage in constant re-evaluation of our work and remind ourselves of our responsibilities to the wider world. Only then will we follow through on the trajectories that will ensure that we have helped to make things matter and have made a genuine contribution to countering the regressive and repressive forces that currently challenge us all.

Acknowledgements

The *Why We Post* project was funded by the European Research Council, ERC-2011-Advanced Grant no. 295486. The ASSA project was funded by the European Research Council, ERC-2016-Advanced Grant no. 740472. The authors are hugely grateful to all the other members of the two project teams, since all the work reported here comes equally from the work of those teams. We are also grateful to our research participants from all over the world whose time and support have provided the evidence for these projects.

Notes

1 141 videos are available to watch on the *Why We Post* YouTube channel: www.youtube.com/channel/UCeaS-I5mRr7Hqf41wO5ntSQ.
2 www.ucl.ac.uk/why-we-post.
3 https://about.jstor.org/news/the-top-10-most-popular-oa-ebooks-in-jstor/.
4 https://openedu.ru/course/hse/WEPOST/.
5 Oxford Cambridge and RSA (OCR) is an awarding body that provides qualifications for people of all ages and abilities.
6 https://sociology.hku.hk/schools-resources/impact-social-media/.
7 www.ucl.ac.uk/anthropology/assa.

12 Prophetic pictures

Or, What time is the visual?

Christopher Pinney

Siegfried Kracauer opens his study of frivolity and catastrophe in nineteenth-century France with an arresting vignette of the Paris Salon of 1831. He describes how crowds gathered each day around Delacroix's *Liberty Leads the People*, a celebration of the July Revolution of 1830. The subject matter was dramatic (a 'half-naked young woman' holding a musket and waving a tricolour, a 'new Joan of Arc') and raised the question of whether she was 'a terrestrial being or a supernatural apparition' (Kracauer 1938: 3). Kracauer surmises that the attraction of the picture may have reflected the suspicion in

> the minds of some of those who came to gaze at it that this picture was not just a graphic representation of the three glorious days of July, but that it also lifted a corner of the veil that hid the future.
>
> (1938: 3–4)

Kracauer, whose work consistently demonstrates a concern with 'uncontemporaneous sedimentation' (Koch 2000: 120) – that is, material resistances to a singular temporality – here dramatises a startling way of viewing images, suggesting a popular understanding and desire for pictures to point to what is yet to be, rather than merely objectify what has already happened. This is perhaps startling only to post-Durkheimians, weighed-down by the idea that representations are after-effects. It may have been much more palatable in an age of Romanticism when Percy Bysshe Shelley could resonantly proclaim that 'Poets are [...] the mirrors of the gigantic shadows which futurity casts upon the present' (Shelley, *A Defence of Poetry* 1821, pub. 1840).

This chapter attempts to anchor this dialectic between the representation of antecedent signifiers and futurity in the specific context of a corpus of illustrated nineteenth-century astrological almanacs before then striking out to attempt to explore the unlikely prophetic echoes apparent in photography. The hypothesis proposed in the first section of the chapter is that compared to language, the visual often displays a relative indeterminacy, a multitude of potential interpretations, and that this makes it the perfect vehicle for predictions. The relevant opposition here is what Lyotard describes as that between 'figure' and 'discourse,' latent potentialities found across all forms of representation, but which in this specific context are differentially situated in 'image' and 'text.' For Lyotard, 'figure' resists the 'linguistic-philosophical closure' of 'discourse.' The 'figural,' a space where 'intensities are felt' is 'relatively free of the demands of meaning' (cited by Carroll 1987: 26–29). Lyotard argued that certain forms of poetry were more 'figural' than certain forms of image (such as a diagram, which might be highly 'discursive'). In general terms this is undoubtedly true and the great merit of Lyotard's contribution is that it liberates us from stale image/text dichotomies. And yet, I hope to persuade the reader that, in these astrological almanacs, 'figural' images do not make themselves hostage in the same way that 'discursive' language does. Rather, figural images offer a flexible ground that can be finessed and sculpted by the linguistic claims of the next year's almanac. The argument here is indebted to Bernadette Bucher's observation that the visual cannot 'negate' in the same way that language can (she notes that Montaigne could say, when looking at images of the Tupinamba 'cannibals' 'What! They're not wearing breeches'; however, 'it is impossible to portray a thing [visually] by what it is not, it is present or absent, and if it appears it is always positively, in a certain shape' (1981: 35). The visual and language have different kinds of power.

In the concluding section of the chapter a parallel yet distinct argument about the 'positivity' of the visual is explored in relation to photography and the suggestion made that the camera opens up a future-oriented performative and 'proleptic' space. These different modes of image-futurity both offer a challenge to social theories that stress representations as mere receptacles of past actions and as end points of social processes.

Raphael's Prophetic Messenger

The most important of these almanacs is *Raphael's Prophetic Almanac*, although others, such as *Zadkiel*'s, also offered images as signs of the future. Raphael's almanacs were striking for the prominence given to a 'prophetic hieroglyphic' published as a frontispiece to each annual issue. This nineteenth-century usage echoes Walter Benjamin's sense of a 'picture-writing' descended from the great Renaissance text *Hypnerotomachia Poliphili*. Large, fold-out, hand-coloured copperplate engravings and lithographs, offered visual predictions of what would happen in the year ahead. They also served as marketing tools, for the reader would only be able to decode the 'truth' of these prognostications if they bought the subsequent year's publication (where 'all was explained'). These images, of which there is a significant corpus, have never been discussed in any detail (they feature fleetingly in histories of European astrology).

The hieroglyphic usually features in history as emblematic of the difficulties but, most importantly, the possibilities of the translation of the visual into language. The history sketched here emerges from a moment when the deciphering of the Rosetta Stone seems to open up the utopian possibility of translatability in general. Indeed, it has been argued that the Rosetta Stone's mystical attraction reflects a desire for the translatability of everything, including material forms. In this reading the Stone is to be understood as an object that provides its own caption (Beard 1992; Ray 2007: 5).

In occult and apocalyptic literature, such as William Cunninghame's *Apocalypse* of 1817 (2nd ed., date of first edition not known) hieroglyphics are 'seals' of a future to come. The seven seals of Revelation are to be understood as akin to locks, guarantors of the closure of the text (it was secret knowledge known only to God – hence its un-polluted authority). The history sketched here, however, points in the opposite direction towards a practice in which the image is produced precisely because of its ability to escape the syntagmatic[1] certainty of language.

Astrology, whose appeal in the Middle Ages had been confined to court circles found new audiences in the seventeenth century through works such as William Lily's *Merlin Anglicus Junior, The English Merlin Revived.* Sales of almanacs were greater than those of the Bible (O'Connell 1999: 21) and, by the end of the eighteenth century (at a time when the population of England was less than ten million) half a million almanacs were sold each year (O'Connell 1999: 22). A proliferation of titles gave way at the beginning of the nineteenth century to the supremacy of Francis Moore's *Vox Stellarum*, aka *Old Moore's Almanac*, which had since the French Revolution 'introduced illustrations symbolizing millennial ideas and promoting political radicalism' (O'Connell 1999: 22).

Robert Cross Smith, a self-educated plebeian from near Bristol, who chose to cloak himself in Cabbalistic mystery, would issue the first of *Raphael's Prophetic Messenger* in 1826 (Curry 1992: 47 and 52). The issue rapidly sold out, necessitating a reprint and, by 1831 he was able to claim sales of 8,500 at a time when Moore's was selling 270,000 per year (Curry 1992: 52). Curry suggests that the audience for Raphael's 'overblown occult romanticism' would have been what he calls 'semi-erudite,' and quite distinct from the rural labourers and urban working classes who enjoyed Moore's publication. Curry identifies Raphael's likely consumers as an audience in retreat from the 'successes of secularism, whether as political radicalism, philosophical utilitarianism or science' (1992: 53).

Raphael (Robert Cross Smith) died at the age of 36 in 1832 and control of the *Prophetic Messenger* passed briefly to the astrologer 'Dixon' before passing to a further five 'Raphaels' (Curry 1992: 58). Sales would steadily rise to about 100,000 by mid-century and (depending on your sources) either 150,000 or almost 200,000 by 1900 (Curry 1992: 60).[2]

The choice of 'Raphael' as an authorial device suggests a genuflection towards the Kabala, and the 'East' in general performed a crucial role in the projection of the *Prophetic Messenger*'s authority: Raphael was also advertised, from the 1830s through to the end of the century, as the author of *Pythoness of the East*, which claimed to be based on a found text 'formerly in the possession of Her Imperial Majesty the Empress Josephine' (Raphael 1894).

The Pythoness of the East, and Raphael's nominalist appropriation of a Jewish identity, points to the Prophetic Messenger's role as an agent of cultural critique. This echoes the earlier role of Confucius and China in eighteenth-century cultural critique as argued by Wittkower:

> Sinomania in 18th-century Europe allows some insight into the nature of this kind of quest. Thinkers of the Enlightenment embraced Confucius' moral philosophy which, based upon reason and tolerance, seemed to offer a better foundation for a harmonious communal life than a revealed religion with its fanaticism, obscurantism and intolerance.
>
> (Wittkower 1977: 14)

The Orientalist linkage between the East, prophecy, and the future, was most clearly stated in the 1854 tract *A Plea for Urania,* written in opposition to Justice Leatherhead's proposal that astrology be outlawed. In defence of his science, the anonymous author extols 'the East' as the location in which astronomy and astrology have only recently begun to separate, and as the 'source of all laws, religions, sciences, and modes of government.' Most significantly for our present purposes the East offered a door into futurity: 'the East has attractions for all. Its fascination is made up of the past, the present, and what is probably *to come.*'

Raphael fused an Eastern mystical futurity with a form of visual conjecture (prefiguring Carlo Ginzburg's [1988] usage) with the hieroglyph as a device: 'We have instanced the Egyptian hieroglyphic as the root or source of all pictorial devices of which the signification is obscure, or conjectural.' (Raphael, 1845: 46).[3]

'Ominous' hieroglyphics

Raphael's *Prophetic Messenger* stands apart from its competitors for the lavish illustration that accompanied each issue. The 'Hieroglyphic for the Eventful Year [insert relevant year]' was a large fold out hand-coloured plate (usually with six or nine scenes) predicting, pictorially, the events in the coming year. The title page for the 1830 almanac – the earliest copy I have managed to obtain – proclaimed (in capital letters) the presence of 'a singularly ominous hieroglyphic for 1830 on a large copper-plate, carefully coloured.'

Given that Raphael's hieroglyphics appeared a few years after Champollion's 1822 translation of the Rosetta Stone (building on Thomas Young's earlier work on the demotic passages – Wood 1954: 206ff.) we might be forgiven for assuming that Raphael's usage reflected popular enthusiasm for the mysteries of Egyptian picture-writing. In fact, Raphael's usage owed more to the Renaissance fantasy of picture-writing descended from Francesco Colonna's *Hypnerotomachia Poliphili* first published by Aldus Manutius in 1499 in Venice.

Walter Benjamin invokes the hieroglyphic several times in his writings. In 'The Antinomies of Allegorical Exegesis,' written in 1925, he uses the term to describe the baroque emergence of the visual as a mode of revelation. 'If script is to be granted a sacred character [...] then it will press toward complexes, towards hieroglyphics' (2008: 176). In 'A Glimpse Into the World of Children's Books,' an essay written in the following year, Benjamin describes the *Hypnerotomachia Poliphili* as providing the 'patent of nobility' for Renaissance hieroglyphics whose nineteenth-century offspring were the rebuses, the picture-writing (*Bilderschrift*) or 'puzzle pictures' for children, which so interested him (2008: 230).

Raphael's 1830 illustration (Figure 12.1) divides its pictorial space into three loosely structured visual registers. A non-exhaustive description would include a cherub unfurling a cartouche adorned with mystical signs, an exploding volcano, ecclesiastical buildings on fire, a notice concerning the eclipse of the moon in September 1830, a funeral cortege, a great naval battle,

Figure 12.1 Hieroglyphic for the Eventful Year 1830, Handcoloured copperplate engraving.

Source: Private Collection.

a storm that appears to be wreaking havoc with agriculture, a naval calamity, a crowd demonstrating their support for 'Reform,' and a learned astrologer/scientist with a telescope and globe who points to a vast book recording that 'Furious Mars, Warns of Danger From afar!'

The astrologer/scientist wears a cummerbund adorned with the name of 'Raphael' and bears considerable similarity to the figure captioned 'Raphael in Italy,' who appears in the elaborate frontispiece to *Raphael's Witch or Oracle of the Future* (1831). In this image, titled 'The Tablet and Questions & The Cabalistical Tablet of the Stars,' Raphael is seated in front of a vast telescope surrounded by the tools of his astrological and alchemical trade.

Witch has an intriguing etymology, being 'derived,' John Brand notes in his 1810 annotations on Henry Bourne's earlier text,

> from the Dutch Witchelen, which signifies *whinnying* and neighing like a Horse: In a secondary sense, also to *foretell* and *prophecy*; because the Germans as Tacitus informs us, used to divine and foretell Things to come by the *whinnying* and *neighing* of *their Horses*.
>
> (Brand 1810: 353)

Raphael's raison d'être was 'witchery.' As he wrote in the 1830 almanac:

> 'I behold (in a dark vista of the future, which science illumines) the sea foaming and raging with fury; the earth quaking; rivers overwhelming their bounds; torrents roaring; the winds of heaven let loose to work his work of vengeance' (1829: 17).

'Imagality': the interpretability of the visual

The beautiful annual Hieroglyphics gave Raphael's publications their unique character and were central to a clever piece of marketing. They were offered as hostages to the future, pregnant spaces of interpretability whose precise meanings would be revealed after the event in the following year's almanac. Of course, the readers were obliged to purchase the following year's publication if they wanted to benefit from such revelation.

The 1830 almanac announced the 'remarkable fulfilment of the predictions prefigured in the hieroglyphic for 1829,' an image that sadly it has not been possible to source.[4] Raphael's exegesis provides ample testimony of both the extent to which certain simmering political events could be safely predicted and also of the way in which the visual offered a productive indeterminacy, a field of interpretative possibility available for sculpting once the actual events that they supposedly prefigured was known. Hence Raphael claimed a presentiment of 'The Catholic Bill and its contingencies' in

> the symbol of *a monk wearing a mask, with a flag in one hand, having thereon the words* RELIGIOUS LIBERTY; in the other hand, a sword: while at

a distance off *a lamb is seen, advanced in growth and pawing in triumph*; at his feet a serpent, monk's cowl, and the various insignia of Popery – plainly prefiguring *the fashionable apostacy* of the times, and the *mask* under which the Catholics obtained Government to sanction their menacing petitions.

(Raphael 1930: 7)

Raphael's description of the success of his 'Omens relative to Spain' illustrates a spectrum of interpretability from the denotative to the connotative. A 'celestial figure [...] seen holding a banner, with the word "*Hispaniola*," the appellation of Spain, first *pointing to a tomb*, which denotes the death of the Spanish Queen that took place,' seems impressively precise (although of course an ailing monarch is quite likely to die) whereas the claim that the banner on which is written '*Resurgam* [...] denote[s] the efforts now made by Spain to assume her former dominion in the New World – witness the famous *Mexican expedition* now approaching the shores of the South American regions' seems more contentious. Nevertheless, Raphael proclaims his belief that 'the *literality*[5] of these portentous omens are too obvious to be explained away on any other principle' before then drawing attention to his wonderful escape clause: 'There are also others which I leave the reader to decipher by *the events that have yet to follow*.' In the 1843 issue this escape clause is presented more poetically: 'the remaining events [...] are still hidden in the womb of time' (1842: 71). Because each new almanac went to press between September and November in each year a significant number of unfulfilled prophecies could be assumed to prefigure events which were yet to pass.

The Hieroglyphic for 1831 (Figure 12.2) featured in its lower register a striking skeleton waving a 'reform' and astrological nativity flag astride a coffin inscribed 'Lo The Time is Come.' The top register features an urban conflagration and enormous ships alongside a cartouche predicting that 'MONARCHS TREMBLE NATIONS MOURN, Ocean[s] rage and Cities burn, Gazing with a prophet's eye, THUS HATH RAPHAEL READ THE SKY.'

The following year's almanac, published in November 1831 (sic), presented the previous edition's Hieroglyphic as a prediction 'relative to the March of the fearful Cholera Morbus.' Reading the image retrospectively with the benefit of a largely unfolded year, Raphael urges the reader to

take into thy notice [...] the fearful Signs in my Hieroglyphical Omens for the year 1831. The Trio of Coffins, Enthroning of Death, &c., and thou wilt be enabled to establish Astrological foresight beyond the shadow of a doubt. Alas! Gentle Reader! TOO TRULY has thy friend Raphael's Predictions, in this instance, been fulfilled!

What makes these hieroglyphic frontispieces much more than a matter of antiquarian interest is their foregrounding of questions of interpretability. Clearly, the challenge to the creator of the hieroglyphic involved the production of images that were both seemingly predictively specific – mobilising the

Figure 12.2 Hieroglyphic for the Eventful Year 1831, Handcoloured copperplate engraving.
Source: Private Collection.

concrete certainty of mimesis – but which were *also* sufficiently vague that they would not become hostages to fortune.

We can see further evidence of the ambivalence of the visual, and the reliability of various probable events in Raphael's 'Explanation of the scenes in the hieroglyphic of 1841' published in the 1842 *Prophetic Messenger* (Figure 12.3). Rather unusually, Raphael is able to rationalise most of the elements of the hieroglyphic (excepting the central motif and the vignette to its left). The building site at the top left is explained as a presentiment of the number of new Catholic buildings erected; the top centre vignette depicting courtiers flanking a veiled throne is explained as 'emblematic of the recent occasions which have rendered it necessary for our Court to go into mourning.'

The top right scene requires a more elaborate exegesis. Described as a 'lion and cock, in a menacing attitude,' these are, Raphael continues, emblems of France and England and 'no one who has read the public journals, detailing the warlike preparations in both countries [...] can doubt the application of this part of the hieroglyphic.' The eagle on the rock, he continues, represents Russia and 'shows the wily policy of that country, ready to seize upon the slightest opportunity afforded by either country.' The scene below this, which one might suppose to depict the poor desperately catching stray grains of wheat (the protectionist Corn Laws, which inflated the price of wheat, would

Figure 12.3 Hieroglyphic for the Eventful Year 1841, Handcoloured copperplate engraving.

Source: Private Collection.

not be repealed until 1846), is said, after the event, to represent 'an English porter, carrying out bales of gold to foreigners; otherwise it is typical that immense sums of gold should be drained out of this country and expended on foreign shores.'

After a brief explanation of the right-hand corner scene, Raphael concludes his tour by noting that

> At the bottom we observe Mars and Bellona in their war chariots followed by Disease and Death. In the perspective we observe ships of war engaged in action; while on the margin of the sea are seen factories and the implements of commerce. This has reference to our warlike proceedings with China, arising from the treacherous proceedings of the Chinese with our countrymen, respecting our commercial relations with the Celestial Empire.
>
> (Raphael 1841: 67)

The First Opium War (1839–1842) was already unfolding when the original hieroglyphic was composed and the modern viewer beholding that quayside scene at the bottom left of the image is likely to find nothing either denotatively, or even connotatively, 'Chinese' there.

Raphael's 1832 almanac described the contents of the hieroglyphic as 'Remarkable Events, Celestial and Terrestial, Podigies, Revolutions, Insurections, Outrages, Convulsions of Nature, Political Occurrences, Remarkable Deaths etc. etc.' This list encompasses many of the topics that feature as staple ingredients of modern 'news.' However if we are nevertheless able to detect what would later be extracted and secularised as 'horizontal' narratives we should be clear that in Raphael's worldview the 'Terrestial' is always causatively linked to the 'Celestial.' This is an interrelation, an aspect of providentialist cosmology that is perfectly captured by Auerbach:

> In this conception an occurrence on earth signifies not only itself but at the same time another, which it predicts or confirms, without prejudice to the power of its concrete reality here and now. The connection between occurrences is not regarded as primarily a chronological or causal development but as a oneness within the divine plan.
>
> (Auerbach 1953: 490)

Raphael's hieroglyphic frontispieces are the origin of the news, a space where the messianic and the contingent jostle together in a 'oneness within the divine plan' that is also alert to contingency and the onward flow of history.

In this respect the astrology of the early and mid-nineteenth century imagines a very different addressee from the one vividly sketched by Theodor Adorno in his 1930s study of astrological columns in Los Angeles newspapers. Raphael's addressee is not yet individuated in the manner that Adorno describes ('The standard image is that of a young person or one in his early thirties, vigorous in his professional pursuits, given to hearty pleasures' Adorno 1994: 83). Raphael's readers are imagined as a collective with highly messianic and nationalistic concerns and suggest a striking contrast with their twentieth-century North American successors ('the striking feature [is] the almost complete absence of any reference to the major mostly solemn speculations about the fate of mankind at large' – Adorno 1994: 66). By the early twentieth century, however, Raphael's addressees appear increasingly individuated. The 1913 Prophetic Messenger is prefaced by an advertisement for Raphael's *Horary Astrology* featuring a list of questions which this new publication would answer. Several reflected anxieties about impending conflict ('Will two armies fight?'; 'Shall I return from the War?') but many more were focused on individual financial and marital success ('Shall we be successful in business?'; 'When shall I marry?'; 'Shall I be happy in marriage?' – Raphael 1912: 2).

Social theory and temporality

This chapter opened with Shelley's claim that 'Poets are [...] the mirrors of the gigantic shadows which futurity casts upon the present.' This was a common Romantic ambition (one also thinks of William Blake's insistence, c. 1808, in his annotations to Joshua Reynolds *Discourses on Art* that 'Empire follows Art and Not Vice Versa as Englishmen suppose' – Blake 1927: 970). The rise of Durkheimian explanations did much to reverse this causal order. Since Durkheim, Latour has famously written: 'To become a social scientist is to realise that the inner properties of objects do not count, that they are mere receptacles for human categories' (1993[1991]: 52). Objects and images in this account are terminal deposits for earlier thoughts: their embrace of futurity, their prophetic potential, is extinguished. In this spirit Durkheim announced that 'the totem is not only a name; it is an *emblem*, a virtual coat of arms whose resemblance to the heraldic coat of arms has often been noted' (Durkheim 2001: 94, emphasis added), 'emblem' suggesting the antithesis of hieroglyphic, being backward looking rather than shadowed by futurity.

It would be misleading, however, to see Latour's critique as marking a new and previously unanticipated break from Durkheim: many other thinkers and writers – broadly post-structuralist in temperament – have proposed more radical ideas. Jacques Attali, for instance, makes a striking argument that sight's predictive abilities have 'dimmed': 'it no longer sees into our future, having constructed a present made of abstraction, nonsense, and silence.' 'By listening to noise,' by contrast 'we can better understand where the folly of men and their calculations is leading us' (Attali 1985: 3). Attali adumbrates a sonic echo of the hieroglyphic suggesting that 'noise [...] constitutes the audible waveband of the vibrations and signs that make up society' (1985: 4). Like Adorno and Lyotard, though noticeably less so than Kracauer, Attali envisages both a progressive aesthetics of futurity[6] and the stale nostalgia of the 'popular.'[7] (Kracauer's work was for the large part dedicated 'anthropologically' to the erosion of such stark political adjudications). Attali in certain respects inverts Durkheim's model, appropriating Nietzsche's claim that art was a Dionysian mirror of the world and explaining that it 'is a mirror, because as a mode of immaterial production it relates to the structuring of theoretical paradigms, far ahead of concrete production' (1985: 9). Attali suggests here that effervescence and objectification have different materialities and different 'weights.' It is precisely music's 'code,' its 'speed' and immateriality, that allows it to operate in advance of the material world. As he puts it

> Music is prophecy. Its styles and economic organisation are ahead of the rest of society because it explores, much faster than the material reality can, the entire range of possibilities in a given code. It makes audible the new world that will gradually become visible.
>
> (1985: 11)

Might we not make a similar, if less dramatic, claim for Raphael's Hieroglyphics? Might it be that their visual code, their ambivalent 'imagality' (both indeterminate and capable of conjuring ex-post-facto visual certainty) enabled them to seduce the shadow of futurity?

Conclusion: the photograph, a small window on the future

Inspired by our ruminations on Raphael, the same claim can surely also be made for the photograph, or at least certain iterations of it. Roland Barthes's contrast between 'the civilised code of perfect illusions' (the stale and used-up way in which photography is normally approached) and the photograph's 'wakening of intractable reality' (1981: 119), for which he argues so passionately, is nowhere more evident, and more illuminatingly present than in the work of the French sociologist and ethnographer Pierre Bourdieu. In Bourdieu's case the conflict lies in his claims for the perfect (we might say 'emblematic') illusions that French vernacular photography delivers (illusions he reproduces in his own analysis of these practices), and conversely the intractable reality that his own fieldwork photography in late 1950s Algeria was forced to confront (for a fuller discussion of which see Pinney 2016).

Bourdieu's ethnographic study of French photography is exemplary of what we might think of as pre-Latourian anthropology – that is, a form of analysis which is Marxist in intent, but indebted to Emile Durkheim. Bourdieu's social science, while politically radical in its critique, is conventional in its basic modality: he treats photographs as crystallisations of sociality, terming them 'solemnisations,' something akin to Durkheimian 'collective representations,' like the totemic kangaroo glimpsed in the twilight of the desert, running away (Cladis 2001: xix). It is against this backdrop that we can place Barthes's sense of the 'oddness' that fellow scholars had not noticed photography's potential for 'disturbance' (1981: 12), for in the work of those such as Bourdieu the apparent work of the photograph was to merely serve (as Latour would later phrase it) as a screen onto which the cinema of society was projected.

In *Un art moyen* (Bourdieu 1996[1965]; translation *Photography: A Middle-brow Art*, 1990) Bourdieu is very interested in how as he puts it the 'Portrait Gallery has been democratised' and photographers have become their own 'historiographers' (1990: 30), suggesting that photography as a practice is not without transformative power. However, overwhelmingly his stress is on the manner in which photography acts as a mechanism of 'integration' (1990: 19), a 'solemnisation' (1996[1965]: 27) *after the fact* that can be read as a 'sociogram' (1990: 23). The family photographic album objectifies 'social memory' and has, as he says in a memorable metaphor, 'all the clarity of a faithfully visited gravestone' (1990: 30–31). Strikingly, in Bourdieu, the 'community' or 'group' always pre-exists the photographic act whose destiny is to further integrate that group. The photograph is always a 'reaffirm[ation]' (1990: 29). When he writes of acts which 'must be photographed because it realises the image that the group seeks to give of itself as a group' (1990: 24), the camera serves only to make manifest a kind of visual echo of what the group has *already* achieved. As Bourdieu further states in a characteristically tautological manner, 'The photograph itself is usually nothing but the group's image of its own integration' (1990: 26). There is no room here for contingency, or the unexpected. Indeed photographic activity appears predestined and ultimately meaningless: 'one may only photograph what one must photograph' and these images become a sort of 'ideogram or allegory,' signs 'to which one does not have the key' (1990: 36–7). *Un art moyen* is a work lacking in surprises and typifies the kind of analysis attacked by Lyotard in which the 'aesthetic' is granted visibility only in order to demonstrate its ideological function (Carroll 1987: 26).

Walter Benjamin's writing on photography offers a stark alternative to Bourdieu. Recall his question: 'Isn't it the task of the photographer – descendant of the augurs and haruspices – to reveal guilt and point out the guilty in his pictures?' For Benjamin, as for Bourdieu, photographs are allegories, but they are portents of a future for which there might be a key. Like Kracauer, Benjamin searches for the future in the archaic. Both Kracauer and Benjamin (and perhaps the Raphael of *The Prophetic Messenger*) would doubtless have agreed with Henri Focilon's insight that 'If the time of a work of art were the time of all history, and if all history progressed at the same rate, these questions would never need to be asked' (Focilon 1992: 141). It is these unsettled temporalities that the study of images forces us to confront. What time is the visual?

Notes

1 What Umberto Eco termed 'syntagmatic concatenation imbued with argumentative capacity' (cited in Burgin 1982: 38).
2 Heywood, writing in 1900 gives the figures during the previous five years of 158,000 to 162,000 (Heywood 1900: n.p.).
3 Article in the 1845 *Prophetic Messenger* entitled 'On Hieroglyphical Devices, with Illustrations from Rare Examples' 46.
4 The British Library catalogue lists Raphael's almanacs from 1827, but many issues seem to have been lost and many are very damaged.
5 Raphael claims a kind of ocular self-evidence for his predictions: in the 1841 issue, commenting on the previous year's hieroglyphic's predictions he writes that 'An explanation of the plate is almost casting a doubt upon the powers of observation of our readers' (1841: 67). The term 'literality' used by Raphael (and derived from Biblical hermeneutics) in fact points to a 'seeing and believing' which might more properly be termed 'imagality.'
6 Recall Lyotard's insistence that 'Something is always happening in the arts [...] that incandesces the embers glowing in the depths of society' (cited by Carroll 1987: 28).
7 'The monologue of standardised, stereotyped music accompanies and hems in a daily life in which in reality no one has the right to speak any more' (1985: 8).

13 Held in Amma's light

The enchantment and political efficacy of *gopurams* in Tamilnadu

Jill Reese

An enduring and central tenet of material culture studies (MCS) is the necessity to critically engage with objects, images, and other (im)material forms frequently dismissed or rendered invisible by their omnipresent, prosaic, or marginal positioning. This study similarly endeavours to bring into focus a specific form of Tamil spectacle, *gopurams*, or light structures, locally defined by their peripherality to the main events of religious festivals and *vizhaas* – annual celebrations such as those recognising the birthday or death anniversary of a venerated political leader. Gopurams are large-scale towers made of bamboo, curved to form the outline of a figure or symbol that is decorated with strings of lights. They are temporarily installed for use at night to serve as illuminants to attract devotees or participants toward the centre of the *vizhaa* through their affective enchantment. They capture beholders and draw them into the broader spectacle.

This chapter is situated within the particular ethnographic context of the then state chief minister's birthday celebrations and draws upon empirical research in urban Tamilnadu, India, with politicians and artists. It is informed by learning how to make a gopuram, by observing viewers' interactions with towers at numerous festivals and events and, of course, by examining the aesthetic and material qualities of gopurams themselves. The chapter examines the enchantment of gopurams in a discussion that extends material culture articulations of agency, iconicity, and corporeal vision to meet the comparatively nascent 'anthropology of luminosity' (Bille 2019; Bille and Sorenson 2007) by considering the material qualities of light within politically charged a/effective 'atmospheres' (Böhme 2017). I argue that the enchantment of gopurams rests upon a constellation of material affordances, processes of making, image-based semiotics, and local sacro-civic notions of vision and corporeality. Further, when operating within a broader image regime carefully constructed to appeal to constituents as devotees, such as that by Chief Minister Jayalalithaa Jayaram[1] and her political party, gopurams are not only enchanting and affective but also politically efficacious.

Figure 13.1 Jayalalithaa as *gopuram* to celebrate her birthday (*pirandanaal*). Madurai 2014.

Source: Photo by author.

Political lineage as luminance

In Madurai, Tamilnadu, three celestial figures tower above the playing fields at Thamukkam Grounds, a public events and sports arena. The figures bathe hundreds of spectators in multi-coloured light as they watch multiple *kabadi*[2] matches taking place simultaneously. These towering figures are incarnations of a powerful political lineage that begins with C.N. Annadurai, founder of the Dravidian Nationalist party,[3] the DMK (Dravida Munnetra Kazhagam). In 1967, the DMK was the first party in the country to oust the Indian National Congress[4] from governmental power at a state level. The second gopuram figure is that of M.G. Ramachandran, affectionately known as MGR, the former film star-turned-politician who rebelled against M. Karunanidhi, the successor to DMK leadership after Annadurai's death. MGR founded the AIADMK (All-India Anna Dravida Munnetra Kazhagam) and by incorporating 'Anna'[5] into the name, claimed himself a direct political descendant of Annadurai. The final figure whose radiant light most prominently manifests above the fields, and whose image is iterated in various print and painted forms across Thamukkam Grounds, is MGR's former co-star and political heir, Chief Minister Jayalalithaa (Figure 13.1). It is the 23rd of February 2014, and the event is a celebration of her 66th birthday (*pirandanaal*) the following day, although she is present only through her multivalent imagery, and most spectacularly through her towering figure of light.

The many gopurams are impossible *not* to see – each was a beacon of light extending across the civic landscape to grab the attention of those far away and draw them forth into the action. They are typically placed at or near the entrance to an event, such as the 'pendant' gopurams of MGR and Jayalalithaa above the entrance to the kabadi arena (Figure 13.2) under which spectators must pass. This is in similarity to their namesake – the vibrant and colourful gopurams, or entrance towers, of South Indian Hindu temples that rise many storeys above their temple complexes. Those specifically surrounding the kabadi fields are distinct in that they enclose the field of action rather than attract those outside toward it, and their frontal directionality means that they are illuminated for the eyes of those within, rather than approaching, Thamukkam Grounds.

I begin with this scene to foreground the aesthetic and material dimensions that make gopurams enchanting, affective, and efficacious within a particular political context to inform the sections that follow on in the processes of making the material affordances of gopurams and the anticipated ways viewers see them, all from the perspectives of their creators. The chapter will then return to gopurams in the field – or rather, above it – as they cultivate an affective atmosphere enclosed within the multiple gopurams operating within the AIADMK's image regime seeking to legitimise Jayalalithaa as an amman – a mother, a goddess or a mother goddess.

Making, aesthetics, and affect

In the course of conducting fieldwork, I was drawn to gopurams due to their near ubiquitous instantiations at the major political, civic, and religious events held throughout Madurai and its surrounding villages. During a walk to document daily changes in streetside political imagery, I came across the

Figure 13.2 MGR and Jayalalithaa as *gopurams* at entrance to sports arena. Madurai 2014.

Source: Photo by author.

outdoor workspace of Alagupandy, the man widely considered to be the best gopuram maker in Madurai, if not all of Tamilnadu. He and those in his team refer to themselves as 'electricians' rather than 'artists,' in contrast to those who make other kinds of political imagery. Alagupandy was expected to become a general labourer like his father, but after a primary school teacher deemed him a talented artist, he went on to make a living painting signboards and gradually taught himself to make gopurams. Over the years, he has built a stellar reputation within the city and beyond, and his works are displayed as far away as the neighbouring states of Kerala and Karnataka. Alagupandy trained his son, Srinivasan, in the craft and anticipates Srinivasan will take over the business in the years to come. Alagupandy also created his own competition through apprenticing others. He claims that his work is nonetheless distinguished because his gopurams are 'clean' (*suttam*) – the outlines of the figure are clearly demarcated and fluid, and the lights do not muddle together. It was through conversations about what differentiates his work from others that elicited Alagupandy's reflections on the aesthetic and material decisions he makes, in addition to his perceptions about what it is that gopurams 'do,' and what he anticipates their beholders experience – all of which are inextricable from their enchantment and efficacy.

Gopurams are comprised of a series of panels decorated with thin strips of bamboo. The strips are carefully bent and secured with small nails to a wooden grid and adorned with strings of lights. Each bulb is wrapped in coloured cellophane secured with a rubber band. Placement of the strings of lights along the wooden strips creates the illuminated outline figure of a deity, politician, filmstar, or symbol. A gopuram begins with an image – usually a photograph or print – of the person or deity whose form will be manifested in the tower. Sometimes a client provides the image, but usually it is Alagupandy who selects it. He stressed that the image must be immediately recognisable to its audience, even when the person or deity's form is reduced to an outline. For deities, Alagupandy chooses prints in which the deity's associated iconography is evident and clear, and he incorporates colours and items easily recognisable to their audience. For all large-scale gopurams, whether political or religious, the frontality of the deity or figure must be accessible, that is, the full body of the god, goddess, or politician directly faces the beholder.

The images he chooses of politicians in particular tend to be photographs continuously circulated in multiple media formats, such as murals, party posters, vinyl banners, and newspapers. As seen in Figures 13.1 and 13.3,[6] Jayalalithaa raises a hand with two fingers in the shape of a 'V.' This is in reference to the two-leaf symbol of the AIADMK's flag, also appearing as gopurams in Figure 13.1, and is a gesture she and other party leaders frequently make when appearing in public, especially before large audiences. The two-leaf symbol is iterated in numerous party posters, appears in car decals and stickers, and is embroidered into many garments and even pieces of jewellery worn by party members.

Alagupandy also explained that he often chooses pink cellophane wrapping for Jayalalithaa's face and hands because she is often complimented for having a pinkish hue to her skin (Figure 13.3). Gold, or rather yellow, is essential to accurately evoke MGR, who was popularly praised for having a golden skin tone. Such aesthetic choices appear to be common among those creating political imagery. Preminda Jacob (2009) shares similar comments and design choices for these figures when painted by artists as large-scale hoardings installed in Chennai. A mural painter in Madurai also provided me with a similar explanation for his selection of hues when painting Jayalalithaa, adding that he uses more red tones in her cheeks than those visible in her photographs in order to attract the attention of people as they pass his murals.

Alagupandy selects images from the popular archive that are either iconic, or in the process of reaching an iconic fixedness, to the degree to which such a state is attainable. In following Roland Barthes (1972 [1957]) these specific kinds of images of a person or deity have already accrued heavy symbolic density within their narrative, even to become to a large extent 'sedimented' (Ghosh 2011). This level of iconic stability is crucial because it enables their immediate recognition without active cognition on the part of the beholder. As Bishnupriya Ghosh (2011) suggests, icons operate as a mediatory apparatus

Figure 13.3 Jayalalithaa as *gopuram* along street outside Thamukkam Grounds. Madurai 2014.

Source: Photo by author.

between subject and image object, allowing the beholder to 'enter' a then and there, or place and time, that is projected or contained by the icon.[7]

The mechanics of such human–thing relationships are predicated upon the ability of images to work upon the body through affect – a corporeally based zone of intensities, a resonance experienced at a pre-social level, as described by Brian Massumi (1995, see also Mazzarella 2009). Thus, the image that Alagupandy chooses is one rife with affective potential. For Tamil audiences, it is immediately recognisable and is thus intended to stir, at least at a preconscious level, a resonance within the beholder of the embodied experiences of their countless interactions with the icon – as either a person or deity *and* of that particular visual representation. However, it is not only through resonance with the iconic image that gopurams are efficacious, as the following sections will demonstrate.

The materiality and affordances of light

When discussing gopurams, the electricians focus heavily on the qualities of light and very little about the difficult process of translating an image to a chalk outline on their asphalt workspace, on the surface of which they will create the panels. Nor do the men elaborate on the skilful technique of curving and securing bamboo strips onto this frame. Instead they emphasise the brightness and clarity of light in the creation and reception of the figure, as well as the distances at which gopurams can be perceived. Their considerable attention to the efficacy and affects of illumination on those who behold them resonates with Mikkel Bille and Tim Flohr Sørenson's (2007; see also Bille 2019) appeal for an 'anthropology of luminosity' to analyse the materiality of light and the impact it has on social relations. The authors herald Tim Ingold's observation – one that on its surface seems obvious but unfolds to reveal greater complexity and implications – that 'light is the experience of inhabiting the world of the visible, and that its qualities – of brilliance and shade, tint and colour, and saturation – are variations, upon this experience'[8] (Ingold, cited in Bille and Sørenson 2007: 264). The authors extend Ingold's insights by advocating that the agency of light 'works as a significant constituent of experience' by examining 'how light is used socially to illuminate places, people and things [...] in different cultural settings' (265). Bille (2019) later articulates the multivalent ways in which light impacts emotions and senses of being, of place, and of nostalgia – all this despite the seeming, if not paradoxical, invisibility of light due to its omnipresence in human and nonhuman life and experience.

It is not just that gopurams are luminous – the aesthetic qualities of the light as afforded by the materials used to create them can enable or hinder their efficacy for enchantment. Alagupandy and his team lament the rise in popularity of LED (light emitting diode) bulbs over traditional incandescent bulbs for several reasons, including their higher upfront cost despite the fact that LEDs are less expensive over time because they consume less electricity

Figure 13.4 Comparison of LED and incandescent bulbs. Madurai 2013.
Source: Photo by author.

and operate far longer than incandescent lights.[9] As the electricians explained, LEDs are not as bright, in part due to the sizes available in the area at the time, as demonstrated in Figure 13.4. Further, LEDs are unidirectional and most often project light up to 180 degrees, in contrast to the omnidirectionality of incandescent bulbs that can emit light 360 degrees. Gopurams with LED lights are, therefore, visible from shorter distances and from a narrower range of perspectives in relation to a viewer's position. The level of brightness and the more limited directionality of light from LEDs also tend to create a blurring of a figure's lines, which is further exaggerated with distance.

The 'electricians' consider LEDs inferior in quality and assume customers think they are better because the technology is newer and more expensive. Unfortunately, customers attribute the comparative loss of brightness, clarity, and perspective, not to the LEDs, but to the makers' craftsmanship, a complaint to which Alagupandy is particularly sensitive given the reputation for his skill creating clear and clean outlines. This criticism of the kind of bulbs used more broadly reveals an understanding, from the perspective of their makers, about how the affordances (Gibson 2014 [1979]) of materials affect the gopuram's reception and efficacy. Even more so, the kind of light used impacts the extent to which the beholder can identify the deity or politician whose figure is an iteration of an image that is popularly circulated, or whose particular iconography must be clearly discernible to be recognised.

Corporeal vision and the technology of enchantment

The electrician's emphasis on the qualities of light signals a primacy of the visual in that these are, of course, material objects meant to be consumed through sight. A foundational and long-standing tenet of material and visual culture is that 'seeing,' 'looking,' 'gazing' – vision itself – are embodied, synaesthetic activities (Pinney 2004; Babb 1981; Eck 1981) in contrast to a Cartesian perspective that posits a separation of the mind – and its perception through vision – from the body. As demonstrated through innumerable ethnographies, beholders are never disengaged from their corporeal bodies as fields of power, nor detached from affective sensations activated through human–image interactions. We perceive through our bodies, and knowledge, experience, and impressions are transmitted and mutually shared through the interactions between people and things.

In what may initially seem like an unlikely link to gopurams, Eva Hayward (2010) refers to haptic vision between humans, cup corals, and other marine species as 'fingeryeyes' – an evocative term referring to the transmission and reception of impressions through the finger-like tentacles of Pacific cup corals, as they touch and 'sense,' as if the tentacles are eyes. Hayward's term resonates here with the interactions between people and light, particularly with the luminous emissions of gopurams. Gopurams touch and impress upon the beholder through the transmission of light that moves away from the gopuram and reaches out toward the beholder to impress upon their eyes.

Luminous brightness, and in some cases dazzling animation, are at the heart of the mechanisms by which gopurams enchant. Christopher Pinney proposes a somatic approach to the visual combined with the efficacy of design with his term 'corpothetics,' a neologism for 'embodied, corporeal aesthetics' (2004: 8). Corpothetics stresses the efficacy of an image over what it 'looks' like to a viewer within Indian visual practices, particularly with regard to a *murthi* – an image, statue, or other manifestation of the divine in which a deity is embodied or present, and with which a devotee can take *darshan*, the reciprocal interaction between deity and devotee of simultaneously seeing and being seen. Thus, a *murthi*'s efficacy is predicated upon its aesthetics, which must be designed such that it not only allows for recognition of the deity, but also facilitates devotion through vision. Although gopurams are often images of deities created by light, they are rarely objects of veneration in the same regard as a murthi. This instead speaks to the material and visual dimensions of the design and overall aesthetics of gopurams that attract the beholder, as well as the beholder's corporeal vision.

Similar to the Hindu temple towers for which they are named, gopurams are positioned on the periphery of an event, often at its entrance, rather than being the central focus. They play a significant role during *vizhaas*, however, by performing the crucial function of drawing people in by beckoning with their warm light. They are bright, colourful, large-scale and can be seen from great distances. They are attractive *and* they attract. In Alfred Gell's terms, light towers are a 'technology of enchantment' (1988, 1992) through which their seductive qualities of luminosity and, in some cases animation, pull their beholders toward and abduct them, like a trap, and prime them for the engagement at the event's centre, whether it is to venerate a deity on procession, to gather for the performance of politics at a rally, or to demonstrate their patronage, respect, and collective identity, as is the case during Thevar Jayanthi, an annual caste celebration in recognition of freedom fighter Muthuramalinga Thevar, a hero of the Thevar caste who opposed British rule. Every year, Alagupandy and his team build an immense gopuram, often depicting a throne or *mahal* (palace) with regal iconography, placed behind the freedom fighter's statue and the two levels of stairs people climb to reach a platform at the statue's head (Figure 13.5). The gopuram not only guides celebrants travelling from the districts south of the city, who can see the unnatural brightness at a great distance, but also illuminates their actions of pouring milk, garlanding with flowers, or offering *prasadam*[10] to the Thevar hero.

It is not simply that there is light, but that there are specific material aspects to the light, which, in combination with the aesthetics of their designs, makes them enchanting. Other than the fact that they are bright and colourful, the more dazzling and revered gopurams include animation effects created through the timing of individual light circuits. Even while speaking in Tamil, Alagupandy and Srinivasan use the English word 'blink' to describe the sequential changes in imagery created by multiple electric circuits. A simple example is the animation of an arm to create the impression that a politician is waving. In more complex gopurams, there are multiple layers of panels, each with its own lights, whose figures will be shifted every few seconds. A relatively common gopuram for amman (goddess) *vizhaas* is of the Tridevi containing the goddesses Lakshmi, Parvati, and Saraswati, who individually and sequentially appear with each cycle of the alternating electric currents, recognised by their unique iconography during each transition, but who share the same outlined face and body. When Srinivasan explained to me how this kind of gopuram works, he said the name of the goddess and then 'Blink!' using both his hands and eyes to punctuate the change in image, or blinking, before saying the name of the next goddess in the cycle followed by 'Blink!' again. The lights blink, but so do our eyes during these transitions.

Even more dazzling is the effect of movement across space through binary animation. When I asked Alagupandy about which gopuram he was most proud, he described a twenty-meter wide train with multiple railcars that appeared to move across a horizontal landscape as the light circuits shifted from panel to panel. Not only did he create the impression of linear movement, but also the railcars' wheels and animal occupants each moved or changed position from frame to frame. He detailed the joy and excitement he saw on the faces of the crowds drawn to the gopuram, with children pulling parents by the hand to get closer.

Figure 13.5 Large-scale *gopuram* installed for Thevar Jayanthi. Madurai 2014.
Source: Photo by author.

Luminosity, then, becomes an entry point into an understanding of the material, aesthetic, and iconographic dimensions through which gopurams affect and enchant beholders, as well as the work they do within a broader event, such as a temple festival or political rally. The efforts, intentions, and anticipation of the electricians who guided the previous sections now inform an examination of the gopurams situated in the ethnographic setting that opened this chapter.

Image regimes, morality, and 'Amma'

Chief Minister Jayalalithaa's *pirandanaal* is a particularly significant date for the AIADMK party, given that the birthdays of leaders are always marked occasions, before and after their deaths. However, in Tamilnadu, a chief minister's birthday is recognised through a variety of events in cities throughout the state, organised by the party and local civic leaders. Most commonly, these include public gatherings at the statues of the chief minister or of the party founder, as in Jayalalithaa's case, since no statues of her had been installed in Madurai at the time. Party members extol her for the many accomplishments they attribute to their leader, and then they express their loyalty – first to the top of leadership and then to those above themselves in the party's hierarchy. Such events also include forms of giving to poorer communities, usually food but also items of clothing such as *saris* for women and *lungis* or *dhotis*[11] for men, often decorated with party iconography or colours. There are often small exhibitions, held in railway stations or at major municipal buildings, of photographs and posters depicting the initiatives introduced or completed under the chief minister's direction. Such events were similarly held in Madurai to mark Jayalalithaa's birthday but, unlike previous years or under the other party's administrations, there was also a kabadi tournament that drew a far more diverse crowd than would normally attend a political event.

The kabadi tournament around which the many gopurams were placed is part of what I have previously glossed as Jayalalithaa's 'nutrition schemes' (Reese 2017). Expanded here as 'sustenance schemes' to better capture the breadth of ways her political persona and image regime sought political efficacy, they are a means to understand her electoral promises of a broad range of material goods if the party were to be elected – what her critics understandably denigrated as 'freebies' equivalent to bribes for votes. The dozens of material goods included household fans, food grinders, bicycles, rice, milk, flour, goats, and even laptops for every university student. The sustenance schemes grew out of her carefully crafted transformation through public imagery from a female filmstar to a respectable politician, beginning as her political predecessor's divine consort in the early 1990s and eventually to that of 'Amma' (mother), an amman or mother goddess for the Tamil polity in the 2000s.

When Jayalalithaa first contested a general election[12] in opposition to the DMK – the only other party with viable political authority in the state – her claim to ascendancy was visually conveyed through specific choices of images depicting her with MGR, as well as through a particular use of juxtaposition, scale, rhetoric, and framing which, in combination, constitutes an image regime.[13] As proposed by MSS Pandian and V. Geetha (1989), the campaign imagery played upon the popular assumption that the unmarried and childless Jayalalithaa was MGR's mistress. The image regime repackaged the relationship as that of Jayalalithaa as divine consort to MGR as god, an effective positioning given MGR's continued popularity and even widespread veneration of him as a deity in Tamilnadu. Campaign posters presented the pair standing side by side and facing the viewer, or included a photograph of MGR appearing to bestow a blessing upon a younger Jayalalithaa by bending towards her with his hands in supplication. Jayalalithaa, wearing a white *sari* with accents in party colours, gazes up at him with a coquettish smile and slightly bowed head, her hands likewise pressed together. Even more evocative were posters with a photograph in which MGR appears to be giving Jayalalithaa an ornamental sceptre to visually convey MGR's handing of the throne onto Jayalalithaa, which was paired with the campaign slogan that she would 'bring back MGR's rule' (Jacob 1997: 143). Such framing of their relationship was her claim to power, but it also provided Jayalalithaa a method of assuaging concerns regarding her morality as a woman in the public sphere by providing an alternative and benign persona.

As an actress-turned-politician, Jayalalithaa was morally problematic first as an actress who, by virtue of her profession, performed in front of a camera or before a live audience for whom she 'willingly expose[d] herself to the gaze of many unfamiliar men' (Seizer 2002: 120). Her 'publicness' was exacerbated by multiple on-screen romances with men, including MGR (Chinniah 2008). Historically, many actresses became mistresses given that such stigma diminished their prospects of marriage since they were, and still are, generally considered not meeting the criterion of a 'good' wife,[14] and therefore are eliminated as a potential bride for arranged marriages, which remain overwhelmingly the norm in Tamilnadu.

The traditional Tamil ideal for womanhood is that of a *cumankali*, an auspicious and morally chaste married mother of at least one son (Reynolds 1980; Lakshmi 1990). She is a nurturing upholder of Tamil culture and mostly relegated to the domestic sphere, where she is submissive to her husband. Despite being an ideal, it remains one that continues to be valourised in contemporary Tamilnadu, and to which women are frequently measured, even if it is achieved only to varying degrees. A common refrain repeated by both men and women at the women's college where I lived and worked was that 'a woman must first obey her father, and then her husband.'[15]

Jayalalithaa asserted herself as MGR's political successor but, as a woman, she faced multiple stigmas of being an unmarried actress who was now in politics – a very public position counter to the morally appropriate married woman safely protected in the private sphere of the home. Such a stigma was addressed by an inversion or rebranding of the assumed romantic relationship between Jayalalithaa and MGR, in that she became MGR's divine consort appointed to continue his vision.

For a woman without children (like Jayalalithaa), as long as she remains an unmarried virgin upholding chaste womanly virtues in all aspects of her life, she could eventually be considered an amman, which literally means 'mother,' but in this context it refers to a type of unmarried and childless goddess whose *sakti* (power) is unrestrained by a male partner (Egnor 1980; Reynolds 1980; Ganesh 1990). In traditional Tamil society, the amman was the only auspicious option for women who never married. Thus, the next incarnation of Jayalalithaa's image regime began after she achieved electoral power as chief minister in 1991. Jacob (2009) documents the marked change in aesthetics of Jayalalithaa's political posters in which she claims a majority of the image space rather than sharing equally with MGR. Jayalalithaa's forward-facing figure is solid, with arms close to the body, with a sari usually wrapped about her shoulders, her hair pulled into a tight knot at the nape of her neck. She named herself 'Amma,' aesthetically and semantically adopting the identity of an amman – the mother goddess whose children are her devotees.[16]

Jacob (2008) proposes that Jayalalithaa had solidified her amman status in imagery to reach a fixed state of iconicity, with her body in the static and rigid frontal position of a religious idol, a half smile on her otherwise vacant face. I argue (2017), instead, that the initial presentation of 'Amma' by the AIADMK's campaign offered the traditional aesthetics of an amman, but failed to resonate with the public because there was not a sincere, corporeal, and devotional link between Jayalalithaa as Amma and the constituents to whom she sought to transform into electoral devotees. Amma was present in image and form but had not also demonstrated loving care for her people and their needs – an exchange of support for devotion expected in Hindu devotional practice. The sustenance schemes were later introduced in an effort to address that failure.

Jayalalithaa's packaging as Amma therefore evolved to become an image regime that pursued a material politics that would resonate within Tamil society. It sought to engage idioms of popular devotion to establish, maintain and reify corporeal and affective registers created through the consumption of items provided through the sustenance schemes. The electoral promises of the impressive range of items were largely kept once she took office, albeit with varying criticisms about their qualities and durability. Despite their purchase through public funds, Amma 'gave' thousands of items through highly publicised ceremonies, the photographs of which were included in scores of large posters pasted along the streets of Tamilnadu's urban centres, including Madurai. Every item given through the schemes was branded with her official portrait, even the desktop image on the laptops the university students received. The desired affect of the simultaneous consumption of her image alongside promised items sought to sediment her persona in the public imagination as the benevolent Amma providing various kinds of sustenance to her devotees.

Amma in light: the political efficacy of gopurams

'Amma All-India Kabadi Tournament,' as the posters with her portrait advertised, was an unprecedented event organised to celebrate her birthday, although it was held in addition to the more quotidian forms of commemoration, such as the garlanding of political portraits or statues during lengthy speeches by local party leadership to valourise her character. A tournament for an Indian sport featuring teams from across the region, most of which were from universities, reflects her broader project to demonstrate divine and maternal care for the bodies of her constituents – in exchange, of course, for political support. In similarity with the material goods branded with her official portrait, the tournament was branded with her iconography in posters, murals, and *as light* through the numerous gopurams. Despite the multitude of forms in which her image manifested throughout the city's sports arena and the streets surrounding it, the most luminous, large-scale, and imposing were those of the light structures that towered above the players and spectators.

At the tournament, the affective enchantment of gopurams operated within an image regime that sought to simultaneously affect the public visually, corporeally, and semiotically by combining the persona of Amma, as mother goddess to the Tamil public, with her portrait and branding, to be consumed along with the material goods she 'gave' to the populace. The intended triggering of devotional and veneration practices of exchange were manifested within the tournament by creating a spectacle of bodies in sporting competition, as well as a spectacle of political luminance as gopurams in the forms of multiple Amma's and her (singular) male predecessors.

In this specific ethnographic context, then, the smaller pendant gopurams were placed in areas to attract and draw people in, as would be the case with nearly all civic or religious *vizhaas*. Of significance here are the far larger gopurams surrounding the fields of play so that spectators were continuously bathed in the light of the oversized figures positioned to face one another while also towering above the comparatively miniature figures in kabadi battle below. Observing people in the crowds during the event reminded me of Alagupandy and Srinivasan, who also taught me about what gopurams *do* by correcting how I saw them. During one of the first days I spent at their workspace, Srinivasan brought out a recently completed gopuram of a colourful mandala that could either be statically illuminated or, with a switch on the electricity circuit box, appear to rotate due to the placement and alternation of the light circuits affixed to the bamboo frame.[17] Before encountering the animated mandala, I immediately approached the gopuram as I would a painting on display by standing about a meter away to take in the entire image, and then stepping in closer to inspect its finer elements. Srinivasan stopped me and directed me to mimic his act of looking. He opened his eyes widely and then quickly squinted. This movement reminded me of the fine adjustments of a camera lens to gain focus on its intended subject. The effect was first a brief blurring of the gopuram's lights, which slightly obscured the figure's form, and then through the quick adjustment of squinting, the lines of light were refined, and the figure of the form was clarified. It was as if, through this directed practice of viewing, the gopuram came into focus. This focus continued as eye muscles relaxed into a gaze before slowly becoming blurred again. This moment of teaching me how to look was my first glimpse of the visual exchanges between gopurams and their beholders, as imagined by Alagupandy, Srinivasan, and the other makers working with them.

During the kabadi tournament, I observed spectators gaze upward at the gopurams and make similar adjustments with their eyes – starting wide and then slightly squinting. Such adjustment was followed, however, by a relaxing of the muscles around the eyes as people continued to gaze, look at another gopuram, and then another, especially during breaks held within or between kabadi matches. It made me consider how people look at illuminated objects but do not usually stare so directly at lights, let alone gaze. Their actions and the luminous glow reflecting on their faces are reminiscent of Hayward's (2010) notion of 'fingeryeyes,' in that the light seemed to touch the eyes of the receptive bodies down below.

Thus, it is not just that gopurams as looming light towers hover at the edges and above the grounds of an event. They are a materialisation of political spectacle implicated in a wider image regime that seeks to operate within a field of relations that resonates with the affective, devotional, and corporeal registers of a Tamil public. Gopurams are not only enchanting and affective but also politically efficacious when they create a radiant atmosphere through the combination of evocation of locally resonant idioms strategically implicated within Jayalalithaa's image regime with their idiosyncratic placement surrounding the central event and in addition to their more traditional places at the periphery and entrances. Spectators at the kabadi tournament beheld, and were held, in Amma's light.

This study draws upon several theoretical and methodological threads within MCS eloquently articulated in this book's introduction that include materiality, agency, relationality, aesthetics, and much more that question and complexify presumed dichotomies between subject and object, human and thing. The case study also resonates with a variety of provocations and concerns expressed throughout the volume and within this section in particular. Gopurams are instantiations of images – whether photographic, iconic, or print – that circulate or 'move' (Spyer and Steedly 2013) to materialise in various media where they have the propensity to become place-objects[18] through which beholders access place, as articulated by Walton, Chapter 16. This is productive along multiple lines of inquiry, but particularly for its assertion that digital photographs can offer a space (broadly conceived), and more specifically a 'place' for ethnographic encounters. This does not mean that such digital photographs become fixed in conceptual or temporal terms, but rather they connect to other evocations within this volume, such as that proposed by Pinney on the image as an 'index of the future' – a potentiality to anticipate and reveal that which is to come.

Thus, an implicit relationship is linked across several chapters between an image, image-object, or object and its possible future as artefact or object of study in museums, archives, exhibits, and collections. Such trajectories of image/object movement may enable discernible individual object biographies, as considered by Drazin, as well as raise both challenges and potential productivity in identifying and connecting provenance, social, and technical practices of making, and materials when there is a slippage or omission of social context and biography for objects in becoming collection artefacts, as examined by Mercier. Gopurams are unlikely candidates for inclusion within collections and exhibits due to their size, however, both the continuing materialisations of their imagery and of the images on which they are based, as well as the numerous political icon-imbued objects 'given' by Amma, have moved (and will continue to move) into exhibit and collection spaces – an example of which is an Amma-branded backpack for the similarly Amma-branded laptop that it once carried. This everyday object-turned-artefact is now used in teaching, but is nonetheless imbued with political claims and the social relations through and across which it materialised. In different but complementary ways, Schacter and Jeevendrampillai and Parkhurst address related complexities in curatorial practices and positionings arising from individual objects and through relationships and juxtapositions between them, as they variously attest, contest or complicate political assertions, values and subjectivities.

As is illustrated across this volume, and in this chapter's case study examining the making, materialisation, and implication of gopurams within Tamil politics, the shifts between lineages and advances nonetheless begins with the premise, as well as promise, of MCS to enable and enrich an understanding of the role of the material in the social (to paraphrase this volume's editors) – not as a myopic enthusiasm for objects, but as an ethnographically grounded approach that generates a greater apprehension and appreciation of humanity. This is the thread carried through the 'undisciplined' legacies of MCS that will extend into future articulations and preoccupations, as well as emerging and alternative object forms, practices, and understandings of the self, the world, and everything in between.

Notes

1 Jayalalithaa Jayaram died on 5 December 2016, while serving her third, non-consecutive, term as state chief minister and leader of the AIADMK.

2 *Kabadi* is a team sport played in India consisting of two sides of four players that alternate attempts to run toward and break through the opposing team's defences in order to eliminate one of the opposing team's players.

3 See Hardgrave (1973), Pandian (1992), and Ramaswamy (1997) for histories of the Dravidian Nationalist political movement.

4 The Indian National Congress, or simply the Congress party, was the dominant national party before and after India gained independence in 1947.
5 *Anna* is the Tamil word for 'elder brother' as well as Annadurai's nickname.
6 To provide a sense of scale, it should be noted that the *gopuram* is taller than the building with four levels at the left side of the photograph.
7 On the relationships between images and/as 'place' using complementary analytical frameworks, see Walton, Chapter 16 for a compelling consideration of images, in this case digital photographs, as 'place-objects.'
8 This resonates with Böhme's (2017) expansive reflections on light, darkness, brilliance and shadow within and in the making of atmospheres – variously conceived and articulated in physical, social, and philosophical terms.
9 The electricians also expressed frustration with the bluer colour temperature of the LEDs that were available in the area at that time. As they did for incandescent bulbs, they selected spherical LED bulbs in order to wrap them in coloured cellophane, which was more economical and offered a greater variety of hues than purchasing a wide range of coloured bulbs to individually place along the strings of lights. They had to make adjustments to achieve their desired shade of 'white' light, however, by wrapping LED bulbs in cellophane to mitigate the bluer temperature, whereas they would simply leave incandescent bulbs bare.
10 Material offering to a deity such as food, flowers, or money.
11 A *lungi* is a traditional garment for men: a long piece of fabric sewn with the ends joined together is wrapped around the torso and secured with careful folds. It is most often made of cotton or cotton blends. It can also be folded up so that the lower legs have greater mobility. A *dhoti*, also known as a *veshti*, is similar in that it is also formed by a lengthy piece of cotton fabric with an embroidered edge, but the ends are not sewn together. It is more formal attire in part because it is not to be folded upward to expose the legs. Over the decades, the *dhoti* has become part of a male politician's uniform when paired with a crisp, white button-up shirt. The result in appearance for both garments is similar to a long sarong or skirt.
12 Jayalalithaa entered politics while MGR was still alive and the party leader in the 1980s. After his death in 1987, Jayalalithaa asserted her claim to AIADMK leadership in opposition to MGR's wife, Janaki. The result was an acrimonious contest between two factions within the party, with Jayalalithaa as the successor. Some of the AIADMK members with whom I conducted research remained reticent to recognise her authority, even decades later. Others who were drawn to the party because of MGR's charisma claimed to support Jayalalithaa *only* because they believed MGR designated her as his successor, but that she had not earned the role by moving up through the party or even by inspiring others to join by believing in her abilities to lead. For similar discussions see Dickey (2008).
13 Building on Martin Jay's (1988) extension of Christian Metz's 'scopic regime.'
14 There are historical associations of actresses with prostitution in the courtly life of medieval kingdoms throughout South India. Such associations have long-standing linguistic ties in the denotations and conflations for the words 'actress' and 'whore' within spoken Tamil (Seizer 2002).
15 Variations of this by Christian colleagues especially emphasised the position of the father and husband in relation to god: 'A woman's god is first her father, then her husband.'

16 See C.S. Lakshmi (1997) on how women in the 1920s Self-Respect Movement in Tamilnadu reconceptualised male party members as family members (brother, father) in an effort to assuage prejudices about their activities in the public sphere and thus justify political involvement.

17 The word 'blink' was also used to describe the appearance of a *mandala* rotating counter-clockwise through a sequence of three different light circuits.

18 See also Sarah Pink (2011a) on photographs as 'place-events' for a different framing as well as attendant applications and implications.

14 A curatorial methodology for anthropology[1]

Rafael Schacter

Curating to the curatorial

The museum has played a central role within the history and development of anthropology. As a discipline that has had, since its earliest days, the 'collection, preservation, exhibition, study, and interpretation of material *objects*' at its very core (Stocking 1985: 4), the sacred site of the museum has acted as a critical pivot through which theories and models of the anthropological subject have been both revealed and refined, a location from which distinct understandings of the so-called ethnographic 'other' have come to be materially instantiated through the display of these others material artefacts. Spearheaded by the figure of the curator, not merely a keeper of objects but an overseer of orientations, the museum has utilised particular modes of presentation and particular narrative techniques in order to *re-present* knowledge garnered in the field (or the armchair, see Buchli, Chapter 2). As such, it has not only brought the social and material world of our 'informants' into visible reach but created specific ways of thinking about these interlocutors themselves. The changing conventions of collection and differing practices of exposition can therefore be understood to disclose as much about our discipline as those set under its disciplinary gaze, the vitrine's and storehouses of our museums not only containing, as Barbara Kirshenblatt-Gimblett notably remarked, the 'artifacts of our disciplines,' but in themselves acting as 'exhibits of those who make them, no matter what their ostensible subject' (1998: 78). A space of accumulation and acculturation, of education and investigation, the museum is thus also a space in which institutional paradigms are materially revealed, a depository of objects and ideologies alike.

This dual mode of revelation can be seen throughout anthropology's long and intimate imbrication with the museum. For example, the linear, teleological, typological arrangement of ethnographic artefacts that General Augustus Pitt Rivers' formed in the late nineteenth century, have retrospectively been seen as a mode of display that implicitly conveyed an 'ethnocentric message of conservative evolutionary gradualism' (Stocking 1985: 8). While attempting to counter the paradigm of degenerationism then dominant (see Jones 1998), the exhibits that Pitt Rivers' curated (notably of weaponry such as sticks or shields) were generated via a method of chronological categorisation paralleling the evolutionary biologism of Darwin – a technique not only placing those from 'elsewhere' into a deep historical past, but containing an entrenched racial hierarchy with Caucasians (especially Victorian ones) at the top. On the other hand, Franz Boas's rejection of this technique as both superficial and 'deceptive,' due to their reliance on 'outward appearance' rather than 'immanent qualities' (1974 [1887]: 66), led to his more contextual, pluralistic, dioramic presentations of tribal life. These scenes, often related to particular practices rather than isolated objects, were an attempt to 'preserve the multiple functions and inner meanings of a given form' (Stocking 1988: 8) and consequently act in a manner as meaningful to those from the communities exhibited as the museum audience. Yet while Boas's work contained a clear 'message of liberal relativism' (Ibid.), it could also, as Michel Verdon argues, be seen to have denied any 'sense of the individual as cultural actor,' creating a depiction of 'quasi-automata acting mostly through habit and imitation' (2007: 447). Moreover, as David Jenkins suggested, Boas's method of classification suppressed 'the historical processes of [the artefacts'] own development,' creating a 'context in which objects exist devoid of their history' (1994: 269); the circumstances of the often highly dubious acquisition of these artefacts – as Boas's assistance in the collection of skeletons, brains, and 'soft parts' of Filipinos who died at the 1904 World Fair in Saint Louis attests to – were hence left either entirely unaddressed or sadly unrecognised.

In a similar way, within the theatrical, immersive *mise en scènes* constructed by Margaret Mead, we can see not only a perpetuation of her mentor Boas's famous life-group displays, but an attempt to provide a more multi-sensorial scenography than previously witnessed, a method whereby viewers would engage through an all-encompassing perceptual experience. Yet the balance between distance and immersion that Mead's curation sought to mediate could itself be seen, as Diane Losche argues, to have disclosed the 'contradictory desires built into modernist ethnology' – the antithetical yearning to both panoptically 'see over – and thereby gain a sense of control over – an area' (2006: 241), while simultaneously attempting to embed the viewer in place and 'recuperate authentic experience' (Ibid.: 224). What's more, Mead's displays disregarded, as Losche continues, the innate 'chasms that separated an actual Pacific environment from words, verbal descriptions, and the overriding minimalist architectural rules of the New York of the 1960s and 1970s' (Ibid.: 238). It ignored the dangerous inauthenticity that these simulacra created, a constructed ambience that, ironically, generated an increased distance rather than proximity with those depicted. Every attempt to renew thus appears only to reveal a new chasm that the discipline has yet to admit.

As such, just as the principal museological paradigm of the last twenty years, the 'contact zone' as engaged in by Anita Herle (2008) and others, has attempted to transcend the technique of similitude that Mead's work epitomised through a mode in which the *subjects* of the ethnographic museum themselves were increasing able to negotiate the collection and narration of their own objects, the discourse of 'consultation' integral to the approach has been seen by some to have in fact bolstered the institution without truly empowering the communities they were purportedly collaborating with. The contact zone helped to form, as Robin Boast reports, an 'asymmetric space where the periphery comes to win some small, momentary, and strategic advantage, but where the centre ultimately gains' an imbalance born from the structure (more than the content) of the museum (2011: 66). The integrally problematic elements of the 'contact zone' as a space, as originally underscored by the creator of the concept, Mary Louise Pratt (1991), in which 'a dominant culture would provide for a 'negotiated' space for certain kinds of cultural exchange [...] *necessary to the maintenance of the imperialistic program*' (Boast 2011: 57, emphasis added), seems to have been perfectly reinforced as the paradigm shifted into the museum. Our current template for best practice can hence be seen as a space in which 'Others come to perform for us, not with us,' in which a 'pragmatic agonism is provided for all, but only to the degree that it returns to, and reinforces, the academy' (Ibid.: 63–64).

The practice of curation within the history of anthropology must, in this way, be seen as integral to both the past disposition and present status of the subject, a technique through which anthropological norms have been reinforced and broken, underpinned and toppled through a material rather than textual mode. It must be understood, not only as a public way of *revealing* our research, an 'impactful' way of displaying 'results,' but as a mode via which concepts are opened up and paradigms initiated, a space from which issues are not only concretised but publicly challenged. As such, curation must be seen as a practice crucial to the anthropological project, a site from which a political and ethical relationship to both our subject and our 'subjects' has been and can be instantiated.

Of course, many contemporary anthropologists have sought to engage the potential that curating gives the discipline, attempting 'exhibition experiments,' as Basu and MacDonald have explored, that 'unsettle accepted knowledge or the status quo' (2007: 17). In the work of Wayne Modest, for example, we find not only an attempt to decolonise the ethnographic museum (Basu and Modest 2013), to utilise it as a site in which issues of social justice can be established (Mears and Modest 2013), but also a method of collaborative curation designed to engage, enable, and emancipate its participants (Modest 2013). In her long-term engagement with Maori communities in the UK and New Zealand, Haidy Geismar has employed a series of innovative digital methods to explore how exhibiting alone – seeing, representing, and even hearing artefacts – could in many cases be less important than 'an understanding of the fields of digital communication as channels for social and spiritual relationships' (Geismar 2015: 313). There is also the creative response to the region of Five Heads in Mongolia, coordinated by Rebecca Empson and Hermione Spriggs, in which ten artist and anthropologist collaborators deployed an 'indigenous principle of exchange' garnered from the field in order to re-imagine the 'aesthetics and practice of anthropology from a minoritarian perspective' (Empson and Spriggs 2018: 7) and engage in 'a 'performative fictioning' of other realities' (O'Sullivan 2018: 147). Equally, we can see the *post-ethnological* or *post-ethnographic* technique pioneered (and coined) by Clementine Deliss, a method meant to 'dislodge' the 'ideological apparatus' of the 'museographic genre' (Deliss 2012: 66) and create 'forms of representation that extend beyond the academic appraisal of past histories and enable one to view the objects in the collection as prototypes for different futures' (Deliss 2013). Each of these differing modes of participation, digitisation, remediation, or decolonisation have today come to place the anthropologist-as-curator in a domain moving with and *beyond* the object. It has placed them, and us, in a domain where they are collaborating not simply collecting, enabling not simply exhibiting.[2]

The curatorial as model

In this chapter, however, I want to explore a particular technique of curating that I have recently explored within my own anthropological practice, a method that can act not merely as exhibit, but that can encompass, research, output, and impact in one. This, framed via the concept of the *curatorial* (as a notion defined in distinction to curat*ing*), is a position emergent from critical art practice rather than contemporary museology per se, one presenting itself as an expanded mode of *research* rather than exhibitory practice. Developing beyond the form of research typically encountered within traditional exhibition practice, however, through a form of 'research-led curating' or method of preliminary critical exploration that ends up housed within a resultant exhibition (in which research is simply a *precursor* to output), this is a mode of research that endures throughout the entire process of each particular project, a mode functioning in a perpetual not preludial manner.

The curatorial, as a conceptual model, can thus be seen to involve a (half) turn away from objects, away, at least, from their overarching hegemony. Echoing the manner in which artists have questioned the dominance of the artefact within contemporary art, something witnessed not only within the field of conceptual art since the 1960s (in which the idea gained dominance over the object) but so too within the more recent fields of post-Internet art (see Quaranta 2013), relational aesthetics (see Bourriaud 2002), and socially engaged art (see Kester 2011), this turn implicitly places the object as a part of a dynamic sequence of events rather than of a static moment of idiosyncrasy. Designating this as a 'post-autonomous' position, Nestor Garcia Canclini (2014) sees this as a zone in which 'art practices based on objects have increasingly been displaced in favor of practices based on contexts' (Ibid.: xviii), in which art moves beyond the 'autonomy of fields (Bourdieu) or of worlds (Becker)' (Ibid.: 11) and is rather understood to be inextricably embedded within social, economic, and political regimes (cf. Becker 1982 and Bourdieu 1993). The object's status as ultimate element of focus is hence substituted (or better reinforced) by a turn toward a more open-ended practice of collaboration, discussion, and collective interrogation, by a practice of close research and site-responsive investigation related to conditions not just spaces. The material procedures of exhibition making are thus here subsumed (or augmented) by the regimes of discourse and dialogue, experiment and enquiry traditionally placed at the backstage. The curatorial moves beyond the exhibition site as a space for the 'unequivocal "utterance" or finalised display' (O'Neill and Wilson 2014: 18) and towards its usage as a space for ongoing possibilities (rather than unchallengeable 'results'), a space to engage developing questions (rather than circumscribed answers), a site of knowledge *production* (rather than presentation). It is thus a move, as Irit Rogoff argues, 'away from illustration, away from exemplification,' and rather toward 'that which we do not yet know or that which is not yet a subject in the world,' towards a practice 'that does not rush to embody itself, does not rush to concretise itself, but allows us to stay with the questions until they point us in some direction we might have not been able to predict' (2006: 3).

Within the curatorial we can thus detect not only a particular emphasis on research, but so too a particular relationship to *time*. The curatorial can be understood as a manner of unfolding rather than concluding, as 'a way station in a process' (Rogoff 2012: 27). While the traditional museological condition has always necessitated time's arrest, freezing the artefact at its moment of collection (DeSilvey 2006), suspending the artwork through ever more complex and potentially dangerous techniques of conservation (Buchli 2008), here the curatorial endeavours to elongate or stretch time, to explore, as Victor Buchli has notably said, 'what happens before and after the artefact,' to explore the '*terms of materiality* rather than material culture itself' (2002a: 19). Johannes Fabian's seminal treatise *Time and the Other* (2014[1983]) can thus, in many ways, provide us with an interesting blueprint shifted from anthropology as text to anthropology as practice. The curatorial can be seen as a method searching for a way beyond what Fabian famously termed 'allochronism' (Ibid.: 32), beyond a mode that spatially and temporally distances our research subjects (or their objects), that relegates our ethnographic partners (or their artefacts) to a distant time, a time other than 'ours,' a time forever still. The classical museum dynamic can hence be seen to almost exactly replicate what Fabian termed the 'schizogenic use of Time' inherent to anthropological discourse (Ibid.: 21). As a space denying any notion of shared temporality, presenting the conditions of knowledge in an abstracted state, functioning through a rhetorical neutrality, the museum, in its archetypal form, embodies the temporal power games and techniques of detachment so trenchantly critiqued by Fabian. True egality in this context can only ever emerge through an emphasis on process, through a focus beyond the moment of display and toward a more 'open-ended, cumulative processes of engagement, interruption, and possibility' (O'Neill 2012: 128). The ambition of the curatorial to 'set a temporal process in motion rather than focussing on the outcome of any final exhibition' (Ibid.: 104), thus becomes a method through which to avoid the 'forces of historical inertia' that museums are 'perhaps inescapably, implicated [within]' (Stocking 1985: 4). More than just engaging themes of contemporaneity or simultaneity, this is about the sharing of *space* with others (cf. Birth 2008), an assertion of the value of ongoing collaboration and cooperation, of process over presentation.

The curatorial in practice

Opening in April 2019 at the University of London's Brunei Gallery, part of the School of Oriental and African Studies (SOAS), *Motions of this Kind: Propositions and Problems of Belatedness*, is a project which, I argue, engages directly with the concept of the curatorial. Emerging out of a three-year British Academy Postdoctoral Fellowship (2014–2017) for which I undertook approximately eight months of fieldwork in Metro Manila, the Philippines, the ethnographic focus of my fellowship was the capital's diverse, highly developed contemporary art scene. Conducting research with artists, gallerists, curators, fabricators, writers, editors, collectors, filmmakers, musicians, producers, coders, designers, publishers, and myriad other individuals connected to the wider art ecosystem, my fieldwork came to centre on the country's much-vaunted return to the Venice Biennale in 2015 (for the first time in 49 years). Already eager to explore the relationships between the local and the global in a famously transnational site, to examine the way that historical and contemporary relations of power were unpacked in a doubly post-colonial location, I thus set out to explore the tension between the lived situation of my research partners in the Philippines and their wider cosmopolitan presence in the world (via international education, international exhibitions, international residency projects, etc., let alone familial ties to Overseas Foreign Workers who make up over 10 per cent of the country's population).

Principally, then, it was the relatively new discourse of 'Global Art' that was the wider paradigm the project lay within (with Venice as the fulcrum that this worked around), and my research came to focus on the ways in which the practices I witnessed and the particulars I was guided towards came to rub up against this new form of Art World hegemony (Schacter and Balaguer 2017). While this new form of Art World hegemony I had engaged in curating within my general academic output in the past, however,[3] producing an exhibition related to this new research project was not an aspiration I took with me to the field (nor one that I discussed with my interlocutors).[4] Nevertheless, at the midpoint of the project I decided to apply to an open call at Brunei Gallery, in the hope that it could help continue my research further (both temporally and theoretically). Accepted in June of 2016 (just under three years in advance of the exhibition), the initial proposal was quite different from the final exhibition itself. Submitting what was, effectively, a proposal for a 'survey' show, a presentation of the latest and most innovatory contemporary art from the Philippines (yet one without a conceptual backbone aside from the regional), once the proposal was accepted the real work of determining the project in fact began.

First and foremost, I felt that curating the project by myself, acting as sole 'impresario' (in O'Neill's terms), would be a clearly inappropriate, if not neo-colonial, move. Seeking to collaborate with local cultural workers whose lived knowledge would always exceed my own, I thus approached two of my most engaging interlocutors, the artist/curator Merv Espina and the writer/curator Renan Laru-an; thankfully, both agreed to participate with me as a three-man curatorial team, and an intense period of planning ensued (both online and in person in Manila).

What became clear early on was that the relationship between where we were working and what we were working on had to be pivotal: The project was not happening in a vacuum (the disinterested white cube is of course a myth), and the site-specific nature of the project needed to be entirely evident. Situated at the *Brunei* Gallery (named after the benefactor of the building, the Sultan of Brunei), was hence a first point of interest. The Sultanate had, for hundreds of years, acted as a key trading and political partner of the Philippines, land commonly interchanging between the two nations (both Palawan and Sulu previously having been part of the Empire of Brunei), as well as many key events (and scandals)[5] occurring between the two neighbours throughout their intertwined historical arcs. Likewise, the location of SOAS and London, where the Brunei Gallery was situated, felt similarly crucial. While a school today famous for its radical left leanings, SOAS was originally founded as 'an instrument to strengthen Britain's political, commercial, and military presence in Asia and Africa,' serving to bolster and support the prolongation of the British Empire through providing 'instruction to colonial administrators, commercial managers, and military officers, but also to missionaries, doctors, and teachers' (Brown 2016: 7). SOAS as a site thus had to act as a central pivot. Added to the fact that the school had recently initiated a new hub, the Philippines Centre (under the guidance of Dr Cristina Juan), and had a huge archive of Filipino material yet to be explored – including nearly 150 boxes of letters, photographs, field-notes, and ephemera from the Ifor Ball Powell Archive – it was the historical and contemporary flows that intertwined and knitted together these three key sites, SOAS, Manila, and Brunei, that we wanted to ensure lay within our sight line.

Yet[6] it was a passage emerging, perhaps unexpectedly, from Sir Isaac Newton's famous treatise of 1687, *Philosophiæ Naturalis Principia Mathematica*, that helped us to solidify this wider desire, a passage emergent from the observations of voyager and astronomer Edmund Halley, in which 'Leuconia' – today known as Luzon, the largest island in the Philippines – is discussed:

> There are two inlets to this port and the neighbouring channels [...] one from the seas of China, between the continent and the island of Leuconia; the other from the Indian sea, between the continent and the island of Borneo.

Laying the foundations for modern science in his unparalleled exposition of the three laws of motion, gravity, and planetary movement, the tidal flows witnessed by Halley left Newton stumped, bewildered by the strange currents surrounding these distant equatorial seas. As he continued,

> whether there be really two tides propagated through the said channels, one from the Indian sea in the space of 12 hours, and one from the sea of China in the space of 6 hours, which therefore happening at the 3rd and 9th lunar hours, motions of this kind add together; or whether there be any other condition imposed by other seas, I leave to be determined by observations on the neighbouring shores.

Inspired by the curious indeterminacy of Newton's analysis, and by the 'motions of this kind' that exposed the limitations of his scientific expertise, we thus set out to explore, not simply 'the rise and fall of the tides,' but, as Filipino historian Ricardo Manapat suggested, the 'historical ebb and flow of ideas' (2011: 292). Newton's work thus led us on an exploration of the turbulent temporal currents flowing between Europe and Southeast Asia, the undertows that both hasten and delay the circulation of knowledge. Placing the theme of *belatedness* as our principal subject matter – as both a concept, reference, and argument – the project aimed to underscore the various 'propositions and problems' that this temporal state, or its accusation, brings. Our emphasis saw belatedness as that in which traces of the past could not only come back to haunt the present, but could act as a tool of resistance, a rupture rejecting the dynamic of primary and secondary (Bhabha 1994). It engaged with a belatedness that asserted itself as a rich terrain, not a linear judgement, a belatedness refusing the strictures of cause and effect, futurity and historicity (as seen in Pinney, Chapter 12), rejecting fixity of people and places. From this conceptual ground zero, however, came the crucial component of the projects enacted by our artist-collaborators themselves. The exhibition commissioned new works and developing ongoing projects by 11[7] artists, working either in or on the Philippines, as well as featuring an archival display from the never-before exhibited Ball Powell archive exploring Britain's informal empire in the Philippines.[8] While each of the nine individual projects produced for *Motions of this Kind* elucidated the key themes and processes brought forth by the exhibition, I want to briefly explore two works – from the artists Cian Dayrit and Mark Salvatus, the artists I collaborated with most closely during the exhibition – whose work can reveal something general about the exhibition through their specificity.

Curating tidal/temporal movements

Like much of his interdisciplinary practice, Dayrit's work – a large embroidered tapestry and a set of archival objects and documents arranged within three museum vitrines – explores notions of authenticity and reproduction, imperialism and indigeneity, power and memory, and the visual apparatus of colonial power. Entitling his project *Northern Conquests in Oriental Soil and Sea*, his tapestry in particular sets out to literally upturn our normative perception of geographic space. Basing the work on a map from 1744 drawn by the renowned English cartographer Emanuel Bowen, Dayrit's rendering both reverses the original orientation of the image (through rotating the north and south, east and west axes) as well as replaces the colonial names delineated in Bowen's 'new and accurate map of the East India Islands' with the collective names of local indigenous groups. As such, while geographically depicting the precise area described by Newton in the passage from *Principia* discussed in our exhibition text – the seas between Leuconia, South China, and Borneo – Dayrit rejects the naturalised efficiency of colonial cartography, acknowledging yet assimilating its power through rerouting the very premises on which it was made.

Yet the capacity of Dayrit's 'counter cartography' not only emerges through its geographic reversal and eponymic renaming, through its ability to interrogate the medium through the medium itself, but so too through its playful anachronism; the contrast of materials, concepts, and time frames within his piece enabled him to 'talk about the issues of today with the language of yesterday' (personal communication), to interweave the colonial histories of American and Spanish domination in the Philippines with the contemporary battles of land reform and ecological extraction that local indigenous and peasant groups currently face. Moreover, while the main body of the tapestry was produced in Metro Manila, final key elements were hand embroidered at the atelier of Hand & Lock, a London-based embroidery house who have provided embellishment services to the Royal Family and Royal Armed Forces since 1767. Imperial and colonial history thus not only exist representationally and conceptually within Dayrit's map, but are woven into its very fabric. Alongside his three vitrine works – displays juxtaposing key documents from the Ball Powell archive with tourist knick-knacks, gift-shop souvenirs, amulets, and talisman, in order to play with positioning and dominance, ideology and belief, in order to 'create a new gaze upon the gaze' (personal communication) – *Northern Conquests in Oriental Soil and Sea* collectively presents a defiant narrative rubbing against the historical grain, refusing the teleological and embracing chronological inconsistency, a simultaneous denial and appropriation of these powerful material forms.

Salvatus's work also addresses the historical record, but that of the more recent colonial past. His installation, *Blue Moon*, an eight-minute video piece installed adjacent to a large vinyl photograph and set of paper and papier-mâché masks, records a masquerade that occurred on 31 December 1910 in Salvatus's hometown of Lucban, a 'simple experiment,' as the local newspaper then reported, that the townsfolk wished to turn into an annual event. Paying 'tribute to the year that had just ended' and intended to provide 'proof,' as the report continued, of the town's 'exquisite taste and love for riots and expansion,' the Lucban Carnival of 1910 was realised just once, however, vanishing

Figure 14.1 Cian Dayrit. *Northern Conquests in Oriental Soil and Sea*, 2019. Tapestry, 215 cm x 238 cm.

Source: Produced in collaboration with Henry Caceres and Karin Beharrell of Hand and Lock. Commissioned with the support of Gasworks. Image © Agnese Sanvito. Courtesy of *Motions of this Kind*.

Figure 14.2 Mark Salvatus. *Blue Moon*, 2019. Video installation in single channel with masks, vinyl, and LEDs. 8', dimensions variable.

Source: Produced by Salvage Projects. Image © Agnese Sanvito. Courtesy of *Motions of this Kind*.

from social memory and remaining entirely unrecognised for over a hundred years. Upon discovering an image of the carnival and the newspaper article in question in Manila's Lopez Museum archives, however, Salvatus decided to re-enact this historical moment in his hometown, a belated reprisal (just 108 years late) reanimating this uncanny event.

Key questions remained for Salvatus, however: Who in fact introduced the masquerade to Lucban? Who influenced its style? At a key transition period between Spanish and American colonisation, between the religious Catholic pageants brought from Iberia and the more secular parades brought from the States (such as the masked Thanksgiving, or Ragamuffin parades today reformulated within the festival of Halloween), the archival images seem to show an intriguing intermeshing of both. Yet the conquistador-style features depicted on many of the masks also reveal a link toward the carnivalesque culture of Mexico, the site from which the majority of colonial administrators came in the Philippines during the rule of Spain. This very local, seemingly parochial, singular event can hence be seen as an opening from which one can enter much wider global networks. Salvatus not only shows the way that colonialism embeds itself into the specificity of social life, but reveals how the history of his hometown and family become intricately, intimately, intertwined with our all-encompassing metanarratives. The quotidian and taken for granted is hence here reclaimed to the status of the truly "historical": the site not of official policy and governmental affairs, but of the citizenship of the everyday.

The work of Dayrit and Salvatus – alongside each of the other nine collaborating artists – intertwined historical and contemporary themes, individual and collective concerns, local and global relations. They took the concept of belatedness as a starting point from which to investigate the relationship between the temporal and the spatial, the colonial, the personal, the political. Yet the focus on the belated so clearly emergent within these works is not, in itself, the only link I want to make toward the curatorial. Instead, it is the project itself as open-ended research method I want to reinstate, the exhibition as a methodological technique, as a space of evolution and unfolding, of speculation over demonstration that I want now to reinforce.

The curatorial as anthropological method

As methodology, the very fact of the exhibition positioned me within the wider political economy of art in a way that pushed me and my research into previously elusive spaces. It gave me access to certain members of the wider art ecosystem (politicians, philanthropists, gallerists),[9] to certain professional practices (into aspects of patronage, processes of delegation and cooperation)[10] I would have otherwise been excluded from. It gave me access to a level of collaborative intensity with my interlocutors that only a joint enterprise could allow, to a period of condensed yet concentrated fieldwork that could only emerge through the fulcrum of a collaborative and shared project. For example, the days spent together in Manila with my co-curators hammering out an initial concept, were in themselves a period of compressed, accelerated fieldwork entirely enabled by the operational structure of a forthcoming exhibition. And this, of course, is without discussing the months (in fact, years) of continuous correspondence and discourse that persisted as we pushed forward as a collective, without discussing the interpersonal relationships engendered and enriched through the mechanism of the wider project. Rather than simply following my research partners as they went about their everyday and professional lives, then, rather than just chatting at exhibition openings, at artist-run spaces, at studio visits or performances, here we were collaborating and discussing ideas together, thinking through concepts in an intrinsically cooperative manner (something also discussed by Geismar in Chapter 6). Without the grounding of a wider enterprise, none of this would have been possible. It would have been an imposition, not a collaboration, an obtrusive demand I would be unwilling to make and they, more to the point, almost certainly unwilling to accept.

As methodology, then, the curatorial enabled me to impel my work upwards (towards people I would otherwise be unable to reach) as well as to burrow it deeper (towards those who would otherwise be unable to afford the necessary time) – to conduct, in George Marcus's famous words, 'ethnography all the way up and down' (1998: 14). It gave me insights into many of the ways the wider art world as a whole functioned, while also creating its own call and response leading to ever more research opportunities and engagements. The project thus generated more ethnographic knowledge and insight, more ethnographic access and output, before the exhibition itself had even begun. It created a temporal trail that set the physical instantiation of the 'show' itself as simply one part of a much larger process, one incorporating an artist and curatorial residency (at Gasworks and the Delfina Foundation), a symposium (held at SOAS), a series of performances and workshops (held at the Koppel Project), two film screenings (at the ICA), as well as a range of public artist talks, academic conversations, and public tours; a second iteration of the exhibition, in Manila in 2021, is presently in its incipient stages.[11] The project, and consequently the ethnographic data, thus extends forward and reaches back far beyond the discrete exhibition moment. It continues as process and open-endedness, to not resolve but remain in continual motion.

Moreover, the very data produced via this technique, due to the long-term relationship and material co-production that the project enabled, meant that the installations and objects produced within the exhibition could be seen to move beyond 'art,' to be understood, in the terms of Hans-Jörg Rheinberger, as 'epistemic objects.' These were each artefacts emergent from a 'poetology of research,' objects stressing both the 'materiality of the research process' as well as their 'capacity to surprise the researcher, to defeat his or her expectations, to dwarf his or her powers of anticipation with their own revelatory richness' (2014). Based upon in-depth fieldwork, on close readings, on long-term engagement, the works of the 11 artists featured in *Motions of this Kind* must therefore be encountered as primary data not secondary reflections, as epistemic forms containing arguments and reasoning yet remaining figurally boundless. The reduction of ethnography to the anecdote can here thus be supplanted (or supported) by the material forms co-produced between curator and artist, ethnographer and research partner, by forms that must be understood as ethnographic data in and of themselves.

As outlined above, many contemporary anthropologists/curators – such as Basu, Deliss, Geismar, Spriggs, and Modest, amongst numerous others – have sought to face the challenges at the root of the museum as colonial record. In this shift from curating to the curatorial, however, the anthropologist moves from solely the custodian of artefacts to simultaneously a 'co-producer [...] critic or agent provocateur' (O'Neill and Wilson 2014: 14–15), a role in which the aim is not merely to *explain* the other but to embed them in a collaborative relationship. Projects taking place here are thus ones in which knowledge is *generated* not simply reproduced, in which exhibitions are understood not simply as a 'medium for representation' but, rather, 'a medium for "enactment"' (Basu and MacDonald 2007: 2, 12). Curation becomes a model in which the anthropologist engages in experimental, speculative, long-term processes wherein we can speak together with our interlocutors, mediating not controlling their own ways of seeing.[12] Rather than forming finalities, here we can create new knowledge (rather than display already confirmed truths); we can engage in an ethnographic process that reveals itself to the world in the midst of the process itself, one that reflexively opens itself up prior to its authorised, constructed result.[13] Here we thus move toward knowledge production as co-creation, a space in which ethnography stretches beyond the space of the preparatory alone. Here we embrace a post-autonomous practice moving with and beyond objects, with and beyond the traditional conception of the ethnographic itself.

As method, the curatorial can thus enable a level of access and temporal engagement we would otherwise be denied, while simultaneously elevating the entire process of the exhibition to a space of active research: It thus refuses the freeze-framing 'museum as mausoleum' first described by Theodor Adorno (1955), the docile museum, in Giorgio Agamben's terms, that presents the 'exhibition of an impossibility of using, of dwelling, of experiencing' (2007: 84): As method, the curatorial can hence eschew the dominance of the autonomous artefact and instead focus on the potential for collaboration and collective practice: This does not mean, of course, that objects are abandoned, but that, as Peter Weibel and Bruno Latour have shown, they become part of a 'performative turn' in which the 'aesthetic product is replaced by an artistic practice that can be object-based or object-free but nevertheless expands the scope for enactment' (2007: 107). As method, the curatorial can act as a site of the *speculative*, the subjunctive (as seen in Pinney, Chapter 12), as a field of evolution rather than exposition: We can hence engage with works that provide not 'doctrinaire answers or programs,' not 'propositions or conclusions,' but, rather, enable us to experience 'the pathways and enigmas of knowledge' (Gárcia Canclini 2014: 16, 28). The curatorial methodology thus moves beyond the idea of exhibition as space of *impact* alone and rather sees it as a space in which research, output and impact are in constant and mutual constitution, in which the project acts as platform for investigation, as site of exposure, as nexus for access. It must therefore be seen a practice holding huge potential for the contemporary anthropological project in itself; not merely a mode of best-practice within the museum, but one in which the relationship to our subjects and their objects is reassessed, one in which the political framing of anthropology as practice can be reformed and reactivated.

Notes

1 Thanks to the editors and two anonymous reviewers for their comments on this chapter. Particular thanks to Ludovic Coupaye for his enlightening insights.

2 Geismar (2015) has, however, made very clear the potential pitfalls housed within the art/anthropology relationship, in particular when art enters into the ethnographic museum but is not critically unpacked as a discourse with heavy baggage in itself and, instead, simply used "as a vehicle for overriding other categories and values surrounding the objects on display" (ibid.: 184).

3 See Schacter 2019 for a discussion of my prior practice of curation in which research-led curating was at the forefront.

4 My research collaborators were used to receiving global members of the artworld in Manila, generally on one-week excursions with numerous daily studio visits. These curators and gallerists were there for a distinct purpose, to bring back booty (or artists) from foreign lands and return them to global. While this is of course a touch cynical (and something I could of course be judged to have perpetrated myself), I hoped that the length of fieldwork undertaken would surpass the issues that these short term, helicoptered in trips can (though not always do) engender. Moreover, had I gone to Manila with the aim of curating an exhibition, I believe my research would have been hampered through falling into a potentially instrumental relationship with my interlocutors from the outset. At the least, my positionality in the field would have shifted from "researcher" to "curator", something that I simply did not want.

5 One of the most notorious scandals, known as the "Brunei Beauties" and concerning Filipina sex workers in Brunei, in fact prompted a congressional enquiry in the Philippines (Law 2000).

6 It is important to note that the ensuing two paragraphs draw heavily from the general exhibition concept for *Motions of this Kind*, written as a collaboration between Merv Espina, Renan Laru-an, and myself.

7 These being Yason Banal, Jon Cuyson, Liza May David, and Gabriel Rossell Santillan, Cian Dayrit, Eisa Jocson, Michelle Dizon, Amy Lien and Enzo Camacho, Kat Medina, and Mark Salvatus.

8 Curated by Dr Cristina Juan of the Philippines Centre, and the curator of the UCL Ethnography Collection, Delphine Mercier.

9 As an independently initiated project, fundraising was part of our remit as 'curators,' a role that brought me into contact with numerous organisations that not only helped to bring the project into fruition, but acted as sites for ongoing ethnographic research.

10 Färber's superb critique of the mode of 'cultural entrepreneurship' necessary to curate exhibitions today highlights the deep 'entanglements of experimental reflexivity in the workings of late modern capitalism' (2007: 219). While I hope to have avoided many of the pitfalls she highlights, it serves as a stark and invaluable warning.

11 This return exhibition, due to take place in December 2021, will centre around an exhibition and attendant conference but will also be the start of a wider 3-year series of programming and infrastructural support between the UK and the Philippines. Again, this will not only help to continue my own research in the Philippines, but will act as a material benefit for my interlocutors and the wider art community in the country.

12 Here I think its also worth thinking about Coupaye's (2018) work on analytical and vernacular categories within the museum, and the way in which the play of multiple vernacularities can reveal the inherently vernacular nature of the museum itself.

13 Here, Schnedier and Wright's (2010) discussion about the potential housed within th "incompleteness" of art over the "forcible completion" of anthropology is very pertinent (ibid.: 19–20).

15 Data aesthetics

Antonia Walford

The age of data is upon us (so we are told). In Euro-America at least, we have been inundated not just with data, but also with images and metaphors of data: from excess, in the form of data floods and deluges, oceans, and cascades (cf. Seaver 2015a), to analogies with natural resources like 'gold' or 'oil,' to its potential for social change, both positive – as when we are confronted with various data 'revolutions' in healthcare or governance – and negative – as when we are told big data analytics herald the end of democracy as we know it (for example Helbing et al. 2017). Disciplines (data science) and professions (data scientist) have grown up around it, and there are industries devoted to its ongoing valorisation through the ramping up of real-time, continuous quantification of any social or natural process, both a continuation of previous historical obsessions with quantification (Porter 1996; Hacking 1990), and a signal of new forms of calculative capital accumulation (Thrift 2005; Zuboff 2018). Data, at least in Euro-America, is emerging as a discrete social phenomenon (Beer 2016), worthy of anthropological attention.

In this chapter, I develop one possible approach to this emergent social phenomenon by asking: what does data *do,* and how does it do it? My contention is that we can begin to answer this through what I am calling an 'aesthetic' approach to data, in which aesthetics is understood, not as universal judgements about the beautiful and good, but rather as about what constitutes a 'persuasiveness of form' in any cultural or social context (Strathern 1991: 10; cf. Riles 1998). Taking partial inspiration from Alfred Gell (1998)[1] – who, admittedly, eschewed 'aesthetics' as a useful term for the anthropology of art – I want to focus on data's social efficacy, shifting attention from what any given configuration of data *means,* to what it *does*. Although Gell was keen to move away from the culturally universalistic and exclusivist pretensions of traditional and Kantian 'aesthetic' approaches to art objects, I maintain the term 'aesthetics' in this chapter because to do so is to deliberately invoke an idiom rarely employed in critical discussions of data practices (in a way Gell might have appreciated). A data set is not conventionally considered an art object in the way a painting might be. However, applying this aesthetic approach to data is illuminating because, as many critical data scholars have pointed out, data is rarely discussed as something that is itself made; it is rather understood as the raw material from which 'knowledge' is gleaned. It appears repeatedly in Euro-American business and corporate practice (Coopmans 2014), popular discourse (Anderson 2008) and in scientific discourse (Bowker 2013) as a material or resource out there in the world, which is collected, harvested or mined. It thus is framed as an inert background from which other forms (of knowledge, or action) are elicited. Challenging this presumed sui generis status has been the aim of much of the critical work done around data – hence the counter claims that 'there is no such thing as raw data' (Gitelman 2013). As these critical scholars have pointed out, data only exists because judgements have been made by humans, who have agendas and presumptions, about what to collect and what not to collect.

But this critique does not go far enough or, rather, it need not stop here. If the argument is that data is not just discovered in the world but actively crafted in different ways, then we need to develop a set of conceptual tools to engage in how this crafting occurs and what different forms of crafting do (see here for example Ratner and Ruppert 2019). My interest in developing an appreciation of data's aesthetic qualities in this sense is for a more nuanced and rigorous critical engagement with data practices and their effects, that not only focuses on whether data 'gets it right' or not. So, although much of social scientific critical work targets the incapacity of big data to accurately represent the fullness of social life (for example, boyd and Crawford 2012), in this chapter I argue that the capacity to accurately represent the world is one of the aesthetic 'effects' of data that we need to investigate. I argue that accurate representation (a persuasive form, at least in Euro-America), as one possible aesthetic effect of data, is a result of a process of delicate and committed crafting, of eliciting qualities, forms, patterns, and intensities, and managing their relationship to other forms of social relations (Gell 1998: 9).[2] This approach will also allow me to countenance the ways in which data's forms are charismatic, how they capture attention, or command action, in ways other than because of their objectivity or accuracy. The merit in this is that it permits me to sidestep the language of 'bias' that is often invoked as a critique of data, problematic because of its corollary that out there, somewhere, is the perfectly accurate data set (cf. Cramer 2019: 37).

A final introductory point to make is that if a study of data aesthetics involves becoming sensitised to the means through which data exerts the effects it does, then aesthetics here also becomes a question of the political work that data does. As Dourish and Mazmanian have argued, a critical study of information needs to pay attention not just to 'the brute fact of material forms' of information such as cables, infrastructures or hardware (2013: 4), but to what they describe as the 'materialities of digital representations' (Ibid.: 9). That is to say, the organisation of information matters for the effects it can have on people, and the way in which it is used to enact the world (cf. Manovich 1999). As Dourish shows (2014), a relational database can do a kind of political and social work very different to a NoSQL database – in terms of distributing property rights, for example.

In order to begin to flesh out this aesthetic approach to data, I will draw on ethnographic work conducted with environmental scientists and data technicians in Brazil, focusing particularly on the practice of manual data 'cleaning.' Data cleaning is understood conventionally as the process by which correct data is sorted from faulty or incorrect data. Focusing on the interplay between revelation and concealment, partiality and holism, and surface and depth that characterise data cleaning allows me to demonstrate how such a practice crafts a data set as a certain kind of internally coherent, persuasive artefact of knowledge with the capacity for particular effects in the world. I subsequently end the chapter by suggesting how an analogous aesthetic analysis could be used to unpack other increasingly pervasive forms of data practice, such as data mining and algorithmic pattern recognition, when machine learning algorithms are trained to 'see' patterns in large data sets. Here, the 'effect' shifts from that of objectivity, to another form of knowledge that is proving extremely charismatic in contemporary Euro-America.

A feeling for the data

The scientific project I am going to focus on is called the Large Scale Biosphere Atmosphere Experiment in Amazonia (LBA), where I conducted ethnographic fieldwork in 2010. Initiated in 1998, the LBA is responsible for managing, collecting, and making available data from a data collection project that at the time had spanned ten years, with the intention of ascertaining the role of the Amazon forest in the global carbon cycle. It brings together hundreds of different researchers in different scientific disciplines. With help from collaborative partnerships with the US National Aeronautics and Space Administration, and various other Brazilian, European, and North American institutions, the LBA built meteorological towers that stretch up to 20m above the top of the forest canopy, which is almost 70m in places.[3] These towers have a profile of equipment on them that measure a number of meteorological variables and generate a large amount of data. This data arrives at the LBA head office in Manaus as an unmarked mass of undifferentiated numbers, which researchers call 'raw data' (*dados crus*). It is then put through a quality control process, or 'cleaned' (*limpar*), as I was told; another term I heard used was *lapidar,* to cut and polish as one would a gemstone. It is this process of cleaning that I want to concentrate on in this section.[4]

Mindful of Gell's insight regarding art objects, but also drawing on Annelise Riles's (1998) analysis of the drafting of UN policy documents by women's rights lobbyists in Fiji, I want to concentrate on how data cleaning is directed not towards eliciting meaning, so much as towards eliciting the right *form*. Moving away from a critical engagement that seeks to uncover the reality 'below' or 'behind' bureaucratic instruments like documents, Riles makes a case for understanding these artefacts of modern knowledge through their surfaces and patterns. These documents work, she argues, because they oscillate between abstraction and concreteness: 'a simultaneous and mutual apprehension of the document as pattern and the document as an independent object or unit' (1998: 379). Riles draws this insight out through a comparison with Fijian mat-making: mats lend themselves much more readily to being analysed as aesthetic objects, and she uses them to think in turn about how policy documents are crafted and shaped:

> Like mats, we might say, intergovernmental agreements such as the *Pacific Platform for Action* partake in a simple non-representational patterning that is replicated again and again within the document, from one document to the next, and in the mechanics of the conference at which documents are negotiated [... l]ike the given and self-evident geometric patterns of Fijian mats, moreover, the analytical sequence was all on the surface, dictated, and known at the start.
>
> (Ibid.: 385)

Riles uses this aesthetic approach to illuminate a startlingly new perspective on the efficacy of documents. What seem to be pointless norms and empty flourishes of bureaucratic practice are revealed as integral to the aesthetic of the document, and to the scalar political work the documents do, bringing into view both the local and the global at the same time (Ibid.: 394).

In similar fashion, in this section I unpack the way in which LBA scientific data is crafted by those who work closely with it; and how that crafting then permits the data to have certain effects.[5] Although Riles's focus is on the 'surface' of documents, what becomes clear is that data is efficacious when its surface form hides a multitude of relations within it.

The data cleaning process at the LBA had a strict protocol which was followed by Raquel and Joanna,[6] who were responsible for cleaning the LBA tower data. As data cleaning could be quite a nonlinear, complex, and painstaking operation, for the sake of clarity, I have divided it up into three main stages. The first is so mundane as to seem unremarkable initially. It consisted in transferring the raw data – which had arrived on their screens as a block of numbers almost impenetrable to the human eye – into a series of spreadsheets. Spreadsheets, and other tabular forms, are banal and everyday technologies of organisation, but they have important implications for how the data was able to subsequently have effects and travel. The first stage in the data cleaning, then, is when this enormous amorphous body of numbers that is the raw data is copied and pasted into the first spreadsheet table of a nested series of spreadsheets that the data cleaners have on their computers. This spreadsheet has sequential time running down the side, and the different variables the data is meant to be measuring, such as wind speed, rainfall and so on, running along the top. The spreadsheet therefore separates the raw data out into columns of different variables, and times. In order to ensure that the raw data has been copied across into the spreadsheet correctly, it is compared to the first column of the spreadsheet, which is the 'control' column, containing the expected time of measurements by day, month, year, hour, and minute. This is filled out before any data has been copied across. Often, there was a mismatch – for example, the data extended below the edge of the table in the spreadsheet, or the times would not quite match up. One of the data cleaners has what I nicknamed the 'museum of errors,' a document with screenshots displaying examples of all the strange permutations that the raw data can arrive in – with missing sections, repeats, and numbers strangely bunched together, or strangely spaced out.

Although it is apparently unremarkable, banal, and executed extremely quickly, this initial transfer of the raw data into the first spreadsheet is very important. The spreadsheet's structure gives a *shape* to the mass of numbers, differentiating them into columns and rows that have meaning attached to them – specific variables and specific times. It also provides the first clue as to whether something is wrong with the raw data set – which it often is. These deviations are seen as errors in the structure or pattern of the data, *seen as a whole* (Figure 15.1). Casting your eye along the data set, you can see where there might be problems. All deviations from the standard rows and columns have to be corrected, because otherwise, the data will be in the wrong place and, if that occurs, it will be nonsensical. Its meaning depends on its position, on its shape.

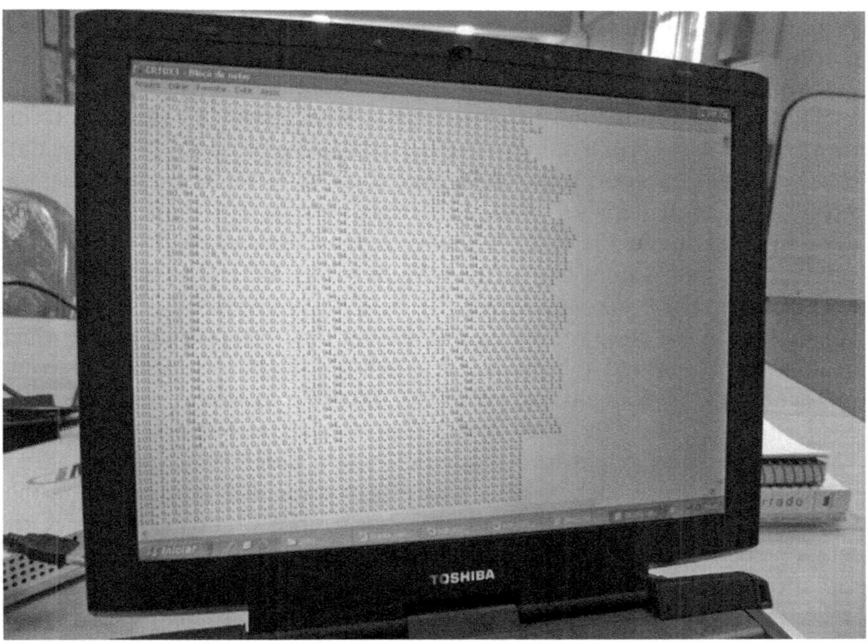

Figure 15.1 Raw data before being copied into the spreadsheet.
Source: Photo by author.

The second stage of the data cleaning process is comparing it to what the data cleaners called 'the range' – a series of values that have been decided upon for each variable, indicating what is and what is not considered possible for a measurement of that variable. When the raw data spreadsheet is checked against it, the range flags any values that are too high or too low, or when the difference between two consecutive measurements is too great. It is considered to be impossible, for example, for there to be more radiation leaving the surface of the earth than arriving – that is just a 'physical law,' I was told, so the 'reflected radiation' value could never exceed the 'incoming radiation' value.[7] Other limits could be given by other standard meteorological patterns, the most basic being that there cannot be radiation after a certain time of day (night). The application of these ranges then fixes the limits and boundaries – the outline, as it were – of the data set.

One of the intriguing characteristic of the ranges that the data cleaners at the LBA use is that they are different for different sites of data collection – in fact, as I was told, 'the range belongs to the place' (*o range pertence ao lugar)*. As a result, the 'ranges are not fixed. The ranges here [at the LBA] have changed several times.' The range for one LBA site was worked out using other data sets, standards, and patterns for one site, and then progressively changed to accommodate other sites that were known to be drier, more humid, or have other characteristics. Developing a range is a slow and painstaking process: 'Experience, literature, meteorological stations – you learn as you go along.' What comes to shape any incoming raw data set, is another, more trustworthy, data set that has evolved out of years of data collection, and into which the new data set might be incorporated.

This leads me to the third stage of data cleaning, which consists in trying to work out why certain values have been flagged as transgressive. The data cleaners spent an enormous amount of their time working out what was wrong with each data set, and then explaining in a document why they decided to maintain the value as data, or remove it as error. Ascertaining whether the data values are 'within the possible,' as they said, proceeds in several ways. There are values that can be quickly dismissed as a gap or an error, because the program that stores the raw data has its own range, which denotes values deemed to be nonsense as NAN ('Not A Number'). This could be because the instrument lost power, for example. However, some cases are harder to work out – the 'grey areas' of possibility, as one researcher put it. For example, there might be a sudden drop in a radiation value, and then a subsequent increase back to what was expected for that time of day in that season. The first way to investigate this occurrence would be to 'check the relations between the measurements,' I was told. In this case, could the low radiation measurement be explained by a sudden rain shower, causing a cloud to briefly block the sun's rays? If so, there should be a spike in rainfall data, an increase in wind speed data, and a change in the pressure data, all at the same time. All these data values would then be compared, and if they demonstrated the right relations to each other then there was good reason to suspect that the value was, in fact, 'possible.'

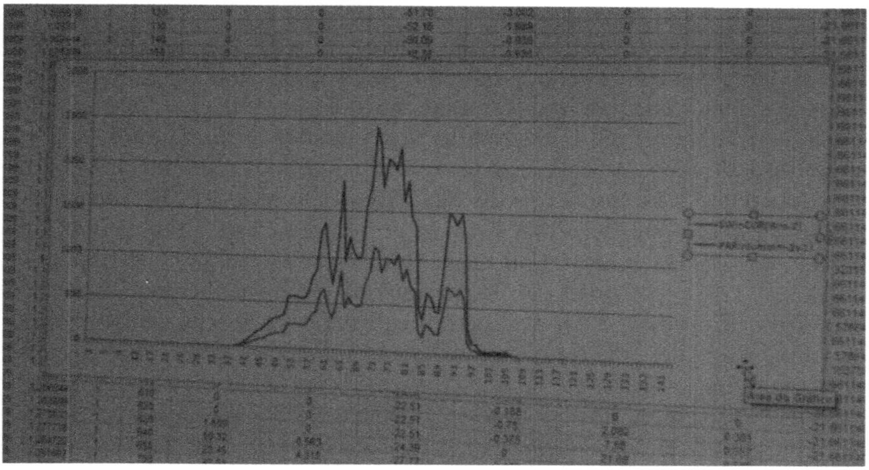

Figure 15.2 Revealing the relations using graphs: here, photosynthetically active radiation and short-wave radiation are meant to follow a certain pattern when plotted against each other.

Source: Photo by author.

This comparison was done visually by using graphs (Figure 15.2): 'there are problems that only emerge when you plot graphs, others that you see immediately' one of the data cleaners tells me. She goes on, showing me: 'if you plot this, you can see it's a problem with the time,' or 'here, on the graph these two should show a relation and they don't.' If they had a suspicious data value over a period of time, they would graph it against other relevant variables, and 'look for the relation.' For example, the latent heat value had 'passed the minimum' several times on one occasion, and so 'when I saw the rainfall spreadsheet, I looked at the same times; when it rains, the temperature drops, and it's cloudy. For several hours, I saw it was very linked to the rain, through a graph that I plotted.' She shows me a graph on which she has plotted all the variables, and traces out the lines with her finger, telling me, 'on this day at this hour, there was a drop in radiation, of short wave, and that's cloud, and then rain – and that is when the sensible heat values went below the minimum – it's because of the sensor, do you see?' Graphs also allowed her to visualise the patterns in the data enough to investigate a suspicious value that has appeared several times. 'We make a graph of the years and compare them. We know that it rains a lot in São Gabriel [...] when we look at the graph and see that this [the rain data] is really low, if it is sustained then there's a problem.' By plotting one variable against another, the data cleaners temporarily reveal the internal patterning within the data. If there is no pattern – if there is too much noise – then there is a problem. This was demonstrated to me one day when one of the data cleaners looked perplexedly at a CO2 profile graph on the screen and told me the 'pattern is wrong.' Even though it was inside the range, it was not exhibiting the right sort of relation to the time of day, and was therefore suspicious.

The capacity to see these patterns comes with what one researcher called getting a 'feeling' for the data. Going through any raw data they did not have source for, they would remark, 'Oh, that one looks like pressure data,' or 'That one looks like battery data.' On another occasion, one of the data cleaners was going through some old data, and told me 'this one looks like a test, it doesn't look like data.' Like the range, the data from each research site has specific patterns and idiosyncrasies that data cleaners have to learn: 'For example, it's hard to get 100 per cent rainfall there. Here, there's humidity – there, there's dust' and, talking of potential infestations of the instruments: 'It's not the bees there, it's the ants!' However, although data cleaners would sometimes go to the field sites (the towers), they would mostly sit in the office in front of a computer, working on the data. That is to say, what they were getting a 'feeling' for, I suggest, was not exactly how the world works (of which the data would just be an encoded representation), but rather how the world of and in the data works.

The data cleaners are cleaning the data so that it can subsequently travel to other researchers and institutions (although the journey is not always uncontentious, see Walford 2012), as 'trustworthy' data that can be used, manipulated and published. Data from the LBA has appeared in countless articles and book chapters and has been responsible for ascertaining several important features of Amazonian climate and ecological systems, theories that have become associated with a number of very prominent Brazilian scientists. That is, the data has certainly been made to mean something beyond its immediate representational capacity as an index of rainfall or wind speed. But this capacity for the data to travel at all, to enter into relation with other data, depends on the careful crafting work of data cleaners to make the LBA data set into a particular form of knowledge object.

This work entails eliciting, and then concealing again, certain relational patterns from the data sets they are presented with. The data cleaners oscillate between looking at the data set as a whole: Does it have the right shape? And decomposing it into its constituent parts: Do these individual variables demonstrate the right pattern? And then reconstituting it again to check the whole. If the data set is an object, then it is an object that has an internal relationality that can be revealed if necessary. However, once it is cleaned, the necessity to decompose it, to reveal its internal patterning, is removed. The data cleaners are only called upon to demonstrate how the data set relates to itself if there is a question mark over the validity of any of the data values. Otherwise, the data set travels as a knowledge object that is presumed to be fully self-relational – that demonstrates relational integrity to itself. The data cleaners' craft is to produce a data set that has a particular totalising form to it: a form that can be looked at from all angles, and that will appear the same from every perspective. This is a perspectival trick that Bruno Latour calls the 'optical devices' of scientific inscriptions (1990: 43). Does the data set look the same from all angles? Do the patterns hold however you look at it? The data set becomes trustworthy, not when it can be shown to accurately represent the world, but when it is made into an object that does not allow for any change in perspective – when it in a sense enforces acquiescence. This is therefore not simply a question of standardisation and enforcing a common way of seeing as a skill developed over time (Carusi 2008), but trying to ensure a common way of seeing is in the data set itself for everyone who comes across it. It is an object that can be decomposed into a set of complex self-referential relations, but that in its composed form is singular, unequivocal, and immutable.[8] What is being crafted so deftly here is the 'form' of objectivity itself.

To return to my initial inspiration, Gell insisted that art is to be understood, not as a symbol, but as a form of agency, a translation or transformation of intentions: 'I view art as a system of action intended to change the world rather than encode symbolic propositions about it' (1998: 6). Such agency works through what Gell called 'abduction,' where the art object becomes an index of social agency – sometimes of the artist or maker – and a form of distributed personhood (Ibid.: 19). This agency is often to 'enchant' (Ibid.: 41), to capture attention, to enforce and elicit behaviours, such as when a Kula canoe prow board's design forces exchange partners to make excessively generous exchanges (Ibid.: 44). As I mentioned earlier, data sets are not conventionally seen as art objects in this sense. But they are crafted with specific intent to persuade others of their veracity, their ability to speak of the world, and this crafting gives them a remarkable power. LBA data cleaning might not seem to evidence the kind of technical virtuosity of a Trobriand prow board carving, nor is the intention of the data cleaner meant to be inscribed into the data set. In fact, data cleaning is assumed to be able to be done by anyone who has the right training, as it is technical work, as opposed to the knowledge production that comes afterwards; data cleaning is not about what the data 'means,' and, as a result, data sets do not carry the name of the person who cleaned them in the way that published papers carry the names of those who turned the data into knowledge. But this is aesthetic work, nonetheless, as it is in data cleaning that the data is given the form of its power to persuade. It is in the data-cleaning process, exactly when the data is understood to become 'true' and 'objective,' that it can come to be what Gell calls a 'secondary agent' – that is, capable of carrying people's (scientists') agency with it. Scientists can no more be scientists without data as soldiers can be soldiers without weapons (1998: 20), because of what the data allows them to *do*. What is erroneous here is to construe data cleaning as 'mere' technical work; it is in fact where the enchantment with and of data begins.

The idea that objectivity is an outcome of what might be glossed as 'subjective choices' is not new to Science and Technology Studies (STS) or the Anthropology of Science, and in fact a concern with the representational strategies of science has been at the heart of STS for decades (see Lynch and Woolgar 1990; cf. Coopmans, Lynch, and Vertesi 2014). There has even been explicit engagement with science as an aesthetic endeavour, particularly around scientific image processing (Lynch and Edgerton 1987; Frow 2014). What I want to draw out of my analysis, however, is that from an aesthetic perspective this is not a question of the subjective versus the objective. The form of objectivity 'persuades' others not because of the idiosyncratic subjective proclivities of the data cleaners; in the same way as there is an 'aesthetic' to universal knowledge evidenced in its various forms – maps, encyclopaedia, archives and databases (Schaffer et al. 2017) – so too is there an aesthetic to objectivity. Scientific data certainly works in a representational register, and it would be false to suggest that it is not considered valuable because it represents the world more or less accurately. But what I want to emphasise is that it is not *only* representational; or even its capacity to represent in the way it does also depends on its functioning in an aesthetic register, which is non-representational, and in which the data's form and its relations to itself (rather than to a world 'out there') come to determine what it can do, and what (scientific) personhood it becomes instrumental in bringing into being.

Algorithmic vision

To finish, I would like to turn to the question of what this analysis suggests for the future study of ever more-ubiquitous data-related activities that seem very different from manual data cleaning in environmental science, such as machine learning, algorithmic pattern recognition, and data mining. These practices are emerging, not just in commercial spheres as part of new capitalist apparatuses of value extraction (Thatcher et al. 2016), but as forms of governmentality (Amoore 2011), as the bedrock of advances in precision medicine (Prainsack 2017) and neuroscience (Rapp 2015); they are pivotal in the automation of the justice system, of transport infrastructures, and medical diagnosis. That these practices are restructuring society is a claim made quite brazenly by their proponents. As Alex 'Sandy' Pentland, an MIT professor recently declared by Forbes as one of the seven most powerful data scientists in the world, states in an interview from 2012:

> There are patterns in those individual transactions that are not just averages, they're the things that are responsible for the flash crash and the Arab spring. You need to get down into these new patterns, these micro-patterns, because they don't just average out to the classical way of understanding society. We're entering a new era of social physics, where it's the details of all the particles – the you and me – that actually determine the outcome.
>
> (Pentland 2012)

The idea that the masses of personal data that is being collected through our communicative and online activities contains within it 'patterns' that just need to be brought to light in order to understand the social whole animates most contemporary data science and is at the heart of the current enchantment of big data. The classic categories of sociological analysis, such as class, race, and gender, are being re-formulated into dynamically shifting 'micro-patterns' and 'vectors' (Cheney-Lippold 2011).[9] Important here is an idea that we have already encountered with the LBA data cleaning: the conviction that (big) data is 'hiding' these patterns within it. In fact, the aesthetic effect of big data – its persuasive power – derives from the idea that all sorts of things might be 'hiding' within it; as Louise Amoore remarked, during a United States House subcommittee hearing in February 2002: 'The invited panel of experts stated clearly that "our enemies are hiding in open and available information" and that, had surveillance and profiling techniques been in place, the events of 9/11 "could have been predicted and averted"' (Amoore 2011: 337). Practices of eliciting or revealing these hidden patterns, insights, threats, and so on, function through logics that are not quite recognisable as the 'objectivity' of the natural sciences. As Orit Halpern points out, we are confronted with a new objectivity: a 'new aesthetic and practice of truth; a valorisation of analysis and pattern seeking that I label "communicative objectivity"' (2014: 15). In another idiom, based on ethnographic work with data analysis in business settings, Catalijne Coopmans has called data mining a form of 'artful revelation' (2014); Amoore, for her part, characterises practices employed by private companies employed by the US government to 'flag' terrorists in biometric data as working through an 'ontology of association' (2011: 27), and the privileging of correlation over causation so infamously claimed for big data analytics is in a sense exactly the celebration of the infinite relationality of big data, its capacity to contain within it countless relational patterns. It is important that these patterns are not on the surface, but concealed. In fact, part of the charisma of these data-driven algorithmic practices is that they are claimed to be so complex that they make decisions based on correlations that humans cannot 'see' directly; that what they are doing is 'rendering perceptible and actionable (almost seeing) that which would otherwise be beyond the threshold of human vision' (Amoore 2018: 12).

But if at the same time as these machine-learning algorithms seem to make patterns and reveal relations 'otherwise unavailable to the senses' (Amoore 2018: 5), they are simultaneously shown to be simply revealing older, equally persuasive, forms. Far from dissolving, classic categories of discrimination are being rendered anew. Racialising and gendering biases, embedded deep into the data sets on which the algorithms are trained, structure these deep patterns, showing them to be nothing more than reflections of human prejudice, and of far older pervasive social patterning that perform and re-perform particular inequalities and violence, particularly racialising ones (Sweeney 2013; McQuillan 2016; Apprich et al. 2019). PredPol, 'predictive policing software' that uses an algorithm to predict where crime will happen based on past crime-rate data, is perhaps one of the most illuminating examples of this. Not only has it come under fire for using biased data sets that rely on data collected disproportionately from racialised populations, and the performative element to its predictive function, which sees these populations subsequently even more heavily policed; but, more recently, it has emerged that the algorithm used by PredPol was originally designed to predict seismic activity, and several data scientists and mathematicians have come forward to say that it is flawed and too simplistic (Haskins 2019a). Nevertheless, it is heralded by many as the forefront in contemporary policing practice and has been adopted in several US states (Haskins 2019b).

So again we might ask: What do algorithms such as PredPol do when they 'see' patterns in data? In her 1993–2008 lectures on Melanesian material culture and decorative practices, published as *Learning to See in Melanesia* (Strathern 1991), Marilyn Strathern makes the pertinent point that we never just 'see'; people show things to us. This then implies that wherever there is some act of revelation, there are also acts of concealment. When looking at Melanesian ceremonial costumes and masks, her question, 'What are we being made to see?' allows her to start to unravel the presumptions woven into what Euro-Americans might think masks 'do.' Whereas we in Euro-America might imagine that masks 'hide' people, in Melanesia they do the opposite: they display people for the other's regard, demonstrating their relational potential: 'When you look at a decorated dancer you see an everted person: what was formerly inside, hidden, is now on display' (1991: 61). In fact, a person in Melanesia is 'hidden' when they are naked, when nothing is being displayed for others to regard, where everything is concealed.

Likewise, we might ask what presumptions of our own shape what we think machine-learning algorithms like that used by PredPol are 'doing' when they 'see' crime. Whether PredPol accurately predicts crime or not is one aspect of a critical anthropological approach, but it is not the only one we need to be developing. No less important is the means of charting the social and cultural histories and imaginaries that are woven through the dynamics of revelation and concealment that make PredPol so persuasive. We must always keep in mind the labour that goes into these acts of displaying and eclipsing: nothing is just there to be seen.[10] How, despite their inaccuracy, are these new forms of algorithmic objectivity so captivating? How do the patterns they reveal hold together both the recognisable and the not-yet imagined? These are the kind of questions, I argue, that an aesthetic approach to data can begin to answer, and in so doing also provide some means of disruption and resistance.

Notes

1 'Partial' because I am not going to follow his schema for understanding the agency of art objects, but would rather make use of his analytical switch from 'art' to 'index,' whereby he focuses on the 'social relations in the vicinity of objects mediating social agency' (1998: 7).

2 This is different from, but potentially complementary to, well-established STS approaches to scientific representation and, indeed, to the agency of objects (Lynch and Woolgar 1990; Coopmans et al. 2014).

3 The LBA has also recently completed building a 300m meteorological tower.

4 There are of course several stages in the production of data that precede this cleaning practice; however, for the purposes of this chapter I am going to concentrate on this specific stage in the data's life cycle.
5 This process of data cleaning is of course highly specific to each scientific discipline and sometimes even to each project, so although all sciences will have protocols around data cleaning and curation, this is not intended as a definitive description of scientific data cleaning.
6 Pseudonyms are used, at their request.
7 Although as these are of different wavelengths ascertaining this is not a simple matter.
8 See here Latour's notion of 'immutable mobiles,' although my analysis here deliberately moves away from a dependence on ideas of correspondence between 'sign' and 'referent' that is crucial to Latour's argument (Latour 1999: 24–77).
9 See Orit Halpern's book *Beautiful Data* (2014) for a history of these practices.
10 My thanks to Rachel Douglas-Jones for her insight here.

16 Place-objects

Anthropology of digital photography/s

Shireen Walton

Introduction

Digital photographs are prominent features of everyday life; seen on smartphones, laptop screens, printed out, and uploaded to storage cloud archives. Rarely, however, have digital photographs/photography been a central focus of anthropological study. This dearth of scholarship is reminiscent of an earlier intellectual moment in visual and material anthropology, whereby (analogue) photographs had been regarded in the discipline, in Elizabeth Edwards' words, as a somewhat 'poorer cousin' to film.[1] A dual problem that digital photographs have posed for anthropologists is that they do not provide some*thing* clear to interact with, nor some*where* 'tangible' to go to, to be with them as part of social research. I explore these two issues in this chapter via my anthropological research on photography in Iran, which was a site of limited geographical access to me as a British passport holder in a politically fraught geopolitical moment in 2012. Focusing on the 'what' and 'where' of digital photography/s, I will suggest, leads us to advancing theories of objects and places, and such thinking and working with what I will call 'place-objects' can have broad relevance for MCS research in places of limited physical access, but which need not preclude the researcher's presence. My attempt to link the (digital) photograph with place follows a definition of place as, in Christopher Tilley's words, 'far more than location' (1994: 15): rather than being of/about a location, places, Tilley writes, are specific 'contexts for human experience, constructed in movement, memory, encounter, and association' (Ibid.). So too are digital photographs bound up in movement, being zones of encounter, memory, and contact. Sarah Pink, in a similarly phenomenological sense, suggests that digital photographs cannot be mapped as static nodes in networks, but can be understood as visual 'place-events' that are 'made, carried, consumed, move forward, and open up potentialities with perceiving embodied persons as part of specific environmental configurations' (2011b: 8). Linking the digital photograph to place along these lines, we see how digital photographs implicate, as do other kinds of physical/digital objects, the places through which they have moved, and the related socialities of each place, in turn, shape their 'objectness' in the world.

The theoretical discussion about objects such as digital photographs as places is also bound to a methodological one. The argument about methodology presented in this chapter is theorised 'backwards' from the field, from the empirical research I undertook on Iranian popular digital-photographic practices between 2012 and 2013, in a range of places online and offline, inside and outside of Iran. In showing how digital photographs are of place, and *em*place, the digital photograph can be seen and handled as a dynamic digital object, moving across social contexts and material registers, while it both reflects and makes place socially, politically, digitally, in this case opening up methodological pathways for MCS research on Iran. The case of my own restricted physical access to Iran provides the backdrop to the chapter's theoretical and methodological advancements, arguing that what might be considered more conventionally as 'problems' of place, such as restricted or unavailable access, can actually *be* accessed by rethinking the relationship between place and object, and taking seriously the proposition that the object becomes a place and a site of field research. In my research on Iran, digital photographic-objects became the places I ethnographically travelled to and with – across a range of geographical, social, and political sites to an extent that superseded the one month I was physically on the ground in the country.

I therefore suggest that a material cultural handling of digital photographs as *place-objects* helps us to understand how digital photographs are not only visual vehicles, carrying representation and aesthetic codes (Campanelli 2010), but, as digital objects, part of what they are as things is embedded with place – and, in turn, the place-making that digital photographs are implicated in creates social space(s) that the ethnographer can enter into and be with/in. This capacity for movement via the place-object, I argue, has broader implications and can matter when research topics touch upon politically precarious geopolitical issues that determine the nature of access in certain contexts of social enquiry. To begin the discussion, a historical overview of where the digital photograph stands in material culture anthropology is required before exploring how digital photographs, as contemporary moving objects, are bound up with place, which altogether forms the notion of place-object as theory and method for MCS.

The digital photograph in MCS: materiality and practice

The intellectual history of working with digital photographs in anthropology comes from a constellation of lineages on photography and the visual, spanning communication studies, visual studies, STS, sociology, and humanities, especially history of art. Two theoretical interests that tie these areas together are *materiality* and *practices*, which I will consider in turn as part of the material culture foundations of the digital photograph.

Materiality

The theoretical understanding and handling of photographs as objects is fundamental in the fields of anthropology and photography. Visual-anthropological scholarship on film and photography had been long focused on images; an approach that privileged analytical frameworks of representation, symbol, and sign.[2] However, in their seminal (2004) volume *Photographs Objects Histories: On the Materiality of Images*, Elizabeth Edwards and Janice Hart marked a certain 'moment' in an anthropology of photography – one that had become object-focused. In moving away from semiotic approaches towards an object-focused framework for photographs based on the materiality of objects, Edwards and Hart and others highlighted the physical properties of photographs as thought to be central to their 'objectness.' In turn, this physical-materiality was found to shape the ways that 'photo-objects,' as they called it, function and have e/affect in discrete ethnographic contexts:

> Materiality can be said here to have a positivistic character, in that it is concerned with real physical objects in a world that is physically apprehendable not only through vision but through embodied relations of smell, taste, touch and hearing.
>
> (Edwards and Hart 2004: 2)

As such, it was the uniqueness of physical objects that constituted objects' 'aura,' an idea espoused by material culture scholarship on photography within this oeuvre. Christopher Pinney, for instance, explicates the social efficacy of photographs in terms of a photograph's material agency (1997: 171). What he terms 'proximal empowerment' denotes the ability of powerful objects (such as photographs) to bring about a sense of proximity on the part of the individual towards sacred beings or deities rooted in the physical materiality of the photograph, predicated on bodily contact with the photographic object. Not altogether departing from this tradition, but rather re-articulating it in a digital context, I aim to illustrate what it is about the digital photograph – whose physical engagements with objects are mediated by computer/phone screens – that constitutes its social efficacy as image *and* object, albeit digital ones. First, this raises the question about the materiality of the digital-photographic object.

That the digital *is* material and vice versa is a founding principal of Digital Anthropology as it was conceived of within UCL Anthropology's material culture section. In Daniel Miller and Heather Horst's seminal 2012 volume, the social order was shown to be a material one, not predicated on abstract systems of societal relations, and this altogether called for a holistic approach to studying the digital (Miller and Horst 2012). The material basis of the digital has since become an established principle, informing the majority of anthropological research on the digital. In their book *Digital Materialities: Design and Anthropology,* Sarah Pink, Elisenda Aredèvol, and

Débora Lanzeni speak to a contemporary world where 'the digital and the material are not separate, but entangled elements' (2016: 1), while in the context of contemporary critical museum studies, Haidy Geismar writes how

> we need to think about the digital not only as material, rather than immaterial, but also in terms of a trajectory of materiality that links our commonplace understandings of the digital to the analogue, information to material, systems to structures, knowledge to form.
>
> (2018: xvii)

Paolo Favero, writing specifically about images in the digital age describes how 'the supposed immateriality of the digital is evidently one of the greatest myths of our epoch' (2018: 78). In the digital context, then, materiality and 'objectness' remains a core framework for understanding digital photographs/photography in social terms, which will be subsequently developed in theorising the place-object.

Practices

In addition to the materiality of photograph*s* is the study of photograph*y* as social practice. This is a distinction Jonas Larsen highlights in developing an 'ethnographic framework' for digital photography:

> [P]hotographing is absent from most theory and research that jumps straight from photography to photographs[, … such studies move] directly to the representational worlds of photographs and skip over their production, movement and circulation. The diverse hybrid practices and flows of photography are rendered invisible.
>
> (Larsen 2008: 143)

Photography as practice, I concur with Larsen, remains fundamental to an anthropological handling of digital photography and photographs – an approach that predates the digital context of the medium. Seminal anthropological scholarship on analogue photography from India (Pinney 1997) and Tibet (Harris 2016) to West Africa (Buckley 2001; Sprague 2003) and Indonesia (Strassler 2010) has examined histories and forms of popular, amateur, and 'vernacular' practices, showing how photography belongs to and/or departs from national visual traditions, image regimes, and wider socio-technological developments. Practice has also been a main framework for multidisciplinary perspective studies of digital photography (Larsen 2008; Gómez Cruz and Lehmuskallio 2016). The emphasis on practice highlights the social dimension of photography as dialogical. In her work on digital amateur photography, Sarah Pink, recalling Pierre Bourdieu's 1992 study of (analogue) amateur photographers in France, notes how 'amateur photography is not necessarily an activity done for its own sake, but is implicated in other activities' (2011a: 95). So too with the popular digital-photographic practice of photoblogging in Iran, as I will discuss, whereby the capacity to dialogue *with* viewers is fundamental in what the practice of photoblogging is (Cohen 2005).

Following Larsen's ethnographic framework for digital photography as 'practices and flows,' I am interested in the myriad practices that can be determined through intentionality and/or 'undetermined' practices (Poster 1999: 16), connected to photographers, digital infrastructures, and the situated and 'interpretative flexibility' (Pinch and Bijker 1984) on the part of viewers. To conceive of digital photography anthropologically, then, in line with material culture perspectives of materiality and practice, is to study photographs as objects and practice, not simply to look at the image as a flat or *finished* representation – though as I will suggest, representation remains a core framework for any study of photographs/y. What is it about the digital-materiality of the digital-photo-object then, and the practices of digital photography, that links it to place? The digital-photographic object creates place via a combination of human and technological processes, practices, and desires – processes that are both human and extra-human, again recalling Christopher Tilley's notion of place as 'both internal and external to the human subject; a personally embedded centre of meanings and a physical locus for action' (1994: 15). With the present chapter I include digital places (and spaces) in addition to a 'physical locus for action' through the theory of the place-object.

Place

Exploring how place is linked with object through digital photography requires reviewing how 'place' has been defined in material culture anthropology and related disciplines. Separate to the notion of 'space,' which Tilley has suggested is a more abstract notion, contingent upon the social construction of it and its reality: 'what space is depends on who is experiencing it and how – it cannot exist apart from the events and activities within which it is implicated' (1994: 15). Places have more 'distinctive meanings and values for persons' and as such, 'are of affective importance[, …] personal and cultural identity are bound up with place' (Ibid.). This phenomenological emphasis on affect and the situated experience of place are important in thinking about the efficacy of the digital photograph as a *place*-object bound up with, in the case of my research, post-1979 revolution Iranian national and transnational identity politics. Crucial to the theoretical argument about the place-object is placing the object in context as part of what geographer Doreen Massey calls 'constellations of processes' (2005: 141) that are unbounded and constituted through movement. With the digital photograph it is the constellation of process and practice(s) that produce the material object and render it visible in discrete digital and social places which, in turn, ends up producing social space. Hence objects, place, and space in this analogy, are not fixed, static, or mappable, but are both objects and environments made through motion.

Representing place: photoblogging in/of Iran

The present moment of material culture is in many ways in a post-representational phase. Material culture scholarship has steadily steered away from linguistic-based semiotic approaches that characterised early and mid-twentieth century approaches, as the spectrum of work detailed in this volume attests. However, with specific regards to photographs as material-culture objects, a capacity for representation remains a fundamental part of how (and why) images exist and are efficacious in contextual places (Pink 2001), and so too as I will argue, in the digital landscape, where 'place-objects' goes some way in explaining the digital photographic object's agency, mobility, and lifespan.

Representation is fundamental in the practice of Iranian photoblogging. What (digital) photographs in these digital and social contexts 'show,' as opposed to the stories or narratives they 'tell' (Grimshaw 2001), holds diverse significance for a range of makers and viewers of the photographs. Anna Grimshaw provides a relevant discussion of 'showing' and 'telling' with regards to film that I find useful in explaining how many Iranian photobloggers I worked with used images. The notion of 'telling,' involves a certain level of didacticism in communicating a narrative message. 'Showing' by contrast, involves a sense of exposition or display of a stand-alone image (or system of images) with a potentially greater discursive capacity, necessitating engagement with an audience. The Iranian photobloggers with whom I conducted research took pride in their capacity, as people with camera phones in everyday life, to show all kinds of images to a range of imagined viewers in order to engage in the historically loaded politics of representing Iran. With their images exhibited online, my informants expressed and demonstrated desires to show (represent) the 'truth' (*haghighat)* and a 'reality' (*vāghey'at*) of life across the country to people inside and outside Iran who may not be familiar with what the country and its visual and sociocultural diversities look like. Amin, in his early thirties and from Tehran, explained how

> photoblogging is my medium of choice if for no other reason than that it involves a camera and the Internet. Even a tiny, barely functional camera phone provides the means to capture what I wanted to show to the world: the truth about Iran.

These 'truths' are subjective visions by the women and men who have the cultural capital and economic and technological means to deploy them.

A typical example of the kind of images seen on Iranian photoblogs can be seen in Figure 16.1. The image is a digital photograph taken on a basic Nokia camera phone in 2008 by the prominent photoblogger who uploaded it to his blog, 'Life Goes on in Tehran' (LGOIT). It depicts Western brands and basic bathroom products, such as 'Nivea' moisturiser, as ostensibly seen in an average home in Iran. LGOIT described their intention of using everyday

Figure 16.1 Mundane 'Iranian' objects of everyday life.
Source: LGOIT. August 2008 archive. Used with permission.

digital images of quotidian objects to visually speak to various international and national narratives about Iran and how Iranians ostensibly live.

These sets of photographs are intended to be viewed as 'normal' images that represent the everyday place(s) of Iran. Such images speak to viewers abroad – including Westerners, and also Iranians who fled or live in exile – about how Iranians inside Iran live in a country that exists ostensibly outside the Western fold, and under heavy US-imposed economic sanctions. A visual discourse of 'ordinariness' and the social salience of ordinary 'stuff' (Miller 2008; 2010) are hereby deployed, as a communicative strategy in these photographs, which are taken and commodified/exhibited to viewers online as representing a 'real Iran.'

The desire for alternative representation of the place of Iran witnessed in Iranian photoblogs sits atop a significant (visual-)ideological history spanning the twentieth century, but which gathered particular traction during and following the revolution of 1979, which saw the founding of the Islamic Republic of Iran. The subsequent decades of social, political, and technological change led to extensive national and geopolitical-strategic campaigning over what Iran was ostensibly becoming – campaigns that were often posited in contradistinction to the United States and UK due to the alleged '*gharbzadegi*' ('westoxification') that the ideological 'nativist' architects of the Islamic regime claimed was a corrosive feature in Iran under the former Pahlavi regime. Images, as a range of scholars have shown, became central to the post-1979 nationalising mission in Iran (Chelkowski and Dabashi 2000; Varzi 2006), and to governments in 'the West' keen to represent the new regime (Adelkhah 2009). State media, photographs, street art, cartoons, paintings, museum exhibitions, and cemeteries were all deployed by the regime to uphold its national imaginary and form a basis by which Iranian subjectivity has been mediated transnationally in recent decades. Iranian photobloggers are acutely aware of these 'representational fields' (Harris 1999),[3] and this visual-narrative history has a tangible impact on contemporary Iranian subjectivity, marking how many research participants expressed how they felt about themselves in the world. As Kiana in Tehran had explained: 'The media can change views about a whole nation. These new online photo-sharing systems (photoblogs) are the same, but this time, they provide *us* with the opportunity to show who we really are.'[4] Kiana conveys how she sees photoblogging as a way to combat reductive and/or stereotypical representations of her country by showing a certain globalised vision of ordinariness that she hopes will have the effect of de-exoticising perceptions of Iran.

Responses to these kinds of banal images from a range of viewers across the world convey how the aesthetic strategy of the mundane employed by photobloggers is considered by some to be an effective means of making people reconsider what they are told by the media about the place of Iran. As one viewer in the US remarked: 'Thank you for your images. I truly enjoy seeing my "enemy" country in the light of non-politics.' The 'non-politics' here equates to the 'harder' aspects of 'top-down' (geo-)political agendas. A study of digital photographs as representational objects thus goes hand in hand with broader political economy of place – recalling in particular what Deborah Poole (1997) has termed the 'visual economy' of a given context that explains the political economic and social matrices in which photographs operate and are deemed valuable,[5] and which pattern their production, circulation, consumption, and possession.

Place-objects beyond representation: digitally (re-)engaging the senses

Beyond representing place, a range of wider characteristics constitute the a/effective qualities of the digital photograph as place-object. This requires what Christopher Pinney and Nicholas Thomas (2001) referred to as looking 'beyond aesthetics' in studying the visual, as it does with Jonas Larsen's (2008) suggestion to move 'beyond representation' in ethnographic approaches to digital photography. In Iranian photoblogs, moving beyond aesthetics and representation points to the senses, to memory, and to wider a/effective engagements with digital objects experienced in digital and physical places. Food photographs are a prominent image category in Iranian photoblogs

Figure 16.2 Majun.
Source: LGOIT. March 2009 archive: www.lifegoesonintehran.com/24_March_2009.html (Accessed: 12/10/12). Used with permission.

and a clear example of this kind of digital-visual work. In one example (seen in Figure 16.2), LGOIT captures a market vendor in South Tehran mixing a popular sweet milk drink, called *ma'jun.* Visceral responses to this image, such as *bah bah* (a colloquial term in Persian and equivalent of 'yum yum' in English) in addition to other posted comments describing the smells, textures, and tastes of the image are tied in with a wider sense and discourses of longing for traditional sights, sounds, and tastes of Tehran, and Iran.

The digital photographs bearing witness to the availability of things like *āb anār* (freshly squeezed pomegranate juice) and *lavāshak* (dried fruit leather, typically sold in Iran) become salient points of connection in viewers' imaginative capacity to foster a sense of being there, between the place *of the* object (in Iran) and the geographical place of viewing the object – often two discrete geographical locations. The digital photograph as place-object hereby forms a material link between the place of the scene and the place of the screen. The image sensorially emplaces people in precise areas of downtown Tehran, based on affective and atmospheric feelings of place (Sheller 2010), recalling Paul Stoller's (1989) earlier argument that print photographs could be essentially 'tasted,' and where digital photographs 'pulse with meaning and affect' (Deger 2016). Such forms of engagement are particularly salient amongst the Iranian diaspora and individuals living in exile. We see here how the place-object engenders a capacity for various forms of emplacement, and this capacity for bringing people somewhere via the place-object has particular salience in contexts of physical or cognitive distance, including contexts of exile, displacement, and/or estrangement with a physical place such as a country or place of origin.

Digital geo-locations

So far, the discussion has highlighted that there is nothing necessarily all that new with the digital context concerning the social efficacies of photographs and photography, and their capacity to emplace. At the same time, a significant part of the *digital* photograph's affordance as a contemporary object is how, once embedded within wider digital infrastructures such as social media platforms and their associated software/affordances, the digital photo-object is able to evidence links to actual physical locations. Again we shall see how the Iranian context of popular digital photography is a salient example of why place-objects 'matter' (see Miller and Haapio-Kirk, Chapter 11), politically and socially to those involved in the practice – and methodologically, too, to the anthropologist conducting research under (geo-)politically sensitive conditions.

Geotagging is a customary practice of using geo-locating techniques such as Google Maps to 'tag' or emplace photographs on social media, websites, and platforms, in physical locations. These techniques are prominent in Iranian photoblogging, adding extra dimensions to individuals' renderings of the 'real' place of Iran. Figure 16.3 shows the geotagged Imamzadeh Zeyd mosque in Tehran, as seen in 2013 by a prominent photoblogger. Such live digital image-making and sharing reflects a unique contemporary socio-technological ability to create 'live postcards of place' (Bell and Lyall 2005), shaking up the past-orientated, 'I was here' analogy of photography of Roland Barthes' (1981) through the socio-technological potential, given the supporting infrastructural conditions, for digital co-witnessing in real-time live-streaming on platforms like Facebook, Twitter, or Telescope. One potential result is a

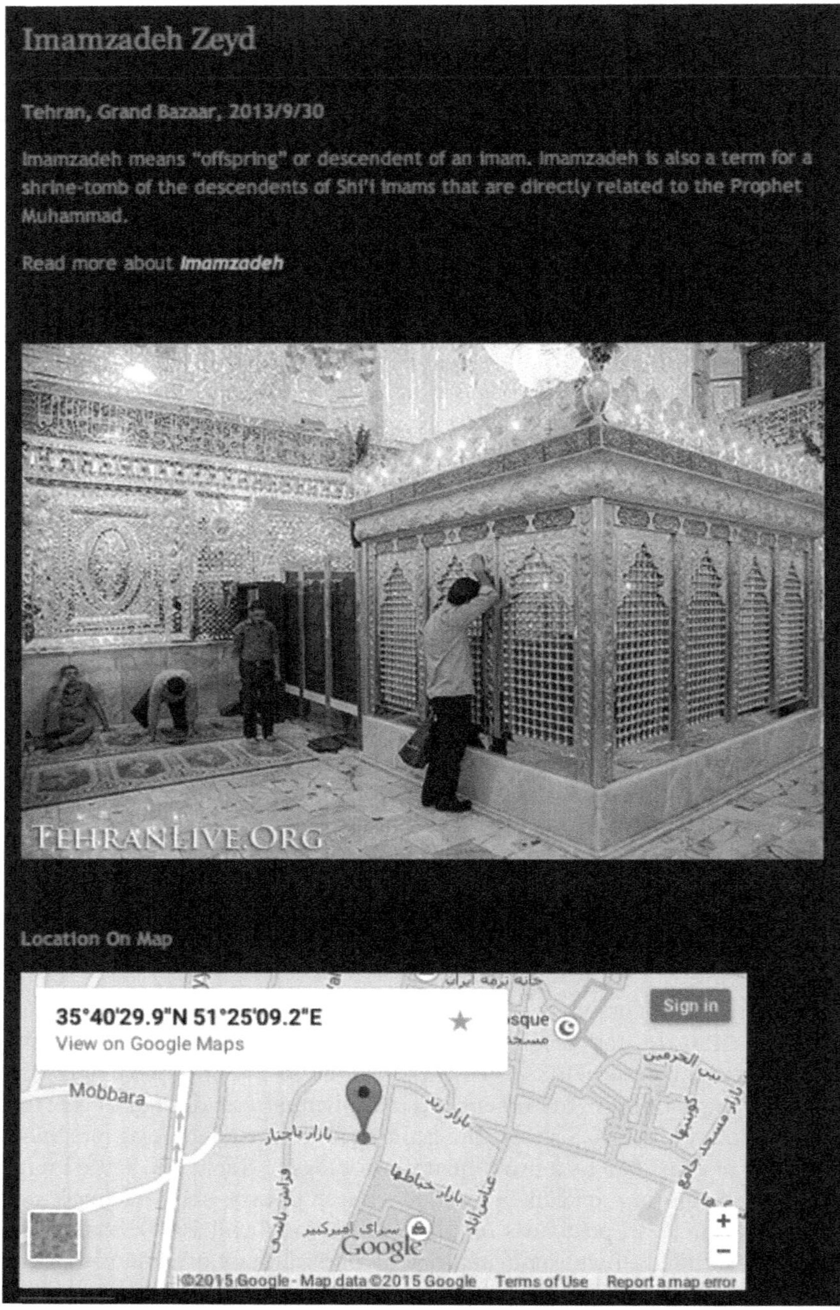

Figure 16.3 Geotagging on Tehran Live photoblog.
Source: 'Imamzadeh Zeyd,' Tehran.

present-rooted, 'I am here' (Bell and Lyall 2005), of digital/social co-presence. Along these lines, Larsen (2008: 152) notes how the digital medium is typified by a 'time-space compression' which, under the necessary conditions, can enable a collective experiencing of 'instantaneous time' (Lash and Urry 1994) in the act of making and viewing images. Geolocative practices seen in online digital image making/viewing thus reconfigures the traditional temporalities of photography by striking real, and even real-time, connections among photographer (image-producer), subject, and viewer.

This potential digital affordance for spatiotemporal presence with the photograph has methodological implications for MCS, and for the digital-ethnographer moving with images in online spaces, where conditions on the ground are politically sensitive and access to a physical field site is restricted – and/or simply that on-the-ground ethnography is entirely unavailable to certain researchers from or in certain countries – both issues directly concerning the anthropology of Iran.[6]

One issue relates to the speed of the moving digital-photographic object. Speed is one part of its agency as a contemporary place-object; it is able to inhabit and traverse multiple places, without the necessarily negative implications signalled by Paul Virilio's 'dromology' (1986; 1991), with which he describes the speed of modern mass communication as destructive of human presence and experience. The digital photographic object, embedded within the digital infrastructures that bring it forth, connects with disparate groups in multiple settings, as seen in Iranian photoblogging in the various sites in Iran and wider contexts across the globe with which it connects. At the same time, this speed is subject to a range of material, economic, and socio-political constraints, all of which make-up the ethnographic 'field' – and its framing for the anthropologist. The creation of *place* by the place-object is embedded with wider flows and conditions, and it is both the digital/wider infrastructures and the human/social agency that determine where the object is *placed* in (online) space, how it is exhibited/seen, by whom and where.

Central in the digital photograph's ontological make-up is its *non-uniqueness* as a thing, able to exist and move within (but not exclusively) digital forms and landscapes. Crucially then, in contradistinction to Walter Benjamin ([1936] 1972) it is the *lack* of a thing's uniqueness as a single entity, but as a potentially replicable and fast-moving object, that has a/effect. Contrary to previous emphasis on the physical qualities of the analogue photograph, the digital photograph's 'aura,' agency, and its e/affective dimensions are therefore constituted through the places and environments through which it moves and leaves traces, which in turn become spaces and places the anthropologist can follow and inhabit as part of contemporary anthropological research. The digital photograph becomes a kind of relic of the 'moment' of capture and its different moments of viewer engagements. The place-object, hence, doubles up as a space-object in the way it constitutes (social) space through place. All the while, the ethnographer may follow 'live photography' and live streaming as part of their online participation-observation strategies.

Considering how place is central in photographic practice reflects in the digital age of geolocative media technologies what some scholars suggest is the need to put movement upfront in thinking about digital photographs' contemporary social contexts. Sarah Pink and Larissa Hjorth (2012) suggest that practices such as geotagging, and its implication of image in place, can be described as 'emplaced visuality' or 'emplaced cartography,' while Paolo Favero (2014) relatedly suggests that geolocative media can reveal a range of aspects about how social actors use and inhabit space. The relationship between media and social environments demonstrated by photobloggers illustrates Lane DeNicola's (2012: 84) claim that 'it is no longer sufficient to talk of landscapes being mediated; what we must recognize are those instances where media have become "landscaped."' In this sense, digital photographs seen here in Iranian photoblogs have been 'landscaped' with Iran and vice versa – the object here has become a place, the place an object, and along the way, social spaces are produced.

Methodological implications for MCS

In the auto-ethnographic sense, place-objects such as photographs demand presence and attention in the present, something that relates to the Object Analysis method in material culture, where there is the sense that one has to sit with and be with the object to comprehend it, to let it in (see Mercier, Chapter 10). The notion of co-presence facilitated by the place-object requires further discussion here. A central tenet of the ethnographic method involves a commitment to being 'in the presence of the people one is studying, not just the texts or objects they produce' (Miller 1997: 72). 'Photography,' Barthes wrote, 'offers an immediate presence to the world – a co-presence' (1981: 84). The digital photograph adds further socio-technological dimensions to these ideas. Internet connectivity and digital practices brings the realms of individual and social experience into 'entanglements' (Hodder 2012) or 'meshworks' (Ingold 2000), recalling the notion of 'quotidian attunement' (Buchli, Chapter 2) that highlights how places (and infrastructures) give rise to social relationships and create spaces in particular ways. The research implications of these new experiential-social realms include digital field sites, as I have further discussed elsewhere (Walton 2017), though it is important to note here that digital co-presence formed the core basis of my research participants' and my shared experience of 'being there together,' online (Schroeder 2010).

This co-presence was not static, but motile, involving movements with and following digital photographic place-objects. The place-object constructs its own field site via its trails and 'traces' (Geiger and Ribes 2011) by which one can 'follow the thing' (Appadurai 1986a; Marcus 1995) within and across a range of social and spatial configurations. Screenshots of digital photographs and of its time/date stamped sociality – such as seen in comment threads in relation to specific images or galleries – helps in this regard, enabling the researcher to be co-present in a digital, phenomenological, and experiential sense with the photo-object long after its original moment of capture/creation. Given my limited access to Iran in a geographical sense, working with photographs as place-objects allowed me to trace the practices and flows of digital photographs exhibited on and channelled by Iranian photoblogs accordingly – including their lifecycles (births and expirations), spaces of production, consumption and circulation – across diverse digital and discursive domains and contexts and publics, inside and outside of Iran. I moved with/in these different digital, visual and material domains and the various media (mobile phones, digital cameras, laptops, PCs, Internet cafes) used by producers and consumers in particular contexts to exhibit and interact with digital photographs on Photoblogs, Flickr, Tumblr, Facebook, Instagram and other platforms. In this manner, I situated myself in what Arnau Monterde and John Postill call (2014) 'mobile ensembles': the intersection of various digital media, participants and issues found in certain instances of mobile technology facilitated social praxis.[7]

Analysis: place-object and their socialities

In the context of analogue photography, Roland Barthes highlighted the 'event' of photography – the moment(s) anchored in time, space, and place that is central to what the photograph is. The digital form and settings of the digital photograph seen in Iranian photoblogging complicates the 'event' in digital photographic practices online, where the time/date stamp of the event is modulated by individual action, by social relations in flux, and by digital infrastructures subject to a range of affordances and constraints. The notion of place-object helps to illuminate the crucial link between digital photographs and place, beyond event(s). Moreover, Barthes' classic theoretical assertion that analogue photography entails an 'I was there' raises the question of 'Where?' *beyond location*, which is central to this chapter. Where, geographically, and socially, was Barthes when he reflected upon a photograph of his mother as a young woman on a beach in his famed 1981 work *Camera Lucida*?

Relatedly, where was I, an anthropologist in the Internet age attempting to maintain my research network and enquiry into Iranian photographic practices after just one month 'on the ground' in the field in Iran? A main step involved exploring the methodological potentials that may be gained in the shift from analogue to digital form, where, with research taking place in multiple physical and digital sites, the 'ground' of fieldwork is ensconced in the 'cloud' of fieldwork. Barthes' dual person- and place-inflected theoretical utterance, '*I* was *there*,' is after all reconfigured in the practice of Iranian photoblogging itself to a present-orientated 'Iran is / we are here' – an act of collective witnessing – that the digital-photograph as place-object both facilitates and declares to a range of actors, including to the anthropologist. The place of Iran meanwhile, enmeshed in desires, practices, and perceptions, is unmoored from geographical rooting and is configured by the place-object, becoming much 'more than location.' The contemporary ethnographer, too, has much more than geographical place to work with and in.

Crucially then, it is the emplaced sociality enmeshed with the digital photograph that forms a significant part of what makes these objects live with action – a notion that Scott McQuire (2013), in the context of museum digital archives and indigenous identity politics and practices, calls an 'operational' quality, denoting active not passive sites where social agency is realised by coming into contact with photographs that relate to collective identities. In a similar vein, Jennifer Deger's (2016) work in the Yolngu Aboriginal Australian context of digital photographic practices examines how the making, sharing, and viewing of elaborated family photographs reaffirms, reconstitutes, and 'thickens' a world of vitality, resonance, and ancestral significance. In Deger's work, as in Iranian photoblogging, it is not just the photographic capture (the creation of the object as a marker of time/space or even of a place 'as is'), but it is the post-production stylising, shaping, curating, and meaning-making in the photography that constitutes a significant component in explaining the efficacy of digital photographs as material objects and as practices.

The digital photograph constantly engaged with, shared, commented upon, celebrated, refuted – in short 'handled' by the public as if it were physically accessible object in a museum or exhibition context – and 'seen' as if representative of the place of Iran. In so doing, it constructs place as 'distinct meanings and values for persons' (Tilley 1994: 15). Indeed, if we take Christopher Tilley's concept of place as 'contexts situatedness in relation to identity and action,' then we can see how the digital photographic object incurs questions of identity and action in relation to place and peoples, photographers and viewers, recalling what philosopher Tim Creswell (2003: 26) calls 'place as an intersection – a particular configuration of happenings in a constant sense of becoming through practice.' Seeing objects as places made up of such 'intersections' helps us to understand the social lives, efficacies, and affordances of objects as themselves places where people congregate within and to and cluster around, for various lengths of time, to various effects – including the ethnographer.

Digital photography/s: towards a new object status

My discussion of digital photographs as place-objects marks a point where old and new practices, theories, and methodologies meet. I have been concerned with the 'oldness' of photographs as material objects, rooted in representation, aesthetics, and incurring phenomenological engagements, while looking simultaneously at the 'newness' of digital photography as platformed practices, shaped through the convergence of different media/platforms and infrastructures through which it exists and operates, and the performances of transnational sociality it engenders. Concurrently, I have examined the 'oldness' of methodological approaches to anthropological enquiry based on physical presence in a place, particularly when access is limited or unavailable, with the 'newness' of engaging with online places and spaces that photographs construct in/for the social sciences. These meetings of 'old' and 'new' present a number of advancements for MCS, including shifts towards real-time, collaborative and networked sociality at-a-distance, while thinking critically about the plurality of objects such as digital photographs in places, and considering what the social/political implications of non-unique objects are – as empirically observed through their efficacies and affordances, attachments and attunements.

The notion of place-object I have posed in the chapter helps to explain how digital photographs can be of place (Iran as a representation), create a sense of place, geographically, through digital-locative techniques – including technological and phenomenological/imaginative capacities – and where places as digital objects, remain ultimately 'contexts for human experience constructed in movement, memory, encounter and association' (Tilley 1994: 15). This theoretical notion has doubled up as a methodological argument, whereby I have suggested that the place-object takes the anthropologist where they need to go for research, including a range of online and offline contexts. Thinking about objects as places, and places through objects, thus has broad implications for material culture scholarship. The ethnographic examples from Iran, and Iranian online environments, reveal how the digital photo-object is 'handled' as a material thing of significance online and/or offline, assembling configurations of peoples, practices, technologies, and desires in places and social spaces, online and offline beyond the geographical context, which may also be inaccessible or of limited physical access. Thinking about the theoretical and methodological potentials of place-objects can offer new pathways for research in and across a range of geographical sites that are politically precarious or otherwise inaccessible to anthropologists.

Acknowledgements

Work on this chapter and the manuscript was undertaken during a postdoctoral fellowship at UCL Anthropology as part of the ERC-funded Anthropology of Smartphones and Smart Ageing (ASSA) project (ERC-2016-Advanced Grant no. 740472). I would like to thank all of the photographers and research participants who took part in the research. I would also like to thank fellow volume editors Timothy Carroll and Antonia Walford for their generous feedback on the various drafts of this chapter.

Notes

1 Ethnographic film being a leading concern of visual anthropology in the UK context during the 1980s and 1990s (D. MacDougall 2005; Banks and Ruby 2011).

2 This move to objects in the study of photography signalled a turn away from the semiotic and linguistic approaches from 1960s anthropology and older lineages from strands of early twentieth-century French sociology (Peirce, Saussure) that

showed a privileging of representation in the study of images as signs, symbols and/or codes.

3 Harris defines the concept of 'representational fields' in relation to her work on visual representations of the idea of Tibet after 1959. She describes this as spaces wherein competing depictions have been mobilised for political agendas.

4 Interview, Kiana (Tehran) 2013.

5 This awareness of how the value of objects is constructed recalls Thompson's 1979 work *Rubbish Theory: The Creation and Destruction of Value.*

6 For an overview of the history and challenges of doing anthropology in/on Iran see S.R. Nadjmabadi (ed.) (2009).

7 Monterde and Postill's use of the term 'Mobile Ensembles' advances an earlier notion of 'media ensembles' (Bausinger 1984) in the digital context, referring to the 'unique set of mobile [and other] technologies that are brought to bear on a specific collective action' (Ibid.: 2).

Bibliography

Abbott, A. (2001) *Chaos of Disciplines*. Chicago: University of Chicago Press.

Abu-Lughod, L. (1991) 'Writing against Culture.' In R. Fox (ed.) *Re-capturing Anthropology: Working in the Present.* Santa Fe: School of American Research Press, 137–161.

Abu-Lughod, L. (1999) 'The Interpretation of Cultures after Television.' In S. Ortner (ed.) *The Fate of 'Culture': Geertz and Beyond.* Berkeley: University of California Press, 110–135.

Ackley, C. (2019) 'The Life Course of Labia: Female Genital Cutting in Somaliland.' In A. Parkhurst and T. Carroll (eds) *Medical Materialities: Toward a Material Culture of Medical Anthropology*. London: Routledge.

Adams, K.M. (2015) 'Back to the Future? Emergent Visions for Object-Based Teaching in and Beyond Classroom.' *Museum Anthropology* 38(2): 88–95.

Adelkhah, F. (2009) 'Islamophobia and Malaise in Anthropology.' In S.R. Nadjmabadi (ed.) *Conceptualizing Iranian Anthropology: Past and Present Perspectives.* New York: Routledge.

Adkins, C. (2011) 'What to Do with a Cartridge Case, a Bullet, or a Primer.' Paper presented at the 38th Annual Meeting of the Alaska Anthropological Association, Fairbanks, AK. [online] Fairbanks AK: University of Alaska, Museum of the North. Available from: www.uaf.edu/museum/collections/archaeo/projects/Adkins2011CartridgeCase.pdf. (Accessed 31 March 2019).

Adorno, T. (1955) 'Valéry Proust Museum.' In D. Crimp, *On the Museum's Ruins*, 1993. Cambridge, MA: MIT Press.

Adorno, T.W. (1994) *The Stars Down to Earth and Other Essays on the Irrational Culture*. London: Routledge.

Adriaans, R. (2019) 'The Humanitarian Road to Nagorno-Karabakh: Media, Morality and Infrastructural Promise in the Armenian Diaspora.' *Identities* 26(1): 69–87.

Africa: Journal of the International African Institute (1973) 'Daryll Forde 1902–1973.' *Africa: Journal of the International African Institute* 43(4): 281–283.

Agamben, G. (2007) *Profanations.* New York: Zone Books.

Ajana. B. (ed.) (2018) *Self-Tracking. Empirical and Philosophical Investigations.* New York: Springer.

Akrich, M. (1992) 'The De-Scription of Technical Objects.' In W.E. Bijker and J. Law (eds) *Shaping Technology / Building Society: Studies in Sociotechnical Change.* Cambridge, MA and London: MIT Press, 205–224.

Alix, C. (2012) 'Introduction: les peuples de l'Arctique et le bois / Introduction: Arctic peoples and wood.' *Études/Inuit/Studies* 36(1): 5–22.

Allen, A. and M. Holbraad (2014) 'Introduction: The Cosmological Frame in Anthropology.' In *Framing Cosmologies: The Anthropology of Worlds*. Manchester: Manchester University Press.
Allington, D., S. Brouillette, and D. Golumbia (2016) 'Neoliberal Tools (and Archives): A Political History of Digital Humanities.' *Los Angeles Review of Books*. 1 May 2016.
Amoore, L. (2006) 'Biometric Borders: Governing Mobilities on the War in Terror.' *Political Geography* 25(3): 336–351.
Amoore, L. (2011) 'Data Derivatives: On the Emergence of a Security Risk Calculus for Our Times.' *Theory, Culture and Society* 28(6): 24–43.
Amoore, L. (2018) 'Cloud Geographies: Computing, Data, Sovereignty.' *Progress in Human Geography* 42(1): 4–24.
Amoore, L., and V. Piotukh (2015) *Algorithmic Life: Calculative Devices in the Age of Big Data*. London: Routledge.
Anderson, C. (2008) 'The End of Theory: The Data Revolution Makes the Scientific Method Obsolete.' *Wired*. March 23. Available at www.wired.com/2008/06/pb-theory/ (Accessed 13 June 2019).
Anon. (1854) *A Plea for Urania: Being a Popular Sketch of Celestial Philosophy with Observations on the Impolicy of the Law Which Is Supposed to Prohibit the Practice of Astral Science in the Present Age*. London: Piper, Stephenson and Spence.
Appadurai, A. (1986a) 'Introduction: Commodities and the Politics of Value.' In Appadurai, A. (ed.) *The Social Life of Things: Commodities in Cultural Perspective*. Cambridge: Cambridge University Press, 3–63.
Appadurai, A. (ed.) (1986b) *The Social Life of Things: Commodities in Cultural Perspective*. Cambridge: Cambridge University Press.
Appadurai, A. (1990) 'Disjuncture and Difference in the Global Economy.' *Theory, Culture and Society* 7(2–3): 295–310.
Appadurai, A. (1996) *Modernity at Large: Cultural Dimensions of Globalisation*. Minneapolis: University of Minnesota Press.
Apprich, C., W.H.K. Chun, F. Cramer, and H. Steyerl (eds) (2019) *Pattern Discrimination*. Minneapolis: University of Minnesota Press.
Arteaga, I. (2019) 'From Attitudes to Materialities: Understanding Bowel Control for Colorectal Cancer Patients in London.' In A. Parkhurst and T. Carroll (eds) *Medical Materialities: Toward a Material Culture of Medical Anthropology*. London: Routledge.
Asad, T. (1973) *Anthropology and the Colonial Encounter*. London: Ithaca Press.
Askew, K. and R. Wilk (eds) (2002) *The Anthropology of Media: A Reader*. Oxford: Blackwell.
Atkinson, M. (2004) 'Tattooing and Civilizing Processes: Body Modification as Self-Control.' *Canadian Review of Sociology* 41(2): 125–146.
Attali, J. (1985) *Noise: The Political Economy of Music*, trans. B. Massumi. Minneapolis: University of Minnesota Press.
Audouze, F. (2002) 'Leroi-Gourhan, a Philosopher of Technique and Evolution.' *Journal of Archaeological Research* 10(4): 227–306.
Auerbach, E. (1953) *Mimesis: The Representation of Reality in Western Literature*. New York: Doubleday Anchor.
Babb, L.A. (1981) 'Glancing: Visual Interaction in Hinduism.' *Journal of Anthropological Research* 37(4): 378–410.
Banerjee, M. and D. Miller (2003) *The Sari*. London: Berg.

Banks, M. and J. Ruby (eds) (2011) *Made to Be Seen: Perspectives on the History of Visual Anthropology*. London: University of Chicago Press.

Banyard, K. (2016) The Dangers of Rebranding Prostitution as 'Sex Work.' *The Guardian*, 7 June 2016. Online at: www.theguardian.com/lifeandstyle/2016/jun/06/prostitution-sex-work-pimp-state-kat-banyard-decriminalisation.

Barad, K. (2003) 'Posthumanist Performativity: Toward an Understanding of How Matter Comes to Matter.' *Signs: Journal of Women in Culture and Society*, 28(3): 801–831.

Barad, K. (2007) *Meeting the Universe Halfway: Quantum Physics and the Entanglement of Matter and Meaning*. Durham, NC: Duke University Press.

Barassi, V. (2017) 'Ethnography Beyond and Within Digital Structures and the Study of Social Media Activism.' In L. Hjorth, H. Horst, A. Galloway, and G. Bell (eds) *Routledge Companion to Digital Ethnography*. London: Routledge, 406–418.

Barnes, A. and R. Lynch (2012) 'From the Classroom to the Museum: Understanding Faculty-designed Assignments in an Academic Museum.' *Museum Management and Curatorship* 27(5): 487–503.

Barry, A. (2013) *Material Politics: Disputes along the Pipeline*. London: Wiley-Blackwell.

Barry, A. and G. Born (2015) *Interdisciplinarity: Reconfigurations of the Social and Natural Sciences*. London: Routledge.

Barthélémy, J.-H. (2012) 'Glossary: Fifty Key Terms in the Works of Gilbert Simondon.' In A. De Boever et al. (eds) *Gilbert Simondon: Being and Technology*. Edinburgh: Edinburgh University Press, 202–231.

Barthes, R. (1972 [1957]) *Mythologies*, trans. A. Lavers. New York: Hill and Wang.

Barthes, R. (1981) *Camera Lucida: Reflections on Photography*, trans. R. Howard. New York: Hill and Wang.

Basu, P. (2013) 'Material Culture: Ancestries and Trajectories in Material Culture Studies.' In J.G. Carrier and D.B. Gewertz (eds) *Handbook of Sociocultural Anthropology*, 370–390. London: Bloomsbury.

Basu, P. and W. Modest (2013) *Museums, Heritage, and International Development*. New York: Routledge.

Batchen, G. (1997) *Photography's Objects*. Catalogue of Exhibition at University of New Mexico Art Museum, Albuquerque, 26 August to 31 October. Albuquerque: University of New Mexico.

Bateson, G. (1972) *Steps to an Ecology of Mind: Collected Essays in Anthropology, Psychiatry, Evolution, and Epistemology*. Chicago: University of Chicago Press.

Bateson, G. (1979) *Mind and Nature: A Necessary Unity*. Toronto: Clarke, Irwin.

Bateson, G. and M.C. Bateson (1988) *Angels Fear: Toward an Epistemology of the Sacred*. London: Bantam.

Battaglia, D. (2012a) 'Arresting Hospitality: The Case of the Handshake in Space.' *Journal of the Royal Anthropological Institute* (N.S.) 18 (s1): S76–S89.

Battaglia, D. (2012b) 'Coming in at an Unusual Angle: Exo-surprise and the Fieldworking Cosmonaut.' *Anthropological Quarterly* 85(4): 1089–1106.

Battaglia, D. (2017) 'Aeroponic Gardens and Their Magic: Plants/Persons/Ethics in Suspension.' *History and Anthropology* 28(3): 263–292.

Battaglia, D., Valentine, D., and Olson, V. (2012) 'Introduction: Extreme Limits and Horizons in the Once and Future Cosmos.' *Anthropological Quarterly* 85(4): 1007–1026.

Battaglia, D., Valentine, D., and Olson, V. (2015) 'Relational Space: An Earthly Installation.' *Cultural Anthropology* 30(2): 245–256.

Bausinger, H. (1984) 'Media, Technology, and Daily Life.' *Media, Culture and Society* 6(4): 343–351.
Bayat, A. (2010) *Life as Politics: How Ordinary People Change the Middle East*. Stanford: Stanford University Press.
Baym, N.K. (2010) *Personal Connections in the Digital Age*. Cambridge: Polity Press.
Beard, M. (1992) 'Souvenirs of Culture: Deciphering (in) the Museum.' *Art History* 15(4): 502–532.
Beaulieu, A. (2010) 'From Co-location to Co-presence: Shifts in the Use of Ethnography for the Study of Knowledge.' *Social Studies of Science* 40(3): 453–470.
Becker, H.S. (1982) *Art Worlds*. Berkeley: University of California Press.
Bedini, S. (1964) 'The Role of Automata in the History of Technology.' *Technology and Culture* 5(1): 24–42.
Beer, D. (2016) 'How Should We Do the History of Big Data?' *Big Data and Society*, January–June 2016: 1–10.
Bell, C., and J. Lyall (2005) 'I Was Here: Pixelated Evidence.' In D. Crouch, R. Jackson, and F. Thompson (eds). *The Media and the Tourist Imagination: Converging Cultures*. London: Penguin.
Bell, J.A. and H. Geismar (2009) 'Materialising Oceania: New Ethnographies of Things in Polynesia.' *The Australian Journal of Anthropology* 20(1): 3–27.
Bender, B. (1998) *Stonehenge: Making Space*. Oxford: Berg.
Benedict, R. (1946) *The Chrysanthemum and the Sword.* Boston: Houghton Mifflin Harcourt.
Benjamin, W. (1972[1936]) 'The Work of Art in the Age of Mechanical Reproduction.' In W. Benjamin (ed.) *Illuminations*. New York: Schocken Books.
Benjamin, W. (2008) *The Work of Art in the Age of Its Technological Reproducibility and Other Writings on Media*. Edited by M.W. Jennings, B. Doherty, and T.Y. Levin. Cambridge, MA: Harvard University Press.
Benjamin, W. and M.W. Jennings (2010) 'The Work of Art in the Age of Its Technological Reproducibility [First Version].' *Grey Room* 39: 11–38.
Bennett, J. (2010) *Vibrant Matter: A Political Ecology of Things*. Durham, NC: Duke University Press.
Benson, S. (2000) 'Inscriptions of the Self: Reflections on Tattooing and Piercing in Contemporary Euro-America.' In J. Caplan (ed.), *Written on the Body: The Tattoo in European and American History*. London: Reaktion Books: 234–254.
Bergson, H., N. Paul, and W.S. Palmer (2004[1911]) *Matter and Memory*. Mineola, NY: Dover Publications, Inc.
Bernstein, E. (2007) 'Sex Work for the Middle Classes.' *Sexualities* 10(4): 473–488.
Bernstein, E. (2010) 'Militarized Humanitarianism Meets Carceral Feminism: The Politics of Sex, Rights, and Freedom in Contemporary Antitrafficking Campaigns.' *Signs: Journal of Women in Culture and Society* 36(1): 45–71.
Bessire, L. and D. Bond (2014) 'Ontological Anthropology and the Deferral of Critique.' *American Ethnologist* 41(3): 440–456.
Beutelspacherm A., E. Martelo, and V. Garcia (2003) 'Does Contraception Benefit Women? Structure, Agency and Well-being in Rural Mexico.' *Feminist Economics* 9(2–3): 213–238.
Bhabha, H.K. (1994) *The Location of Culture.* London: Routledge.
Bialecki, J. (2017) 'After, and Before, Anthropos.' *Platypus: Committee on the Anthropology of Science, Technology and Computing (CASTAC) Blog.* (6 April 2017).

Bialecki, J. (2019) 'Gods, AIs, and Mormon Transhumanism.' *Platypus: Committee on the Anthropology of Science, Technology and Computing (CASTAC) Blog* (14 February 2019).

Bille, M. (2019) *Living with Light: Homely Atmospheres and Lighting Technologies in Denmark*. London: Bloomsbury.Bille, M. and T.F. Sorenson (2007) 'An Anthropology of Luminosity: The Agency of Light.' *Journal of Material Culture* 12(3): 263–284.

Bille, M., and T.F. Sørensen (2016) 'Into the Fog of Architecture.' In. M. Bille and T.F. Sørensen (eds) *Elements of Architecture, Assembling Archaeology, Atmosphere and the Performance of Building Spaces*. London: Routledge.

Bijker, W. (1989) 'The Social Construction of Bakelite: Towards a Theory of Invention.' In W.E. Bijker, T. Hughes and T.J. Pinch (eds) *The Social Construction of Technological Systems*. London: The MIT Press, 159–187.

Bijker, W. (2010) 'How is Technology Made? That is the Question!' *Cambridge Journal of Economics* 34(1): 63–76.

Bijker, W., T. Hughes, and T.J. Pinch (eds) (1987) *The Social Construction of Technological Systems: New Directions in the Sociology and History of Technology*. Cambridge, MA: MIT Press.

Bindel, J. (2016) 'Decriminalising the Sex Trade Will Not Protect Its Workers from Abuse.' *The Guardian*, 13 July 2016. Online at: www.theguardian.com/commentisfree/2016/jul/13/decriminalising-sex-trade-protect-workers-abuse

Bird, D., R. Bird, B, Codding, and D. Zeanah (2019) 'Variability in the Organization and Size of Hunter-Gatherer Groups: Foragers Do Not Live in Small-Scale Societies.' *Journal of Human Evolution* 131: 96–108.

Birth, K.K. (2008) 'The Creation of Coevalness and the Danger of Homochronism.' *Journal of the Royal Anthropological Institute* 14(1): 3–20.

Blake, W. (1927) *Poetry and Prose of William Blake* (ed.) Geoffrey Keynes. London: Nonesuch Press.

Blanchard, M. (1991) 'Post-Bougeois Tattoo: Reflections on Skin Writing in Late capitalist Societies.' *Visual Anthropology Reivew* 7(2): 11–21.

Blanchette, J.F. (2011) 'A Material History of Bits.' *Journal of the American Society for Information Science and Technology* 62(6): 1042–1057.

Blier, S.P. (1987) *The Anatomy of Architecture: Ontology and Metaphor in Batammaliba Architectural Expression*. Cambridge: Cambridge University Press.

Boas, F. (1974 [1887]) The Principles of Ethnological Classification. In *The Shaping of American Anthropology, 1883–1911: A Franz Boas Reader*. New York: Basic Books.

Boast, R. (2011) Neocolonial Collaboration: Museum as Contact Zone Revisited. *Museum Anthropology* 34(1): 56–70.

Boaz, F. (1955) *Primitive Art*. Toronto: Dover Publications.

Boddy, J. (1988) Spirits and Selves in Northern Sudan: The Cultural Therapeutics of Possession and Trance. *American Ethnologist* 15(1): 4–27.

Boellstorff, T. (2007) *A Coincidence of Desires: Queer Theory, Anthropology, Indonesia*. Durham, NC: Duke University Press.

Boellstorff, T. (2008) *Coming of Age in Second Life: An Anthropologist Explores the Virtually Human*. Oxford and Princeton, NJ: Princeton University Press.

Boellstorff, T., B. Nardi, C. Pearce, and T. Taylor (2012) *Ethnography and Virtual Worlds*. Princeton, NJ: Princeton University Press.

Bogost, I. and N. Montfort (2009) 'Platform Studies: Frequently Asked Questions.' *Digital Arts and Culture*, December 12–15. Available at https://escholarship.org/uc/item/01r0k9br (Accessed 13 June 2019).

Böhme, G. (2017) *The Aesthetics of Atmospheres*. Edited by J-P Thibaud. London: Routledge.

Bonilla, Y. and J. Rosa (2015) 'Ferguson: Digital Protest, Ethnography, and the Racial Politics of Social Media in the United States.' *American Ethnologist* 42(1): 4–17.

Bouquet, M. (2000) 'Thinking and Doing Otherwise: Anthropological Theory in Exhibitionary Practice.' *Ethnos* 65(2): 217–236.

Bourdieu, P. (1977) *Outline of a Theory of Practice*. Cambridge: Cambridge University Press.

Bourdieu, P. (1984a) *Distinction: A Social Critique of the Judgement of Taste*. London: Routledge and Kegan Paul.

Bourdieu, P. (1984b) *Homo Academicus*. Cambridge: Polity Press.

Bourdieu, P. (1990) *Photography: A Middle-brow Art*, trans. S. Whiteside. Cambridge: Polity Press.

Bourdieu, P. (1996[1965]). *Un art moyen*. (Photography: A Middle-Brow Art.). Paris: Les Éditions de Minuit.

Bourdieu, P. and R. Johnson (1993) *The Field of Cultural Production: Essays on Art and Literature*. Cambridge: Polity Press.

Bourriaud, N. (2002) *Nicolas Bourriaud: Relational Aesthetics*. Paris: Les Presses du réel.

Bowker, G. (1994) 'Information Mythology and Infrastructure.' In L. Bud-Frierman (ed.) *Information Acumen: The Understanding and Use of Knowledge in Modern Business*. London: Routledge, 231–247.

Bowker, G.C., S.L. Star, and MIT Press (1999) *Sorting Things Out: Classification and Its Consequences*. Cambridge, MA: MIT Press.

Bowker, G.C. (2013) 'Data Flakes: An Afterword to "Raw Data" is an Oxymoron.' In L. Gitelman (ed.) *"Raw Data" is an Oxymoron*. Cambridge, MA: MIT Press, 167–171.

Boyarin, J. (1991) 'Jewish Ethnography and the Question of the Book.' *Anthropological Quarterly* 64(1): 14–29.

boyd, d. (2015) *It's Complicated: The Social Lives of Networked Teens*. New Haven: Yale University Press.

boyd, d., and K. Crawford (2012) 'Critical Questions for Big Data Provocations for a Cultural, Technological, and Scholarly Phenomenon.' *Information Communication and Society: Special Issue: A Decade in Internet Time: The Dynamics of the Internet and Society* 15(5): 662–679.

Boyle, J. (2010) *The Public Domain: Enclosing the Commons of the Mind*. New Haven: Yale University Press.

Braidotti, R. (2013) *The Posthuman*. Cambridge: Polity Press.

Brand, J. (1810) *Observations on Popular Antiquities Including the Whole of Mr. Bourne's Antiquitates Vulgares*. London: Vernor, Hood, and Sharpe.

Bratton, B.H. (2015) *The Stack: On Software and Sovereignty*. Software Studies. Cambridge, MA: MIT Press.

Bräucher, B. and J. Postill (eds) (2010) *Theorising Media and Practice*. New York: Berghahn Books.

Braudillard, J. (1996[1968]) *The System of Objects*. New York: Verso Books.

Bredekamp, H. (2008) Die Fenster Der Monade: *Gottfried Wilhem Leibniz' Theater Der Natur Und Kunst* (ACTA Humaniora). Berlin: Walter de Gruyter.
Brown, B. (2001) 'Thing Theory.' *Critical Inquiry* 28(1): 1–22.
Brown, S.D. (2003) 'Natural Writing: The Case of Serres.' *Interdisciplinary Science Reviews* 28(3): 184–192.
Brown, I. (2016). *The School of Oriental and African Studies: Imperial Training and the Expansion of Learning*. Cambridge: Cambridge University Press.
Brun, J. (1985) 'Biographie de la machine.' *Les ÉtudesPhilosophiques* 1: 3–16.
Bucher, B. (1981) *Icon and Conquest: A Structural Analysis of the Illustrations of de Bry's Great Voyages*. Chicago: University of Chicago Press.
Buchli, V. (2002a) 'Introduction.' In V. Buchli (ed.) *The Material Culture Reader*. Oxford: Berg, 1–22.
Buchli, V. (ed.) (2002b) *The Material Culture Reader*. Oxford and New York: Bloomsbury.
Buchli, V. (2008) 'Material Culture: Current Problems.' In L. Meskell and R. Preucel (eds) *A Companion to Social Archaeology*. Oxford: Blackwell.
Buchli, V. (2010) 'The Prototype: Presencing the Immaterial.' *Visual Communication* 9(3): 273–286).
Buchli, V. (2013) *An Anthropology of Architecture*. London: Bloomsbury Academic.
Buchli, V. (2016) *An Archaeology of the Immaterial*. London: Routledge.
Buchli, V. (2017) 'The Social Condenser: Again, Again and Again–the Case for the Narkomfin Communal House, Moscow.' *The Journal of Architecture* 22(3) 387–402.
Buchli, V. (Forthcoming) 'A Speculative Ethnographers Guide to Low Earth Orbit.' In W. Bracewell et al. (eds) *Anti-Atlas: Towards a Critical Area Studies*. London: UCL Press.
Buchwald, V.F. and G. Mosdal (1985) *Meteoric Iron, Telluric Iron and Wrought Iron in Greenland.* Copenhagen: Kommissionen for videnskabelige Undersøgelser i Grønland.
Buckley, L. (2001) 'Self and Accessory in Gambian Studio Photography.' *Visual Anthropology Review* 16(2): 71–91.
Bunce, L. (2016) 'Appreciation of Authenticity Promotes Curiosity: Implications for Object-based Learning in Museums.' *Journal of Museums Education* 41(3): 230–239.
Burgin, V. (1982) 'Photographic Practice and Art Theory.' In V. Burgin (ed.) *Thinking Photography*. Basingstoke: Macmillan, 39–83.
Burton, J. (2016) *Making Sex Work Work: The Material Practices of Maintaining Sex Work as Work in London*. M.Sc.Social and Cultural Anthropology Dissertation. University College London, unpublished.
Cadena, M. de la. (2010) 'Indigenous Cosmopolitics in the Andes: Conceptual Reflections beyond "Politics."' *Cultural Anthropology* 25(2): 334–370.
Cadena, M. de la. (2015) *Earth Beings: Ecologies of Practice across Andean Worlds*. Durham, NC: Duke University Press.
Campanelli, V. (2010) *Web Aesthetics: How Digital Media Affect Culture and Society*. Rotterdam: NAI publishers.
Campbell, S. (2002) *The Art of Kula*. Oxford: Berg.
Cannon, L. (2014) 'Purpose of the Mormon Transhumanist Association.' Scribd. Mormon Transhumanist Association, 6 March 2014. Web. www.scribd.com/document/210935176/Purpose-of-the-Mormon-Transhumanist-Association Retrieved (Accessed 21 March 2019).

Carrier, J., and D. Miller (1998) *Virtualism: A New Political Economy.* London: Bloomsbury.

Carroll, D. (1987) *Paraesthetics: Foucault. Lyotard. Derrida*. New York: Methuen.

Carroll, T. (2017) 'Textiles and the Making of Sacred Space.' *Textile History* 48(2): 192–210.

Carroll, T. (2018) *Orthodox Christian Material Culture: Of People and Things in the Making of Heaven*. London: Routledge.

Carroll, T. and A. Parkhurst (2019) 'Introduction: A Genealogy of Medical Materialities.' In A. Parkhurst and T. Carroll (eds) Medical Materialities: Toward a Material Culture of Medical Anthropology. London: Routledge, 1–20.

Carroll, T., A. Parkhurst, and D. Jeevendrampillai (2017) 'Introduction: Toward a General Theory of Failure.' In T. Carroll et al. (eds) *Material Culture of Failure.* London: Bloomsbury, 1–20.

Carsten, J. and S. Hugh-Jones (eds) (1995) *About the House: Lévi-Strauss and Beyond.* Cambridge: Cambridge University Press.

Carusi, A. (2008) 'Scientific Visualisations and Aesthetic Grounds for Trust.' *Ethics and Information Technology* 10: 243–254.

Catchpole J. (2008) *The International Space Station: Building for the Future*. Chichester, UK: Springer Praxis Books.

Chatterjee, H. and L. Hannan (eds) (2015) *Engaging the Senses: Object-Based Learning in Higher Education*. Farnham, Surrey: Ashgate.

Chatterjee, H., L. Hannan, and L. Thomson (2015) 'An Introduction to Object-Based Learning and Multisensory Engagement.' In H. Chatterjee and L. Hannan (eds) *Engaging the Senses: Object-Based Learning in Higher Education*. Farnham, Surrey: Ashgate, 1–18.

Chatterjee, H., S. MacDonald, G. Noble, and D. Prytherch (eds) (2008) *Touch in Museums: Policy and Practice in Object Handling*. Oxford: Berg.

Chatterjee, H. and G. Noble (eds.) (2013) *Museums, Health and Well-being*. Farnham, Surrey: Ashgate.

Chelkowski, P.J. and H. Dabashi (2000) *Staging a Revolution: The Art of Persuasion in the Islamic Republic of* Iran. London: Booth-Clibborn.

Cheney-Lippold, J. (2011) 'A New Algorithmic Identity: Soft Biopolitics and the Modulation of Control.' *Theory, Culture and Society* 28(6): 164–181.

Chinniah, S. (2008) 'The Tamil Film Heroine: From a Passive Subject to a Pleasurable Object.' In S. Velayutham (ed.) *Tamil Cinema: The Cultural Politics of India's Other Film Industry*, 29–43. New York: Routledge.

Chladek, J. (2017) *Outposts on the Frontier: A Fifty Year History of Space Stations.* Lincoln: University of Nebraska Press.

Chrysostomos, A. (1981) *Orthodox Liturgical Dress.* Brookline: Holy Cross Orthodox Press.

Cladis, M.S. (2001) 'Introduction to Durkheim, E.' *The Elementary Forms of Religious Life*, trans. C. Cosman. Oxford: Oxford University Press.

Clarke, A. (1998) 'Window Shopping at Home.' In D. Miller (ed.) *Material Cultures.* Chicago: Chicago University Press, 73–99.

Clifford, J. (1988) *The Predicament of Culture: Twentieth-century Ethnography, Literature, and Art*. Cambridge, MA: Harvard University Press.

Clifford, J. (1997) *Routes: Travel and Translation in the Late Twentieth Century.* Cambridge, MA: Harvard University Press.

Clifford, J. (1999[1997]) 'Museums as Contact Zones.' In D. Boswel and J. Evans (eds) *Representing the Nation: A Reader. Histories, Heritage and Museums*. London and New York: Routledge, 435–457.

Clifford, J., and G. Marcus (1986) *Writing Culture: The Poetics and Politics of Ethnography: A School of American Research Advanced Seminar*. Berkeley: University of California Press.

Clynes, M.E. and N.S. Kline. (1960) 'Cyborgs and Space.' *Astronautics* 26 (September): 24–25.

Coates, J. (2017) 'So Hot Right Now: Reflections on Virality and Sociality from Transnational Digital China.' *Digital Culture and Society* 3(2): 77–98.

Cohen, K.R. (2005) 'What Does the Photoblog Want?' *Media, Culture and Society*, 27(6): 883–901.

Coleman, G. (2012) *Coding Freedom: The Ethics and Aesthetics of Hacking*. Princeton, NJ: Princeton University Press.

Collier, S. (2011) *Post-Soviet Social: Neoliberalism, Social Modernity, Biopolitics*. Princeton, NJ: Princeton University Press.

Collini, S. (2018) 'Kept Alive for Thirty Days.' *London Review of Books* 40(21), 8 November 2018.

Colonna, F. (2005) *Hynerotomachia Poliphili: The Strife of Love in a Dream*, trans. J. Godwin. London: Thames and Hudson.

Connell, C. (2011) *An Archaeology of Materials: Substantial Transformations in Early Prehistoric Europe*. London: Routledge.

Coopmans, C. (2014) 'Visual Analytics as Artful Revelation.' In C. Coopmans, J. Vertesi, M. Lynch, and S. Woolger (eds) *Representation in Scientific Practice Revisited*. Cambridge, MA: MIT Press, 37–55.

Coopmans, C., M. Lynch, and J. Vertesi. (2014) *Representation in Scientific Practice Revisited*. Cambridge MA: MIT Press.

Cornell, C. (2011) *An Archaeology of Materials: Substantial Transformations in Early Prehistoric Europe*. London: Routledge.

Costa, E. (2016) *Social Media in Southeast Turkey*. London: University College London Press.

Couldry, N. and J. van Dijck (2015) 'Researching Social Media as If the Social Mattered.' *Social Media + Society* 1(2): 1–7. DOI: 10.1177/20563051 1560417.

Coupaye, L. (2013) *Growing Artefacts, Displaying Relationships: Yams, Art and Technology amongst the Nyamikum Abelam of Papua New Guinea*. Oxford and New York: Berghahn Books.

Coupaye, L. (2015) Chaîne opératoire, transects et théories: quelques réflexions et suggestions sur le parcours d'une méthode classique. In P. Soulier, ed. *André Leroi-Gourhan « l'homme tout simplement »*. Paris: Éditions de Boccard –Travaux de la MAE – Maison de l'Archéologie et de l'Ethnologie, René-Ginouvès, 69–84.

Coupaye, L. (2018) 'Realising Fantasies: Objects as Contexts, Processes and Presence.' In P. Cordez et al. (eds) *Object Fantasies: Experience and Creation*. Berlin: De Guyter, 209–222.

Coupaye, L. (2021). 'Making "Technology" Visible: Technical Activities and the Chaîne Opératoire.' In M.H. Bruun and A. Wahlberg (eds) *The Handbook of Anthropology of Technology*. New York: Palgrave Handbooks.

Coupaye, L., P. Pitrou, F. Provost (eds) (2016) Des êtres vivants et des artefacts. L'imbrication des processus vitaux et des processus techniques, 9–10 April 2014, Musée du quai Branly. Paris: Musée du quai Branly Jacques Chirac.

Cramer, F. (2019) 'Crapularity Hermeneutics: Interpretation as the Blind Spot of Analytics, Artificial Intelligence and Other Algorithmic Producers of the Postapocalyptic Present.' In C. Apprich, W.H.K. Chun, F. Cramer, and H. Steyerl (eds) *Pattern Discrimination*. Minneapolis: University of Minnesota Press, 23–58.

Crawford, K. and T. Paglen (2019) 'Excavating AI: The Politics of Images in Machine Learning Training Sets.' www.excavating.ai. 19 September 2019. www.excavating.ai. (Accessed 28 October 2019).

Cresswell, T. (2003) 'Theorizing Place. Thamyris/Intersecting: Place, Sex and Race, Mobilizing Place.' In G. Verstraete and T. Cresswell (eds) *Placing Mobility: The Politics of Representation in a Globalized World*. Leiden: Brill, 11–31.

Cunninghame, W. (1817) *A Dissertation on the Seals and Trumpets of the Apocalypse and the Prophetical Period of the Twelve Hundred and Sixty Years*. London: T. Cadell and W. Davies.

Curry, P. (1992) *A Confusion of Prophets: Victorian and Edwardian Astrology*. London: Collins and Brown.

Dalakoglou, D. (2017) *The Road: An Ethnography of (Im)Mobility, Space, and Cross-border Infrastructures in the Balkans.* Manchester: Manchester University Press.

Damon, F. (2016) *Trees, Knots and the Outrigger: Environmental Knowledge in the Northeast Kula Ring*. Oxford: Berghahn.

Daniels, I. (2010) *The Japanese House*. Oxford: Berg.

Daniels, I. (2019) *What are Exhibitions For?* London: Bloomsbury.

Das, V. (1995) 'On Soap Opera: What Kind of Anthropological Object Is It?' In D. Miller (ed.) *Worlds Apart: Modernity Through the Prism of the Local.* London: Routledge.

David, D.-A. and C. Craven (2016) *Feminist Ethnography: Thinking Through Methodologies, Challenges and Possibilities*. New York: Rowman and Littlefield.

de Beaune, S.A. (2004) 'The Invention of Technology: Prehistory and Cognition.' *Current Anthropology* 45(2): 139–162.

de Castro, E.B.V. (2004) 'Exchanging Perspectives: The Transformation of Objects into Subjects in Amerindian Ontologies.' *Common Knowledge* 10(3): 463–484.

Deacon, A.B. (1934) *Malekula: A Vanishing People in the New Hebrides*. C.H. Wedgewood (ed.). London: Routledge.

Deacon, A.B. and C. Wedgewood (1934) 'Geometric Drawings from Malekula and Other Islands of the New Hebrides.' *Journal of the Royal Anthropological Institute* 64: 129–175.

DeCerteau, M. (1984) The *Practice of Everyday Life*. Berkeley, Los Angeles, and London: University of California Press.

Deger, J. (2016) 'Thick Photography.' *Journal of Material Culture* 21(10): 111–132.

Dekker, R., G. Engbersen, J. Klaver, and H. Vonk (2018) 'Smart Refugees: How Syrian Asylum Migrants Use Social Media Information in Migration Decision-Making.' *Social Media + Society* 4(1): 1–11.

Deleuze, G. (1983) *Cinema 1: The Movement Image* (*Cinéma 1, L'Image-Mouvement*). Paris: Les Editions de Minuit.

Deleuze, G., and F. Guattari (1987) *A Thousand Plateaus: Capitalism and Schizophrenia*. London: Continuum.

Deliss, C. (2012) 'Performing the Curatorial in a Post-ethnographic Museum.' In M. Lind (ed.), *Performing the Curatorial: Within and Beyond Art*. Berlin: Sternberg Press.

Deliss, C. (2013) 'Trading perceptions in a post-ethnographic museum.' Retrieved on 31.4.19 from http://theatrum-mundi.org/library/trading-perceptions-in-a-post-ethnographic-museum/.(Accesed 29 August 2020).

DeNicola, L. (2012) 'Geomedia: The Reassertion of Space within Digital Culture.' In H. Horst, and D. Miller (eds) *Digital Anthropology*. London and New York: Berg.

DeSilvey, C. (2006) 'Observed Decay: Telling Stories with Mutable Things.' *Journal of Material Culture* 11(3): 318–338.

Dickey, S. (2008) 'The Nurturing Hero: Changing Images of MGR.' In S. Velayutham (ed.) *Tamil Cinema: The Cultural Politics of India's Other Film Industry*, 111–123. New York: Routledge.

Douglas, M. (1991) 'The Idea of Home: A Kind of Space.' *Social Research* 58(1): 287–307.

Douglas, M. (2002[1966]), *Purity and Danger*. London: Routledge.

Douglas, M. and B. Isherwood (1978) *The World of Goods*. London: Allen Lane.

Dourish, P. (2014) 'NOSQL: The Shifting Materialities of Database Technology,' Computational Culture. November. Available at: http://computationalculture.net/no-sql-the-shifting-materialities-of-database-technology/ (Accessed June 13 2019).

Dourish, P. (2016) 'Rematerializing the Platform: Emulation and the Digital-Material'. In S. Pink, E. Ardèvol, and D. Lanzeni (eds) *Digital Materialities: Design and Anthropology*. London: Bloomsbury Academic.

Dourish, P. (2017) *The Stuff of Bits*. Cambridge, MA: MIT Press.

Dourish, P. and Mazmanian, M. (2013) 'Media as Material: Representations as Material Foundations for Organizational Practice.' In D. Nicolini, P.R. Carlile, and H. Tsoukas (eds) *How Matter Matters: Objects, Artefacts and Materiality in Organization Studies*. Oxford: Oxford University Press, 92–118.Dow Schull, N. and M. Ruckenstein (2017) The Datafication of Health. *Annual Review of Anthropology* 46: 261–278.

Downey, G., and J. Dumit (eds) (1997) *Cyborgs and Citadels: Anthropological Interventions in Emerging Sciences and Technologies*. Santa Fe and Seattle: School of American Research Press; Distributed by the University of Washington Press.

Downey, G.L., J. Dumit and S. Williams (1995) 'Cyborg Anthropology.' *Cultural Anthropology* 10(2): 264–269.

Drazin, A. (2013) 'Design Anthropology: Working on, with and for Digital Technologies.' In H. Horst and D. Miller (eds) *Digital Anthropology*. London: Bloomsbury, 245–265.

Drazin, A. (2015) 'To Live in a Materials World.' In A. Drazin and S. Küchler (eds) *The Social Life of Materials*. London: Bloomsbury.

Dudley, S. (2011) 'Encountering a Chinese Horse. Engaging with the Thingness of Things.' In S.H. Dudley (ed.) *Museum Objects: Experiencing the Properties of Things*. London: Taylor and Francis Group.

Durkheim, E. (2001) *The Elementary Forms of Religious Life*, trans. C. Cosman. Oxford: Oxford University Press.

Eastham, J. (2016) 'A Radical Moment for Britain's Workers.' *The Guardian*. www.theguardian.com/global/2016/jul/04/sex-workers-commons-inquiry-prostituion-brothels-soliciting-legalised-womens-safety-stunning-victory. (Accessed 9 September 2016).

Eberbach, C. and K. Crowley (2005) 'From Living to Virtual: Learning from Museum Objects.' *Curator* 48(3): 317–338.

Eck, D.L. (1981) *Darsan: Seeing the Divine in India*. Chambersburg: Anima Books.

Editorial (2016) 'The Guardian View on Criminal Policy: Sex, Money and the Long Arm of the Law.' *The Guardian*, 7 April 2016. Online at: www.theguardian.com/commentisfree/2016/apr/07/the-guardian-view-on-criminal-policy-sex-money-and-the-long-arm-of-the-law. (Accessed 29 August 2020).

Edwards, E. (1994) *Anthropology and Photography*. New Haven: Yale University Press.

Edwards, E. (2011) 'Tracing Photography.' In M. Banks and J. Ruby (eds) *Made to Be Seen: Perspectives on the History of Visual Anthropology*. Chicago and London: University of Chicago Press.

Edwards, E. and J. Hart (eds) (2004) *Photographs Objects Histories: On the Materiality of Images.* London and New York: Routledge.

Egnor, M. (1980) 'On the Meaning of *Sakti* to Women in Tamil Nadu.' In S.S. Wadley (ed.) *The Powers of Tamil Women*, 1–34. Syracuse: Maxwell School of Citizenship and Public Affairs.

Empson, R. and H. Spriggs (2018) 'Introducing Five Heads: The Artist as Antanae.' In H. Spriggs (ed.) *Five Heads (Tavan Tolgoi): Art, Anthropology, and Mongol Futurism*. Berlin: Sternberg Press.

Erden, Y. (2013) 'Neural Implants, Human Identity, and Perceptions of Illness.' *RiCT*, June 17, 2013.

Eriksen, T.H. (2001) *Small Places, Large Issues*. London: Pluto Press.

Eriksen, T.H. (2006) *Engaging Anthropology*. Oxford: Berg.

Eriksen, T.H. (2008) 'The Otherness of Norwegian Anthropology.' In A. Boskovic (ed.) *Other People's Anthropologies*. Oxford: Berghahn, 169–185.

Eriksen, T.H. (2016). 'Public Anthropology in the 21st Century, with Some Examples from Norway.' *Český Lid* 103: 23–36. doi:http://dx.doi.org/10.21104/CL.2016.1.01.

Fäber, A. (2007) 'Exposing Expo: Exhibition Entrepreneurship and Experimental Reflexivity in Late Modernity.' In S. Macdonald and P. Basu (eds) *Exhibition Experiments*. London: Blackwell, 219–239.

Fabian, J. (2014 [1983]) *Time and the Other: How Anthropology Makes Its Object*. New York: Columbia University Press.

Färber, A. (2007). 'Exposing Expo: Exhibition Entrepreneurship and Experimental Reflexity in Late Modernity.' In S. Macdonald and P. Basu (eds) *Exhibition Experiments,* 219–239. Oxford: Blackwell.

Farman, A. (2012) 'Re-enchantment Cosmologies: Mastery and Obsolescence in an Intelligent Universe.' *Anthropological Quarterly* 85(4): 1069–1088.

Fassin, D. (ed.) (2017) *If Truth Be Told: The Politics of Public Ethnography*. Durham, NC: Duke University Press.

Favero, P. (2014) 'Learning to Look Beyond the Frame: Reflections on the Changing Meaning of Images in the Age of Digital Media Practices.' *Visual Studies* 29(2): 166–179.

Favero, P. (2018) *The Present Image: Visible Stories in a Digital Habitat.* London: Palgrave Macmillan.

Feenberg, A. (1999) *Questioning Technology.* London and New York: Routledge.

Feld, S. (1994) 'From Schizophrenia to Schismogenesis: on the Discourse and Commodification Practices of "World Music" and "World Beat".' In C. Keil and S. Feld, *Music Grooves*. Chicago: University of Chicago Press, 257–289.

Ferreira, V.S. (2011) 'Becoming a Heavily Tattooed Young Body: From a Bodily Experience to a Body Project.' *Youth and Society* 46(3): 303–337.

Flannery, T. (1994) *The Future Eaters: An Ecological History of the Australasian Lands and People*. Chatswood, N.S.W: Reed Books.

Flexner, J.L. (2016) 'Archaeology and Ethnographic Collections: Disentangling Provenance, Provenience, and Context in Vanuatu Assemblages.' *Museum Worlds* 4(1): 167–180.

Flowers, T.H., M.J. Smith, and J. Brunton (2019) 'Colouring of Pacific Barkcloths: Identification of the Brown, Red and Yellow Colourants Used in the Decoration of Historic Pacific Barkcloths.' *Heritage Science* 7(2).

Focilon, H. (1992) *The Life of Forms in Art*. New York: Zone Books.

Forde, C.D. (1963[1934]) *Habitat, Economy and Society: A Geographical Introduction to Ethnology*. New York: E.P. Dutton.

Forty, A. (1986) *Objects of Desire: Design and Society since 1750*. London: Thames and Hudson.

Foucault, M. (1970) *The Order of Things: An Archaeology of the Human Sciences*. London: Tavistock Publications/Routledge.

Foucault, M. (1977) *Discipline and Punish: The Birth of the Prison*. New York: Pantheon Books.

Foucault, M. (2007) *Security, Territory, Population: Lectures at the Collège de France, 1977–1978*. New York: Palgrave Macmillan.

Foucault, M. (2008) *The Birth of Biopolitics: Lectures at the Collège de France, 1978–1979*. New York: Palgrave Macmillan.

Foucault, M., and C. Gordon (1980) *Power/Knowledge: Selected Interviews and Other Writings, 1972–1977*. London: Harvester Wheatsheaf.

Fowles, S. (2016) 'The perfect subject (postcolonial object studies).' *Journal of Material Culture*, 21(1), 9–27.

Fox, N., and P. Alldred (2018) 'New Materialism.' In *SAGE Encyclopedia of Research Methods*, edited by P.A. Atkinson, S. Delamont, M.A. Hardy, and M. Williams. London: Sage.

Fox, R. and A. Gingrich (eds) (2002) 'Introduction.' *Anthropology, by Comparison*. London: Routledge.

Freis-Bryce, A. (2016) 'Jeremy Corbyn Is Right: Decriminalising the Sex Industry is the Way Forward.' *The Guardian*. 4 March 2016. Online at: www.theguardian.com/voluntary-sector-network/2016/mar/04/jeremy-corbyn-right-decriminalising-sex-industry-way-forward.

Friedman, J. and M. Rowlands (eds) (1977) *The Evolution of Social Systems*. London: Duckworth.

Frow, E.K. (2014) 'In Images We Trust? Representation and Objectivity in the Digital Age,' in C. Coopmans, M. Lynch, and J. Vertesi (eds) *Representation in Scientific Practice Revisited*. Cambridge, MA: MIT Press, 249–267.

Frow, J. (2001) 'A Pebble, A Camera, A Man Who Turns into a Telegraph Pole.' *Critical Inquiry* 28(1): 270–285.

Fuller, M. (2008) *Software Studies: A Lexicon*. Cambridge, MA: MIT Press.

Furness, W.H. (1910) *The Island of Stone Money, Yap of the Carolines*. JB Lippincott.

Gabrys, J. (2011) *Digital Rubbish: A Natural History of Electronics*. Ann Arbor, MI: University of Michigan Press.

Ganesh, K. (1990) 'Mother Who Is Not a Mother: In Search of the Great Indian Goddess.' *Economic and Political Weekly* 25(42/43): WS58–WS64.

Gárcia Canclini, N. (2014) *Art Beyond Itself: Anthropology for a Society without a Story Line*. Durham, NC: Duke University Press.

Geertz, C. (1973) *The Interpretation of Cultures*. New York: Basic Books.

Geertz, C. (2004[1966]) 'Religion as a Cultural System.' In M. Banton (ed.) *Anthropological Approaches to the Study of Religion*. London: Routledge, 1–46.

Geiger, R.S., and Ribes, D. (2011) 'Trace Ethnography: Following Coordination through Documentary Practices.' Proceedings of the 44th Annual Hawaii International Conference on Systems Sciences.

Geismar, H. (2011) '"Material Culture Studies" and Other Ways to Theorize Objects: A Primer to a Regional Debate.' *Comparative Studies in Society and History* 53(1): 210–218.

Geismar, H. (2012) 'Museum + Digital =?' In Miller, D and Host, H. (eds) *Digital Anthropology*. London and New York: Berg, 266–287.

Geismar, H. (2013) 'Defining the Digital.' *Museum Anthropology Review*, 7(1–2): 254–263.

Geismar, H. (2015) 'Post-Photographic Presences, or How to Wear a Digital Cloak.' *Photographies* 8(3): 305–321.

Geismar, H. (2018) *Museum Object Lessons for the Digital Age*. London: UCL Press.

Geismar, H. and W. Mohns (2011) 'Database Relations: Rethinking the Database in the Vanuatu Cultural Centre and National Museum.' *Journal of the Royal Anthropological Institute*, 17(S1): S133–S155.

Geismar, H., D. Miller, S. Küchler, M. Rowlands, and A. Drazin (2014) 'Material Culture Studies: Eine kurze Geschichte der Materiellen Kultur am UCL.' In *Handbook Materielle Kultur: Bedeutungen, Konzepte, Disziplinen*. Edited by Herausgegeben von Stefanie Samida, Manfred K.H. Eggert, and Hans Peter Hahn. Stuttgart: Verlag J.B. Metzler, 309–315.

Gell, A. (1985) 'How to Read a Map: Remarks on the Practical Logic of Navigation.' *Man, New Series, Royal Anthropological Institute* 20(2): 271–286.

Gell, A. (1988) 'Technology and Magic.' *Anthropology Today* 4(2): 6–9.

Gell, A. (1992) 'The Technology of Enchantment and the Enchantment of Technology.' In J. Coote and A. Shelton (eds) *Anthropology Art and Aesthetics*. Oxford: Clarendon Press, 40–63.

Gell, A. (1993) *The Anthropology of Time*. London: Athlone Press.

Gell, A. (1998) *Art and Agency: Towards an Anthropological Theory*. Clarendon Press, Oxford.

Gershon, I. (2010) *The Breakup 2.0: Disconnecting Over New Media*. Ithaca, NY: Cornell University Press.

Gershon, I. (2017) *Down and Out in the New Economy: How People Find (or Don't Find) Work Today*. Chicago: University of Chicago Press.

Ghosh, B. (2011) *Global Icons: Apertures to the Popular*. Durham, NC: Duke University Press.

Gibson, J. (2014[1979]) *The Ecological Approach to Visual Perception*. Houghton Mifflin Harcourt: Boston.

Gille, B. (1966[1964]) *The Renaissance Engineers*. London: Lund Humphries.

Gille, B. (1986[1978]) *The History of Techniques*. New York: Gordon and Breach Science Publishers.

Ginzburg, C. (1988) 'Morelli, Freud and Sherlock Holmes: Clues and the Scientific Method.' In U. Eco and T.A. Sebeok (eds) *The Sign of Three: Dupin, Holmes, Peirce*. Bloomington: Indiana University Press.

Ginzburg, C. (1989) 'From Aby Warburg to E.H. Gombrich.' In *Clues*. Baltimore: Johns Hopkins University Press.

Ginzburg, C. (2001) *Wooden Eyes: Nine Reflections on Distance*. New York: Columbia University Press.

Gitelman, L. (2006) *Always Already New: Media, History, and the Data of Culture*. Cambridge, MA and London: MIT Press.

Gitelman, L. (2013) *'Raw Data'* Is an Oxymoron. Cambridge, MA: MIT Press.

Goddard, E. (2015) *The Bionic Self: Neural Implants and Threats to Identity: Implications for Selfhood and Social Relations*. Doctoral thesis, University of Tasmania.

Godelier, M. (1972) *Rationality and Irrationality in Economics*. London: New Left Books.

Gómez Cruz, E. and A. Lehmuskallio (eds) (2016) *Digital Photography and Everyday Life: Empirical Studies on Material Visual Practices*. Abingdon: Routledge.

Goold, I. and H. Maslen (2014) 'Must the Surgeon Take the Pill? Negligence Duty in the Context of Cognitive Enhancement.' *The Modern Law Review* 77(1): 60–86.

Gorman, A. (2007) 'La Terre et l'Espace: Rockets, Prisons, Protests and Heritage in Australia and French Guiana.' *Archaeologies* 3(2), 153–168.Gorman, A. (2009a) 'Cultural Landscape of Space.' In A.G. Darrin, and B.L. O'Leary (eds) *Handbook of Space Engineering, Archaeology and Heritage*. Boca Raton: CRC Press, Taylor and Francis Group.

Gorman, A. (2009b) 'Heritage of Earth Orbit: Orbital Debris – Its Mitigation and Cultural Heritage.' In A.G. Darrin, and B.L. O'Leary (eds) *Handbook of Space Engineering, Archaeology and Heritage*. Boca Raton: CRC Press, Taylor and Francis Group.

Gorman, A. (2009c) 'The Gravity of Archaeology.' *Archaeologies* 5(2): 334–359.

Gorman, A. (2017) 'Pale Blue Dot: Everyday Material Culture on the International Space Station.' www.dayofarchaeology.com/pale-blue-dot-everyday-material-culture-on-the-international-space-station/ (Accessed August 28, 2020).

Gorman, A. and Walsh, J. (forthcoming) 'Archaeology in a Vacuum: Obstacles to and Solutions for Developing a Real Space Archaeology.' In H. Barnard and R. Boytner (eds) *Archaeology Out-of-the-Box*. Los Angeles: Cotsen Institute of Archaeology Press.

Graeber, D. (2001) *Toward an Anthropological Theory of Value: The False Coin of Our Own Dreams*. New York: Palgrave.

Graeber. D. (2013) *Debt: The First 5,000 Years*. Reprint ed. New York: Melville House Publishing.

Graeber, D. (2015) 'Radical Alterity Is Just Another Way of Saying "Reality": A Reply to Eduardo Viveiros de Castro.' *HAU: Journal of Ethnographic Theory* 5(2): 1–41.

Graham, M. (2014) *Anthropological Explorations in Queer Theory*. London: Routledge.

Grimshaw, A. (2001) *The Ethnographer's Eye: Ways of Seeing in Modern Anthropology*. Cambridge: Cambridge University Press.

Guidieri, R. and F. Pellizi (1981) 'Nineteen Tableaux on the Cult of the Dead in Malekula, Eastern Melanesia.' *RES*: 2: 1–97.

Haapio-Kirk, L. (2017) 'Why We Post.' *Teaching Anthropology* 7(1).

Hacking, I. (1990) *The Taming of Chance*. Cambridge: Cambridge University Press.

Haddon, A.C. (1890) 'The Ethnography of the Western Tribe of Torres Straits.' *Journal of the Anthropological Institute of Great Britain and Ireland* 19(1890): 297–444.

Hallowell, A.I. (1955) *Culture and Experience*. Philadelphia: University of Pennsylvania Press.

Hallowell, A.I. (2002[1960]) 'Ojibwa Ontology, Behavior, and World View.' In G. Harvey (ed.) *Readings in Indigenous Religions*. London: Continuum, 17–49.

Halpern, O. (2014) *Beautiful Data: A History of Vision and Reason Since 1945*. Durham, NC: Duke University Press.

Hampp, C. and S. Schwan (2014) 'Perception and Evaluation of Authentic Objects: Findings from a Visitor Study.' *Museum Management and Curatorship* 29(4): 349–367.

Handler, R. (1993) 'An Anthropological Definition of the Museum and its Purpose.' *Museum Anthropology* 17(1): 33–36.

Hannan, L., R. Duhs and H. Chatterjee. (2013) 'Object-Based Learning: A Powerful Pedagogy for Higher Education.' In A. Boddington, J. Boys, and C. Speight (eds) *Museums and Higher Education Working Together: Challenges and Opportunities*. Farnham: Ashgate Publishing.

Haraway, D. ([1985] 2000) 'A Manifesto for Cyborgs: Science, Technology, and Socialist Feminism in the 1980s.' In F. Hovenden, L. Janes, G. Kirkup, and K. Woodward (eds) *The Gendered Cyborg: A Reader*. London and New York: Routledge, 50–57.

Hardgrave Jr., R.L. (1973) 'Politics and the Film in Tamilnadu: The Stars and the DMK.' *Asian Survey* 13(3): 288–305.

Harman, G. (2015) 'Object-Oriented Ontology.' In M. Hauskeller, T.D. Philbeck, and C.D. Carbonell (eds) *The Palgrave Handbook of Posthumanism in Film and Television*. Palgrave Macmillan: London: 401–409.

Harman, G. (2018) *Object-Orientated Ontology: A New Theory of Everything*. London: Pelican Books.

Harris, C. (1999) *In the Image of Tibet: Tibetan Painting after 1959*. London: Reaktion Books.

Harris, C. (2001) 'Politics and Personhood of Tibetan Icons.' In C. Pinney, and N. Thomas (eds) *Beyond Aesthetics: Art and the Technologies of Enchantment*. Oxford and New York: Berg.

Harris, C. (2012) *The Museum on the Roof of the World*. Chicago: University of Chicago Press.

Harris, C. (2016) *Photography and Tibet*. London: Reaktion Books.

Harris, M. (1968) *The Rise of Anthropological Theory: A History of Theories of Culture*. Walnut Creek, CA: AltaMira Press.

Hart, K. (1982) 'On Commoditization.' In E. Goody (ed.) *From Craft to Industry: The Ethnography of Proto-Industrial Cloth Production*. Cambridge: Cambridge University Press, 38–49.

Hart, K. (2007) 'Marcel Mauss: In Pursuit of the Whole. A Review Essay.' *Comparative Studies in Society and History* 49(2): 473.

Hart, K., and C. Hann (2011) *Economic Anthropology*. Cambridge: Polity.

Harvey, D. (2005) *A Brief History of Neoliberalism*. Oxford: Oxford University Press.

Harvey, P. (2005) 'The Materiality of State Effects.' In Christian Krohn-Hansen and Knut G. Nustad (eds) *State Formation: Anthropological Perspectives*. London: Pluto Press.

Harvey, P. and H. Knox (2012) 'The Enchantments of Infrastructure.' *Mobilities* 7(4): 521–536.

Harvey, P., C.B. Jensen, and A. Morita (2017) *Infrastructures and Social Complexity: A Companion*. London: Routledge.

Harvey, P., E. Casella, G. Evans, H. Knox, C. McLean, E. Silva, N. Thoburn, and K. Woodward (2013) *Objects and Materials: A Routledge Companion*. London: Routledge.

Haskins, C. (2019a) 'Academics Confirm Major Predictive Policing Algorithm is Fundamentally Flawed.' *Vice*. February 14. Available at www.vice.com/en_us/article/xwbag4/academics-confirm-major-predictive-policing-algorithm-is-fundamentally-flawed. (Accessed 13 June 2019).

Haskins, C. (2019b) 'Dozens of Cities Have Secretly Experimented with Predictive Policing Software.' *Vice*. February 6. Available at www.vice.com/en_us/article/d3m7jq/dozens-of-cities-have-secretly-experimented-with-predictive-policing-software. (Accessed 13 June 2019).

Haynes, N. (2016) *Social Media in Northern Chile*. London: University College London Press.

Hayward, E. (2010) 'Fingeryeyes: Impressions of Cup Corals.' *Cultural Anthropology* 25(4): 577–599.

Hebdige, D. (1988) *Hiding in the Light: On Images and Things*. London and New York: Routledge.

Hegel, G. (2008) *Outlines of The Philosophy of Right*. Oxford: Oxford University Press.

Heidegger, M. (1993) *Basic Writings from 'Being and Time' (1927) to 'The Task of Thinking' (1964)*. London: Routledge.

Helbing, D., B.S. Frey, G. Gigerenzer, E. Hafen, M. Hagner, Y. Hofstetter, J. van den Hoven, R.V. Zicari, A. Zwitter (2017) 'Will Democracy Survive Big Data and Artificial Intelligence?' *Scientific American*, February 25th. Available online: www.scientificamerican.com/article/will-democracy-survive-big-data-and-artificial-intelligence/ (Accessed 13th June 2019).

Henare, A., M. Holbraad and S. Wastell (eds) (2006) *Thinking Through Things: Theorizing Artefacts Ethnographically*. London: Routledge.

Henning, M. (2007) 'The Pig in the Bath: New Materialisms and Cultural Studies'. *Radical Philosophy* 145: 11–19.

Herder, J.G. (2002[1778]) *Sculpture: Some Observations on Shape and Form from the Pygmalion's Creative Dream*, trans. J. Gaiger. Chicago: University of Chicago Press.

Herle, A. (2008) 'Relational Objects: Connecting People and Things through Pasifika Styles.' *International Journal of Cultural Property* 15(2): 159.

Herle, A. and S. Rouse (eds) (1998) *Cambridge and the Torres Strait: Centenary Essays on the 1898 Anthropological Expedition*. Cambridge: Cambridge University Press.

Hetherington, K. (ed.) (2019) *Infrastructure, Environment and Life in the Anthropocene*. Durham, NC: Duke University Press.

Heywood, A. (1900) *English Almanacs During the First Third of the Nineteenth Century*. Manchester: Privately Printed.

Hicks, D. (2010) 'The Material-Culture Turn: Event and Effect.' In D. Hicks and M.C. Beaudry (eds) *Oxford Handbook of Material Culture Studies*, 25–98. Oxford: Oxford University Press.

Hicks, D. and M.C. Beaudry (2010) 'Introduction: Material Culture Studies: A Reactionary View.' In D. Hicks and M.C. Beaudry (eds), *Oxford Handbook of Material Culture Studies*, 1–24. Oxford: Oxford University Press.

Hindess, B. and P. Hirst (1975) *Pre-Capitalist Modes of Production*. London: Routledge and Kegan Paul.

Hodder, I. (1989) 'This Is Not an Article about Material Culture as Text.' *Journal of Anthropological Archaeology* 8(3): 250–269.

Hodder, I. (2012) *Entangled: An Archaeology of the Relationships between Humans and Things*. Oxford: Wiley-Blackwell.

Holbraad, M. and M.A. Pedersen (2017) *The Ontological Turn: An Anthropological Exposition*. Cambridge: Cambridge University Press.

Horst, H.A. (2006) 'The Blessings and Burdens of Communication: The Cell Phone in Jamaican Transnational Social Fields.' *Global Networks: A Journal of Transnational Affairs* 6(2): 143–159.

Holt, J. and P. Vondereau (2014) '"Where the Internet Lives": Data Centers as Cloud Infrastructure.' In L. Parks and N. Starosielski (eds) *Signal Traffic*. Chicago: University of Illinois Press.

Horst, H. and D. Miller (2012) *Digital Anthropology*. London: Bloomsbury.

Horst, H., and D. Miller (2006) *The Cell Phone: An Anthropology of Communication*. Oxford: Berg.

Howatson, M.C. (1997) *Oxford Companion to Classical Literature*. Oxford: Oxford University Press.

Hughes, T. (1986) 'The Seamless Web: Technology, Science, Etcetera, Etcetera.' *Social Studies of Science* 16(2): 281–292.

Hui, Y. (2012) 'What Is a Digital Object?' *Metaphilosophy* 43: 4.

Hui, Y. (2013) 'Lecture: On Latour and Simondon's Mode of Existence.' Lecture given at a workshop on Latour at Denkerie, 28th of January 2013. http://digitalmilieu.net/?p=289 (Accessed 11 April 2019).

Hui, Y. (2015) *Simondon et la question de l'information. Cahiers Simondon, 6*. Paris: L'Harmattan.

Hui, Y. (2016) *On the Existence of Digital Objects*. Minneapolis: The University of Minnesota Press.

Humphrey, C. (2005) 'Ideology in Infrastructure: Architecture and Soviet Imagination.' *Journal of the Royal Anthropological Institute (N.S)* 11(1): 39–58.

Ihde, D. (1993) Postphenomenology. Evanston: Northwestern University Press.

Ihde, D. (1998) *Expanding Hermeneutics: Visualism in Science*. Evanston: Northwestern University Press.

Ihde, D. (2008) 'Introduction: Postphenomenological Research.' *Human Studies* 31(1): 1–9.

Ihde, D. (2010) *Heidegger's Technologies. Postphenomenological Perspectives*. New York: Fordham University Press.

Ingold, T. (2000[1988]) 'Tools, Minds and Machines: An Excursion in the Philosophy of Technology.' In T. Ingold, (ed.) *The Perception of the Environment: Essays on Livelihood, Dwelling and Skill*. London and New York: Routledge, 294–311.

Ingold, T. (2007) 'Materials against Materiality.' *Archaeological Dialogues* 14(1): 1–16.

Ingold, T. (2008) 'When ANT Meets SPIDER: Social Theory for Arthropods.' In C. Knapett and L. Malafouris (eds) *Material Agency. Towards a Non-Anthropocentric Approach*. New York: Springer, 209–215.

Ingold, T. (2010) 'The Textility of Making.' *Cambridge Journal of Economics* 34(1): 91–102.

Ingold, T. (2012) 'Toward an Ecology of Materials.' *Annual Review of Anthropology* 41: 427–442.

Ingold, T. (2013) *Making: Anthropology, Archaeology, Art and Architecture*. London: Routledge.

Ingold, T. (2015) *The Life of Lines*. London: Routledge.

Ingold, T. (2016) 'Tim Ingold on the Future of Academic Publishing.' Allegra Lab [Online]. Retrieved from http://allegralaboratory.net/interview-tim-ingold-on-the-future-of-academic-publishing/ (Accessed 9 September 2020).

Jacob, P. (1997) 'From Co-star to Deity: Popular Representations of Jayalaitha Jayaram.' In Vidya Dehejia (ed.) *Representing the Body: Gender Issues in Indian Art*. New Delhi: Kali for Women, 140–165.

Jacob, P. (2008) 'Tamil Cinema in the Public Sphere: the Evolving Art of Banner Advertisements in Chennai.' In S. Velayutham (ed.) *Tamil Cinema: The Cultural Politics of India's Other Film Industry*. New York: Routledge, 95–110.

Jacob, P. (2009) *Celluloid Deities: The Visual Culture of Cinema and Politics in South India*. Lanham: Lexington Books.

Jakubowski, D. (2016), *Expandable Space Aesthetics: International Space Station Social Media as an Expandable Democratic Surround*. Unpublished MA Thesis, Department of Anthropology, UCL.

Jasanoff, S. (2002) 'New Modernities: Reimagining Science, Technology and Development.' *Environmental Values* 11(3): 253–276.

Jay, M. (1988) 'Scopic Regimes of Modernity.' In H. Foster (ed.) *Vision and Visuality*. Seattle: Bay Press, 3–23.

Jeffrey, I. and A. Doron (2013) *The Great Indian Phone Book: How Cheap Mobile Phones Change Business, Politics and Daily Life*. London: Hurst.

Jenkins, D. (1994) 'Object Lessons and Ethnographic Displays: Museum Exhibitions and the Making of American Anthropology.' *Comparative Studies in Society and History* 36(2).

Jones, G. (1998) 'Contested Territories: Alfred Cort Haddon, Progressive Evolutionism and Ireland.' *History of European Ideas* 24(3): 195–211.

Jones, A.M. and B. Alberti (2016) 'Archaeology after Interpretation.' In B. Alberti, A. Jones and J. Pollard (eds) *Archaeology after Interpretation: Returning Materials to Archaeological Theory*. London: Routledge.

Joyce, R.A. and J. Pollard (2010) 'Archaeological Assemblages and Practices of Deposition.' In D. Hicks and M. Beaudry (eds) *The Oxford Handbook of Material Culture Studies*. Oxford: Oxford University Press, 291–309.

Kapferer, B. (1988) *Legends of People, Myths of State*. Washington, DC: Smithsonian Institution Press.

Kaplan, L. 'Inuit or Eskimo: Which Name to Use?' University of Alaska, Fairbanks: Alaska Native Language Center. Online Available at: www.uaf.edu/anlc/resources/inuit_or_eskimo.php (Accessed 27 August 2020).

Keane, W. (1997) *Signs of Recognition: Powers and Hazards of Representation in an Indonesian Society*. Berkeley, Los Angeles, and London: University of California Press.

Keane, W. (2003) 'Semiotics and the Social Analysis of Material Things. Language and Communication 23(3–4), 409–425.

Keane, W. (2005a) 'The Hazards of New Clothes: What Signs Make Possible.' In S. Küchler and G. Were (eds) *The Art of Clothing: A Pacific Experience*. London: Taylor and Francis.

Keane, W. (2005b) 'Signs Are Not the Garb of Meaning: On the Social Analysis of Material Things.' In D. Miller (ed.) Materiality . Durham, NC: Duke University Press.

Keane, W. (2006) 'The Hazards of New Clothes: What Signs Make Possible.' In S. Küchler and G. Were (eds) *The Art of Clothing: A Pacific Experiment*. London: UCL Press, 1–16.

Keane, W. (2013) On Spirit Writing: Materialities of Language and the Religious Work of Transduction. *Journal of the American Institute of Planners* 19(1): 1–17.

Kelty, C. (2008) *Two Bits: The Cultural Significance of Free Software*. Durham, NC: Duke University Press.

Kester, G. (2011) *The One and the Many: Contemporary Collaborative Art in a Global Context*. Durham, NC: Duke University.

Khan-Magomedov, S.O. (2015) *Georgii Krutikov: The Flying City and Beyond*. Barcelona: Tenov Books.

Khomami, N. (2016) 'Decriminalisation of Sex Workers in England and Wales Backed by MPs.' *The Guardian*. 1 July 2016. Online at: www.theguardian.com/society/2016/jul/01/decriminalisation-of-sex-workers-in-england-and-wales-backed-by-mps.

Kirshenblatt-Gimblett, B. (1998) *Destination Culture: Tourism, Museums and Heritage*. Berkeley: University of California.

Kintz, N. and L. Palinkas (2016) 'Communication Delays Impact Behavior and Performance Aboard the International Space Station.' *Aerospace Medicine and Human Performance* 87(11): 940–946.

Knappett, C. (2005) *Thinking Through Material Culture: An Interdisciplinary Perspective*. Philadelphia: University of Pennsylvania Press.

Knauft, B. (2010) *South Coast New Guinea Cultures*. Cambridge: Cambridge University Press.

Knox, H. (2014) 'Footprints in the City: Models, Materiality and the Cultural Politics of Climate Change.' *Anthropological Quarterly* 87(2):405–429.

Knox, H. (2017) 'Affective Infrastructures and the Political Imagination.' *Public Culture* 29(2): 363–384.

Knox, H. and D. Nafus (eds) (2018) *Ethnography for a Data-Saturated World. Materialising the Digital*. Manchester: Manchester University Press.

Knox, H. and A. Walford (2016) 'Is There an Ontology to the Digital?' *Theorizing the Contemporary, Cultural Anthropology Website* (blog). 24 March 2016. https://culanth.org/fieldsights/818-is-there-an-ontology-to-the-digital (Accessed 9 September 2020).

Koch, G. (2000) *Siegfried Kracauer: An Introduction*, trans. J. Gaines. Princeton, NJ: Princeton University Press.

Kohn, E. (2013) *How Forests Think: Toward an Anthropology beyond the Human*. Berkeley: University of California Press.

Kopytoff, I. (1986) 'The Cultural Biography of Things: Commoditization as a Process.' In A. Appadurai (ed.) *The Social Life of Things: Commodities in Cultural Perspective*. Cambridge: Cambridge University Press, 64–91.

Kosut, M. (2000) 'Tattoo Narratives: The Intersection of the Body: Self-Identity and Society.' *Visual Studies* 15(1): 79–100.

Kovtun, E.F. and Douglas, C.D. (1981) 'Kazimir Malevich.' *Art Journal* 41(3): 234–241.

Kracauer, S. (1938) *Orpheus in Paris: Offenbach and the Paris of His Time*, trans. G. David and E. Mosbacher. New York: Alfred Knopf.

Kreps, C. (2015) 'University Museums as Laboratories for Experiential Learning and Engaged Practice.' *Museum Anthropology* 38(2): 96–111.

Krmpotich, C. (2015) 'Teaching Collections Management Anthropologically.' *Museum Anthropology* 38(2): 112–122.

Kubler, G. (1962) *The Shape of Time: Remarks on the History of Things.* New Haven: Yale University Press.

Küchler, S. (2001) 'Why Knot?' Towards a Theory of Art and Mathematics.' In C. Pinney and N. Thomas, *Beyond Aesthetics: Art and the Technologies of Enchantment*. Oxford and New York: Berg, 51–71.

Küchler, S. (2002) *Malanggan: Art, Memory and Sacrifice.* Oxford: Berg.

Küchler, S. (2003) 'Imaging the Body Politic: The Knot in Pacific Imagination.' *L'Homme* 165(Jan–Mar., 2003): 205–222.
Küchler, S. (2005) 'Materiality and Cognition: The Changing Faces of Things.' In D. Miller (ed.) *Materiality*. Durham, NC: Duke University Press.
Küchler, S. (2007) 'The String in Art and Science: Rediscovering the Material Mind.' *Textile: The Journal of Cloth and Culture* 5(2): 124–140.
Küchler, S. (2008) 'Technological Materiality: Beyond the Dualist Paradigm.' *Theory Culture and Society* 25(1): 101–120.
Küchler, S. (2013) 'Threads of Thought: Reflections on Art and Agency.' In L. Chua and M. Elliott (eds) *Distributed Objects: Meaning and Mattering after Alfred Gell*. London: Berghahn.
Küchler, S. (2014) 'Virtual Archives: An Ethnographic Perspective on their Creation and Transmission.' In A. Schnapp (ed.) *World Antiquarianism*. Los Angeles: Getty Research Institute, 106–120.
Küchler, S. (2017) 'Differential Geometry, the Informational Surface and Oceanic Art: The Role of Pattern in Knowledge Economies.' *Theory Culture and Society*. 34(7–8): 75–97.
Küchler, S. and T. Carroll (2021a) 'Art (in the Twentieth Century and Beyond).' In L. Wilkie and J. Chenoweth (eds) *A Cultural History of Objects: Modern Period, 1900 to Present*. Oxford and London: Bloomsbury.
Küchler, S. and T. Carroll (2021b) *A Return to the Object: Alfred Gell, Art and Social Theory*. London: Routledge.
Küchler, S. and A. Eimke (2009) *Tivaivai: The Social Fabric of the Cook Islands*. London: British Museum Press.
Lakoff, G. and M. Johnson (2008[1980]) *Metaphors We Live By*. Chicago and London: University of Chicago Press.
Lakshmi, C.S. (1990) 'Mother, Mother-Community and Mother-Politics in Tamil Nadu,' *Economic and Political Weekly* 25(42/43): WS72–WS83.
Lakshmi, C.S. (1997) 'Bodies Called Women: Some Thoughts on Gender, Ethnicity and Nation.' *Economic and Political Weekly* 32(46): 2953–2962.
Lamb, J. (2011) *The Things Things Say*. Princeton, NJ: Princeton University Press.
Larkin, B. (2013) 'The Politics and Poetics of Infrastructure.' *Annual Review of Anthropology* 42(1): 327–343.
Larsen, J. (2008) 'Practices and Flows of Digital Photography: An Ethnographic Framework.' *Mobilities* 3(1): 141–160.
Lasch, C. (1991) *The Culture of Narcissism: American Life in an Age of Diminishing Expectations*. New York: Norton.
Lash, S. and J. Urry (1994) *Economies of Signs and Spaces*. London: Sage.
Latour, B. (1990) 'Drawing Things Together.' In M. Lynch and S. Woolgar (eds) *Representation in Scientific Practice*. Cambridge, MA: MIT Press, 19–68.
Latour, B. (1991) 'Technology Is Society Made Durable.' In J. Law (ed.) *A Sociology of Monsters: Essays on Power, Technology and Domination*. London: Routledge.
Latour, B. (1993[1991]) *We Have Never Been Modern*, trans. C. Porter. Cambridge, MI: Harvard University Press.
Latour, B. (1993) *An Inquiry into Modes of Existence*. Cambridge, MA and London: Harvard University Press.
Latour, B. (1994) 'On Technical Mediation.' *Common Knowledge* 3(2): 29–64.
Latour, B. (1999) 'On Recalling ANT' in J. Law and J. Hassard (eds) *Actor Network Theory and After*. Oxford: Blackwell.

Latour, B. (2000) 'The Berlin Key or How to Do Words with Things.' In P. Graves-Brown (ed.) *Matter, Materiality and Modern Culture*. London: Routledge, 10–21.

Latour, B. (2005) *Reassembling the Social: An Introduction to Actor-Network-Theory*. Oxford: Oxford University Press.

Latour, B. (2014) 'Technical Does Not Mean Material. Comment on Lemonnier, Pierre. 2012. *Mundane Objects: Materiality and Non-verbal Communication*. Walnut Creek, CA: Left Coast Press.' *Hau: Journal of Ethnographic Theory* 4(1): 507–510.

Latour, B. (2015) *An Inquiry into the Modes of Existence*. Cambridge and London: Harvard University Press.

Latour, B. (as Jim Johnson) (1988) 'Mixing Humans and Nonhumans Together: The Sociology of a Door-Closer.' *Social Problems* 35(3): 298–310.

Latour, B. and S. Woolgar (1979) *Laboratory Life: The Social Construction of Scientific Fact*. London: Sage.

Law, J. (1986) *Power, Action, and Belief: A New Sociology of Knowledge?* London: Routledge and Kegan Paul.

Law, J. (1999) 'After ANT: Complexity, Naming and Topology.' In J. Law and J. Hassard (eds), *Actor Network Theory and After*. Oxford: Blackwell.

Law, J. and J. Hassard (1999) *Actor Network Theory and After*. Oxford: Blackwell.

Law, L. (2000) *Sex Work in Southeast Asia*. London: Routledge.

Layard, J.W. (1942) *Stone Men of Malekula*. Cambridge Expedition to the Torres Straits, 1908. London: Chatto & Windus.

Leach, E. (1954) *Political Systems of Highland Burma: A Study of Kachin Social Structure*. London: London School of Economics and Political Sciences.

Leach, E. (1961) *Rethinking Anthropology*. London: Routledge Kegan and Paul.

Leach E. and J. Leach (eds.) (1983) *The Kula: New Perspectives on Massim Exchange*. Cambridge: Cambridge University Press.

Leinhardt, G. and K. Crowley. (2002) 'Objects of Learning, Objects of Talk: Changing Minds in Museums.' In S.G. Paris (ed.) *Perspectives on Object-Centered Learning in Museums*. Mahwah, NJ: Lawrence Erlbaum Associates, 301–324.

Lemonnier, P. (1986) 'The Study of Material Culture Today: Toward an Anthropology of Technical Systems.' *Journal of Anthropological Archaeology* 5(2): 147–186.

Lemonnier, P. (1992). *Elements for an Anthropology of Technology*. Ann Arbor, MI: University of Michigan Press.

Lemonnier, P. (2012) *Mundane Objects: Materiality and Non-Verbal Communication*. Walnut Creek, CA: Left Coast.

Leonardi, P.M. (2010) 'Digital Materiality? How Artefacts without Matter, Matter.' *First Monday* 15(6–7).

Leroi-Gourhan, A. (1971[1943]) *Evolution et Techniques I: L'Homme et la Matière*. Paris: Albin Michel.

Leroi-Gourhan, A. (1973[1945]) *Evolution et Techniques II: Milieu et techniques*. Paris: Albin Michel.

Leroi-Gourhan, A. (1993[1964]) *Gesture and Speech*. Cambridge, MA and London: MIT Press.

Lessig, L. (2006) *Code: Version 2.0*. 2nd ed. New York: Basic Books.

Levi-Strauss, C. (1966) *The Savage Mind*. Chicago: University of Chicago Press.

Levi-Strauss, C. (1969) *The Elementary Structures of Kinship* (trans. J.H. Belle, J.R. von Sturmer, and R. Needham). Boston: Beacon Press.

Levi-Strauss, C. (1982) *The Way of the Masks*. Seattle: University of Washington Press.

Lévy-Bruhl, L. (1966 [1928]) *The 'Soul' of the Primitive.* Chicago: Henry Regnery Company.

Lewis, J. (2014) 'Making the Invisible Visible: Designing Technology for Nonliterate Hunter-gatherers.' In J. Leach and L. Wilson (eds) *Subversion, Conversion, Development: Public Involvements with Information and Communication Technologies*. Cambridge, MA: MIT Press Infrastructures Series, 127–152.

Lindstrøm, C.T. (2015) 'Agency in "itself": A Discussion of Inanimate, Animal, and Human Agency.' *Archeological Dialogues*, 22(2): 207–238.

Lindstrøm, C.T. (2017) 'Agency: A Response to Sørenson and Ribeiro.' *Archaeological Dialogues,* 24(1): 109–116.

Liu, A. (2012) 'Where Is Cultural Criticism in the Digital Humanities?' 2012. http://dhdebates.gc.cuny.edu/debates/text/20. (Accessed September 9, 2020).

Livingstone, S. and K. Ólafsson (2018) *When Do Parents Think Their Child Is Ready to Use the Internet Independently? Parenting for a Digital Future: Survey Report 2.* Department of Media and Communications, The London School of Economics and Political Science: London.

Lloyd, K. (2016) *Objects of Desire.* London and New York: Routledge. Time Out. 15 July 2016. Online at: www.timeout.com/london/things-to-do/objects-of-desire-

Losche, D. (2006) 'The Fate of the Senses in Ethnographic Modernity: The Margaret Mead Hall of Pacific Peoples at the American Museum of Natural History.' In E. Edwards, C. Gosden, and R.B. Phillips (eds) *Sensible Objects: Colonialism, Museums and Material Culture.* London: Bloomsbury Academic.

Lupton, D. (2015) *Digital Sociology*. London: Routledge.

Lupton, D. (2016) *The Quantified Self.* Cambridge: Polity Press.

Lupton, D. (2018) How Do Data Come to Matter? Living and Becoming with Personal Data. *Big Data and Society* 5(3): 1–11.

Lury, C., and N. Wakeford (eds) (2012) *Inventive Methods: The Happening of the Social. Culture, Economy and the Social.* London: Routledge.

Lynch, M. and S.Y. Edgerton (1987) 'Aesthetics and Digital Image Processing: Representational Craft in Contemporary Astronomy.' *The Sociological Review* 35(1): 184–220.

Lynch, M. and S. Woolgar (eds) (1990) *Representation in Scientific Practice*. Cambridge, MA: MIT Press.

Macdonald, S. and P. Basu (2007) *Exhibition Experiments*. Oxford: Blackwell.

MacDougall, D. (1992) 'Photo Hierarchicus: Signs and Mirrors in Indian Photography.' *Visual Anthropology* 5(2).

MacDougall, D. (2005) *The Corporeal Image: Film, Ethnography and the Senses*. Princeton, NJ: Princeton University Press.

Made in Space (2019) Fiber Optics. Made in Space, website. Available at: https://madeinspace.us/mis-fiber (Accessed 26 August 2020).

Madianou, M. and D. Miller (2012) *Migration and New Media: Transnational Families and Polymedia*. London: Routledge.

Maher, N. (2017) *Apollo in the Age of Aquarius*. Cambridge, MA: Harvard University Press.

Mailhot, J. (1978) L'étymologie de « Esquimau » revue et corrigée. *Études/Inuit/Studies* 2(2): 59–69.

Malaby, T. (2009) *Making Virtual Worlds: Linden Lab and Second Life*. 1st ed. Ithaca, NY: Cornell University Press.

Malafouris, L. (2013) *How Things Shape the Mind*. Cambridge, MA: MIT Press.

Malevich, K. (1920) Suprematizm: 34 Risunka, UNOVIS: Vitebsk (German facsimile edition).

Malinowski, B. (1922) *Argonauts of the Western Pacific: An Account of Native Enterprise and Adventure in the Archipelagoes of Melanesian New Guinea*. London: Routledge.

Malinowski, B. (1988) *Malinowski Collected Works, Vol. 1: Malinowski Among the Magi*. London: Routledge.

Manapat, R. (2011) 'Mathematical Ideas in Early Philippine Society: Posthumous Essay.' *Philippine Studies* 59(3): 291–336.

Mandle, C. (2015) 'Meryl Streep, Kate Winslet and Emma Thompson urge Amnesty International not to decriminalise sex trade.' Independent. 28 July 2015. Online at: www.independent.co.uk/news/people/meryl-streep-kate-winslet-and-emma-thompson-urge-amnesty-international-not-to-decriminalise-sex-10421863.html.

Maniglier, P. (2006) *La vie enigmatique des signes: Saussure et du structuralisme. la naissance*. Paris: Editions Leo Scheer.

Maniglier, P. (2013) 'The Order of Things.' In C. Falzon, T. O'Leary, and J. Sawicki (eds) *A Companion to Foucault*. London and New York: Blackwell Publishing: 103–121.

Manovich, L. (1999) 'Database as Symbolic Form.' *Convergence* 5(2): 80–99.

Manovich L. (2013) *Software Takes Command*. London and New York: Bloomsbury.

Marcus, G. (1995) 'Ethnography in/of the World System: The Emergence of Multi-sited Ethnography.' *Annual Review of Anthropology* 24: 95–117.

Marcus, G.E. (1998) *Ethnography Through Thick and Thin*. Princeton, NJ: Princeton University Press.

Marres, N. (2015) *Material Participation: Technology, the Environment and Everyday Publics*. Basingstoke: Palgrave Macmillan.

Marres, N. (2017) *Digital Sociology: The Reinvention of Social Research*. Cambridge: Polity Press.

Marres, N. and C. Gerlitz (2018) 'Social Media as Experiments in Sociality.' *Inventing the Social*, 253–283. London: Mattering Press.

Marres, N., M. Guggenheim, and A. Wilkie (eds) (2018) *Inventing the Social*. Manchester: Mattering Press.

Martin, E. (1992) 'The End of the Body.' *American Ethnologist* 19(1): 121–140.

Martin, E. (1998) 'The Fetus as Intruder: Mother's Bodies and Medical Metaphors.' In *Cyborg Babies: From Techno-Sex to Techno-Totts*. New York: Routledge, 125–142.

Martin, E. (2006) 'The Pharmaceutical Person.' *BioSocieties* 1(3): 273–287.

Martin, F. (2013) *Money: An Unauthorized Biography*. London: Vintage Books, Random House Publishing.

Martjin, C.A. (1964) 'Eskimo Carving in Historical Perspective.' *Anthropos* 59(3/4): 546–596.

Marx, L. (2010[1997]) '"Technology": The Emergence of a Hazardous Concept.' *Technology and Culture* 51(3): 561–677.

Massey, D. (2005) *For Space*. London, Thousand Oaks, and New Delhi: Sage.

Massumi, B. (1995) 'The Autonomy of Affect.' Cultural Critique 31: 83–109.

Mauss, M. (1954) *The Gift*, trans. I. Cunnison. London: Cohen and West.

Mauss, M. (1973 [1935]) 'Techniques of the Body.' *Economy and Society* 2(1): 70–88.

Mauss, M. (2002[1925, 1967]) *The Gift: Forms and Functions of Exchange in Archaic Societies*. London: Routledge.

Mauss, M. (2003[1909]) *On Prayer*. New York and Oxford: Durkheim Press/Berghahn Books.

Mauss, M. (2006) *Techniques, Technology and Civilization*, N. Schlanger (ed.). New York: Durkheim Press; Oxford: Berghahn Books.
Mazzarella, W. (2009) 'Affect: What Is It Good For?' In S. Dube (ed.) *Enchantments of Modernity: Empire, Nation, Globalization*. New York: Routledge, 291–309.
McAuley, A., B. Stewart, G. Siemens, and D. Cormier (2010) 'The MOOC Model for Digital Practice.' MASSIVE OPEN ONLINE COURSES Digital ways of knowing and learning [Online]. Retrieved from https://pdfs.semanticscholar.org/ad53/b9655587771edcf4ae028d4490a218d87ff2.pdf (Accessed 29 August 2020).
McDonald, T. (2016) *Social Media in Rural China*. London: UCL Press.
McLuhan, M. (1995[1964]) *Understanding Media: The Extension of Man*. Cambridge MA: MIT Press.
McQuillan, D. (2016) 'Algorithmic Paranoia and the Convivial Alternative.' *Big Data & Society*, July–December 1–12.
McQuire, S. (2013) 'Photography's Afterlife: Documentary Images and the Operational Archive.' *Journal of Material Culture* 18(3): 223–241.
Mears, H. and W. Modest (2013) 'Museums, African Collections and Social Justice.' In R. Sandell and E. Nightingale (eds) *Museums, Equality, and Social Justice*. Abingdon, Oxon: Routledge.
Merleau-Ponty, M. (1962) *Phenomenology of Perception*. London: Routledge.
Merry, S.E., and J.M. Conley (2011) 'Measuring the World: Indicators, Human Rights, and Global Governance.' *Current Anthropology* 52(Supp 3): S83–S96.
Mertz, E. (2007) Semiotic Anthropology. *Annual Review of Anthropology* 36: 337–353.
Messeri, L.R. (2011) *Placing Outer Space: An Earthly Ethnography of Other Worlds*, Unpublished PhD Dissertation, MIT.
Messeri, L. (2016) *Placing Outer Space: An Earthly Ethnography of Other Worlds (Experimental Futures)*. Durham, NC: Duke University Press.
Messeri, L. (2017) 'Resonant Worlds: Cultivating Proximal Encounters in Planetary Science.' *American Ethnologist* 44(1): 131–142.
Meyer, B. (ed.) (2009) *Aesthetic Formations: Media, Religion and the Senses*. London: Palgrave Macmillan.
Meyer, B. (2011) 'Mediation and Immediacy: Sensational Forms, Semiotic Ideologies and the Question of the Medium.' *Social Anthropology* 19(1): 23–39.
Meyer, B. (2012) *Mediation and the Genesis of Presence: Toward a Material Approach to Religion*. Utrecht: Universiteit Utrecht. Faculteit Geesteswetenschappen.
Milbrath, S. (2009) 'Archaeoastronomy, Ethnoastronomy, and Cultural Astronomy.' In Darrin, A.G, and B.L. O'Leary (eds) *Handbook of Space Engineering, Archaeology and Heritage.* Boca Raton: CRC Press, Taylor and Francis Group.
Miller, D. (1987) *Material Culture and Mass Consumption*. Oxford: Blackwell.
Miller, D. (1997) Capitalism: An Ethnographic Approach. Oxford: Berg.
Miller, D. (ed.) (1998) *Material Cultures: Why Some Things Matter*. London: UCL Press.
Miller, D. (2005a) 'What Is Best "Value"? Bureaucracy, Virtualism and Local Governance.' In P. du Gay (ed.) *The Values of Bureaucracy*. Oxford: Oxford University Press, 233–256.
Miller, D. (ed.) (2005b) *Materiality*. Durham, NC and London: Duke University Press.
Miller, D. (2008) *The Comfort of Things*. Malden, MA: Polity.
Miller, D. (ed.) (2009) *Anthropology and the Individual: A Material Culture Perspective*. Oxford: Berg.
Miller, D. (2010) *Stuff*. Cambridge: Polity Books.
Miller, D. (2011) *Tales from Facebook*. Malden, MA: Polity.

Miller, D. (2015) 'Photography in the Age of Snapchat.' *RAI Series: Anthropology and Photography*, 1: 1–17.

Miller, D. (2017) *The Comfort of People.* Cambridge: Polity.

Miller, D. and H. Horst (eds) (2012) *Digital Anthropology*. London and New York: Beg.

Miller, D. and H. Horst (2012) 'The Digital and the Human.' In H. Horst and D. Miller (eds), *Digital Anthropology*. London: Bloomsbury.

Miller, D. and H. Horst (Forthcoming, 2020) 'Six Principles for a Digital Anthropology.' In H. Geismar and H. Knox (eds) *Digital Anthropology 2.0*, 2nd. ed. London: Bloomsbury.

Miller, D. and J. Sinanan (2013) *Webcam*. Polity Press.

Miller, D. and D. Slater (2000) *The Internet: An Ethnographic Approach*. London: Bloomsbury.

Miller, D. and C. Tilley (1996) 'Editorial'. Journal of Material Culture 1(1): 5–14.

Miller, D. and S. Woodward (2012) *Blue Jeans: The Art of the Ordinary*. Berkeley, Los Angeles and London: University of California Press.

Miller, D., E. Costa, N. Haynes, T. McDonald, R. Nicolescu, J. Sinanan, J. Spyer, S. Venkatraman, and X. Wang. (2016) *How the World Changed Social Media*. London: UCL Press.

Miller, J. (ed.) (1991) *Soviet Space, an Exhibition in the City of Fort Worth*. Fort Worth: The Fort Worth Museum of Sciences and History Association.

Mintz, S. (1985) *Sweetness and Power*. New York: Viking.Mitchell, T. (2002) *Rule of Experts: Egypt, Techno-politics, Modernity*. Berkeley: University of California Press.

Mitchell, W.J.T. (1996) 'What Do Pictures "Really" Want?' *October* 77: 71–82.

Mitchell, W.J.T. (2005) *What do Pictures Want: The Lives and Loves of Images*. Chicago and London: University of Chicago Press.

Modest, W. (2013) 'Co-curating with Teenagers at the Horniman Museum.' In V. Golding and W. Modest (eds) *Museums and Communities: Curators, Collections, and Collaboration*. London and Oxford. Bloomsbury.

Mosko, M.S. (2013) 'Omarakana Revisited, or "Do Dual Organizations Exist?" in the Trobriands.' *Journal of the Royal Anthropological Institute* 19: 482–509.

Monterde, A. and J. Postill (2014) 'Mobile Ensembles: The Uses of Mobile Phones for Social Protest by Spain's Indignados.' In G. Goggin and L. Hjorth (eds), *Routledge Companion to Mobile Media.* London: Routledge, 429–438.

Moore, A. (2016) 'Anthropocene Anthropology: Reconceptualizing Contemporary Global Change.' *Journal of the Royal Anthropological Institute* 22(1): 27–46.

Morozov, E. (2012) *The Net Delusion*. London: Penguin.

Mumford, L. (1970[1938]) *The Culture of Cities*. London: Harcourt Brace Jovanovich.

Munn, N. (1966) 'Visual Categories: An Approach to the Study of Representational Systems.' *American Anthropologist* 68(4): 936–950.

Munn, N. (1986) *The Fame of Gawa*. Chicago: Chicago University Press.

Myers, F. (2002) *Painting Culture. The Making of an Aboriginal High Art*. Durham, NC: Duke University Press.

Nadjmabadi, S.R. (ed.) (2009) *Conceptualizing Iranian Anthropology: Past and Present Perspectives*. New York and Oxford: Berghahn.

Nafus, D. (2016) 'The Domestication of Data: Why Embracing Digital Data Means Embracing Bigger Questions.' *Ethnographic Praxis in Industry Conference Proceedings*. (1): 384–399.

Nafus, D. (2016) *Quantified: Biosensing Technologies in Everyday Life*. Cambridge, MA: MIT Press.

Nardi, B. (2009) *My Life as a Night Elf Priest: An Anthropological Account of World of Warcraft*. Ann Arbor, MI: University of Michigan Press.

Nardi, B. (2015) 'Virtuality.' *Annual Review of Anthropology* 44(1): 15–31.

Nichols, B. (2000) 'The Work of Culture in the Age of Cybernetic Systems.' In J.T. Caldwell (ed.) *Electronic Media and Technoculture*. New Brunswick, NJ: Rutgers University Press.

Nietzsche, F. (1927) *The Will to Power*. In *The Philosophy of Nietzsche*, trans. A.M. Ludovici. Random House: New York, paragraph 960.

Nietzsche, F. (1974) *The Gay Science*, trans. W. Kaufmann.

Nixon, D. (2017) *International Space Station: Architecture Beyond Earth*. London: Circa Press.

Nussbaum, M. (2003) 'Capabilities as Fundamental Entitlements: Sen and Social Justice.' *Feminist Economics* 9(2–3): 33–59.

Lévi-Strauss, C. (1963). *Do Dual Organizations Exist? In Structural Anthropology*. New York: Basic Books, 128–160.

O'Connell, S. (1999) *The Popular Print in England*. London: British Museum.

O'Neill, P. (2012) *The Culture of Curating and the Curating of Culture(s)*. Cambridge, MA: MIT Press.

O'Neill, P. and M. Wilson (2014) 'An Opening to Curatorial Enquiry: Introduction to Curating and Research.' In *Curating Research*, London: Open Editions/de Appel.

O'Reilly, K. (2012) *Ethnographic Methods*. 2nd ed. London: Routledge.

O'Sullivan, S. (2018) 'Fictioning Five Heads (On the Art-Anthropology Hybrid).' In H. Spriggs (ed.) *Five Heads (Tavan Tolgoi): Art, Anthropology and Mongol Futurism*. Berlin, Sternberg Press.

Oksanen, A. and J. Turtiainen (2005) 'A Life Told in Ink: Tattoo Narratives and the Problem of the Self in Late Modern Society.' *Auto/Biography* 13(2): 111–130.

Oliver, K. (2015) *Earth and World: Philosophy after the Apollo Missions*. New York: Columbia University Press.

Olson, V. (2010) *American Extreme: An Ethnography of Astronautical Visions and Ecology*. Unpublished PhD Thesis. Houston: Rice University.

Olson, V. (2018) *Into the Extreme: U.S. Environmental Systems and Politics Beyond Earth*. Minneapolis: University of Minnesota Press.

Olson, V. and L. Messeri (2015) 'Beyond the Anthropocene: Un-earthing an Epoch.' *Environment and Society* 6(1): 28–47.

Ott, K., D. Serlin, and S. Mihm (eds) (2002) *Artificial Parts, Practical Lives: Modern Histories of Prosthetics*. New York: NYU Press, Chapter 1.

Pandian, M.S.S. (1992) *The Image Trap: M.G. Ramachandran in Film and Politics*. New Delhi: Sage Publications.

Pandian, M.S.S. and V. Geetha. (1989) 'Jayalalitha: "Sworn Heir."' *Economic and Political Weekly* 24(11): 551.

Parikka, J. (2012a) 'New Materialism as Media Theory: Medianatures and Dirty Matter.' *Communication and Critical/Cultural Studies* 9(1): 95–100.

Parikka, J. (2012b) *What Is Media Archaeology*? Cambridge: Polity.

Parkhurst, A. (2012) 'Becoming Cyborgian: Procrastinating the Singularity.' *The New Bioethics* 18(1): 68–80.

Parkhurst, A. (2016) 'The Human Figure and Urban Ground: Cyborgs and the City.' *The New Bioethics* 22(2): 91–103.

Parkhurt, A. and D. Jeevendrampillai (2020) 'Toward an Anthropology of Gravity: Emotion and Embodiment in Microgravity Environments.' *Emotion, Space and Society* 35, Article 100680. https://doi.org/10.1016/j.emospa.2020.100680

Pedersen, M. (2007) 'Talismans of Thought: Shamanist Ontologies and Extended Cognition in Northern Mongolia.' In A. Henare et al. *Thinking Through Things: Theorising Artefacts Ethnographically*. London: Routledge, 141–165.

Pedersen, M.A. and M. Carey (2017) 'Introduction: Infrastructures of Certainty and Doubt.' *Cambridge Anthropology* 35(2): 18–29.

Pellegram, A. (1998) 'The Message in Paper.' In D. Miller (ed.) *Material Cultures: Why Some Things Matter*. London: UCL Press.

Pentland, A. (2012) 'Reinventing Society in the Wake of Big Data.' Edge. August 30. Available at: www.edge.org/conversation/alex_sandy_pentland-reinventing-society-in-the-wake-of-big-data (Accessed June 13 2019).

Pérez-Gómez, A. (2016) *Attunement: Architectural Meaning After the Crisis of Modern Science*. Cambridge, MA: MIT Press.

Pfaffenberger, B. (1992) 'Technological Drama.' *Science, Technology and Human Values*, 17(3): 282–312.

Phillips, R. (2005) 'Dress and Address: First Nations Self-Fashioning and the 1860 Royal Tour of Canada.' In S. Küchler and G. Were (eds) *The Art of Clothing: A Pacific Experience*. London: UCL Press.

Pickles, A. (2009) 'Part and Whole Numbers: an "Enumerative" Reinterpretation of the Cambridge Anthropological Expedition to Torres Straits and its Subjects.' *Oceania* 79: 293–315.

Pietz, W. (1985) 'The Problem of the Fetish, I.' *Res* 9: Spring, 5–17.

Pinch, T.J. and W.E. Bijker (1984) 'The Social Construction of Facts and Artefacts: or How the Sociology of Science and the Sociology of Technology Might Benefit Each Other.' *Social Studies of Science* 14(3): 399–441.

Pinch, T.J. and W.E. Bijker (1989) 'The Social Construction of Facts and Artifacts: Or How the Sociology of Science and the Sociology of Technology Might Benefit Each Other.' In *The Social Construction of Technological Systems*, edited by W.E. Bijker, T. Hughes, and T.J. Pinch. Cambridge, MA and London: MIT Press, 17–50.

Pink, S. (2001) *Doing Visual Ethnography: Images, Media and Representation in Research*. London: Sage.

Pink, S. (2006) *The Future of Visual Anthropology: Engaging the Senses*. London: Routledge.

Pink, S. (2009) *Doing Sensory Ethnography*. London, Thousand Oaks, CA, New Delhi, and Singapore: SAGE Publications.

Pink, S. (2011a) 'Amateur Photographic Practice, Collective Representation and the Constitution of Place.' *Visual Studies* 26(2): 92–101.

Pink, S. (2011b) 'Sensory Digital Photography: Re-thinking "Moving" and the Image' *Visual Studies* 26(1): 4–13.

Pink, S. and L. Hjorth. (2012) 'Emplaced Cartographies: Reconceptualising Camera Phone Practices in an Age of Locative Media.' *Media International Australia* 145(1): 145–155.

Pink, S., E. Aredèvol and D. Lanzeni (2016) *Digital Materialities: Design and Anthropology*. London: Bloomsbury.

Pink, S., H. Horst, J. Postill, L. Hjorth, T. Lewis, and J. Tacchi (eds) (2016) *Digital Ethnography: Principles and Practice*. Los Angeles: SAGE.

Pinker, D. (2018) *Enlightenment Now*. New York: Penguin.
Pinney, C. (1997) *Camera Indica: The Social Life of Indian Photographs*. London: Reaktion Books.
Pinney, C. (2002) 'Visual Culture.' In V. Buchli (ed.) *The Material Culture Reader*. Oxford: Berg, 81–86.
Pinney, C. (2003) 'Notes from the Surface of the Image: Photography, Postcolonialsm and Vernacular Modernism.' In Peterson, N. and Pinney, C. (eds) *Photography's Other Histories*. Durham, NC and London: Duke University Press.
Pinney, C. (2004) *Photos of the Gods: The Printed Image and Political Struggle in India*. London: Reaktion Books.
Pinney, C. (2005) 'Things Happen: Or, From Which Moment Does That Object Come?' In D. Miller (ed.) *Materiality*. Durham, NC: Duke University Press.
Pinney, C. (2008) *The Coming of Photography in India*. London: British Library.
Pinney, C. (2012) 'Seven Theses on Photography.' *Thesis Eleven* 113(1): 141–156.
Pinney, C. (2016) 'Bruises and Blushes: Photography "Beyond" Anthropology.' In E. Balsom and H. Peleg (eds) *Documentary across Disciplines*. Berlin: Haus der Kulturen der Welt and Cambridge, MA: MIT.
Pinney, C. and N. Thomas (eds) (2001) *Beyond Aesthetics: Art and the Technologies of Enchantment*. Oxford and New York: Berg.
Pomian, K. (1990) *Collectors and Curiosities: Paris and Venice, 1500–1800*. Cambridge: Polity Press.
Poole, D. (1997) *Vision, Race, and Modernity: A Visual Economy of the Andean Image World*. Princeton, NJ: Princeton University Press.
Porter, T. (1996) *Trust in Numbers: The Pursuit of Objectivity in Science and Public Life*. Princeton, NJ: Princeton University Press.
Poster, M. (1999) 'Underdetermined.' *New Media and Society* 1(1): 12–18.
Postill, J. (2018) *The Rise of Nerd Politics: Digital Activism and Political Change*. London: Pluto Press.
Povinelli, E. (2016) *Geontologies: A Requiem to Late Liberalism*. Durham and London: Duke University Press.
Prainsack, B. (2017) *Personalised Medicine: Empowered Patients in the 21st Century?* New York: NYU Press.
Pratt, M.L. (1991) 'Arts of the Contact Zone.' *Profession*, 33–40.
Price, D.H. (2011) *Weaponizing Anthropology: Social Science in Service of the Militarized State*. Petrolia and Oakland, CA: CounterPunch; AK Press.
Punjabi, S. and C. Pinney (2012) *Artisan Camera: Studio Photography from Central India*. Chennai: Tara Books.
Pype, J. (2018) 'Smartness from Below: Variations on Technology and Creativity in Contemporary Kinshasa.' In C. Mavhunga (ed.) *What Do Science, Technology, and Innovation Mean from Africa*? Cambridge, MA: MIT Press, 97–116.
Quaranta, D. (2013) *Beyond New Media Art*. Brescia: LINK Editions.
Quiroga, R.Q., S. Dudley, and J. Binnie (2011) 'Looking at Ophelia: A Comparison of Viewing Art in the Gallery and in the Lab.' *ACNR* 11(3): 15–18.
Ramaswamy, S. (1997) *Passions of the Tongue: Language Devotion in Tamil India, 1891–1970*. Berkeley: University of California Press.
Raphael (1829) *The Prophetic Messenger, with Two Almanacs for 1830*. London: William Charlton Wright.
Raphael (1831) *Raphael's Witch or Oracle of the Future*. London: William Charlton Wright.

Raphael (1840) *Raphael's Prophetic Almanac Being the Prophetic Messenger, with Agricultural, Commercial, and Astrological Almanacs for 1841*. London: William Charlton Wright.

Raphael (1841) *Raphael's Prophetic Almanac Being the Prophetic Messenger, with Agricultural, Commercial, and Astrological Almanacs for 1842*. London: William Charlton Wright.

Raphael (1842) *Raphael's Prophetic Almanac Being the Prophetic Messenger, Weather Guide, and Ephemeris, for 1843*. London: William Charlton Wright.

Raphael (1845) *Raphael's Prophetic Almanac Being the Prophetic Messenger, Weather Guide, and Ephemeris, for 1846*. London: William Charlton Wright.

Raphael (1894) *Raphael's Pythoness of the East; Or, Complete Key to Futurity*. Translated from the original ms. of the celebrated *Mystical Divining Book*, formerly in the possession of Her Imperial Majesty the Empress Josephine. London: W. Foulsham.

Raphael (1912) *Raphael's Almanac; or the Prophetic Messenger and Weather Guide for 1913*. London: W. Foulsham.

Rapp, R. (2015) 'Big Data, Small Kids: Medico-scientific, Familial and Advocacy Visions of Human Brains.' *Biosocieties* 11(3): 296–316.

Rappaport, R. (1999) *Ritual and Religion in the Making of Humanity*. Cambridge: Cambridge University Press.

Rasheed, A. (1989) 'Our Bauhaus, Others' Mudhouse.' *Third Text* 3(6): 3–14.

Ratner, H. and E. Ruppert (2019) 'Producing and Projecting Data: Aesthetic Practices of Government Data Portals.' *Big Data and Society* 6(2): 1–16.

Ray, J. (2007) *The Rosetta Stone and the Rebirth of Ancient Egypt*. Cambridge, MA: Harvard University Press.

Reese, J. (2017) 'Amma Provides for All: Popular Devotional Aesthetics, Material Politics and the Amman in Tamil Nadu.' In U. Skoda and B. Lettman (eds) *India and its Visual Cultures: Mapping Visualities*. Delhi and London: SAGE.

Reno, J. (2018) 'Making Time with Amateur Astronomers and Orbital Space Debris: Attunement and the Matter of Temporality.' *Journal of Contemporary Archaeology* 5(1): 4–18.

Reynolds, H.B. (1980) 'The Auspicious Married Woman.' In S.S. Wadley (ed.), *The Powers of Tamil Women*. Syracuse: Maxwell School of Citizenship and Public Affairs, 35–60.

Rheinberger, H.-J. (2014). 'On Epistemic Objects, and Around. Think-Essays.' Available at: https://www.fkawdw.nl/en/review/think/on_epistemic_objects_and_around (Accessed 9 September 2020).

Ribeiro, A. (2016) 'Against Object Agency: A Counterreaction to Sørrensen's "Hammers and Nails"' *Archaeological Dialogues* 23(2): 22–235.

Ribeiro, A. (2018) 'Death of the Passive Subject: Intentional Action and Narrative Explanation in Archaeological Studies.' *History of Human Sciences*, 1(3): 105–121.

Riles, A. (1998) 'Infinity Within the Brackets.' *American Ethnologist* 25(3): 378–398.

Riles, A. (2001) *The Network Inside Out*. Ann Arbor, MI: University of Michigan Press.

Rio, K. (2007) *The Power of Perspective: Social Ontology and Agency on Ambrym Island, Vanuatu*. Oxford: Berghahn Books.

Risam, R. (2019) *New Digital Worlds: Postcolonial Digital Humanities in Theory, Praxis, and Pedagogy*. Evanston, IL: Northwestern University Press.

Rivière, P. (1995) 'Houses, Places and People: Community and Continuity in Guiana.' In J. Carsten and S. Hugh-Jones (eds) *About the House: Lévi-Strauss and Beyond*. Cambridge: Cambridge University Press.

Rode, J. (2011) 'Reflexivity in Digital Anthropology.' In *Proceedings of the 2011 Annual Conference on Human Factors in Computing Systems – CHI 2011*, 123. Vancouver, BC: ACM Press.

Rodriguez, P. and J. Blanco (2017) 'Organization and Information in Simondon's Theory of Individuation.' *Culture and Organisation* 23(1): 34–43.

Rogoff, I. (2006) 'Smuggling – a Curatorial Model.' In V.J. Muller and N. Schafhausen (eds) *Under Construction. Perspectives on Institutional Practice*. Köln: Verlag der Buchhandlung Walther König.

Rogoff, I. (2012) 'Curating/curatorial: A Conversation between Irit Rogoff and Beatrice von Bismarck.' In B. von Bismarck (ed.) *Cultures of the Curatorial*. New York: Sternberg Press.

Rowlands, M. (1983) 'Material Culture Studies at British Universities: University College London.' Royal Anthropological Institute Newsletter 59: 15–16.

Rowlands, M. (1984) 'Objectivity and Subjectivity in Archaeology.' In M. Spriggs (ed.) *Marxist Directions in Archaeology*. Cambridge: Cambridge University Press, 108–113.

Rubin, A.M (1984) 'Ritualized and Instrumental Television Viewing.' *Journal of Communication* 34(3): 67–77.

Sahlins, M. (1976) *Culture and Practical Reason*. Chicago: University of Chicago Press.

Sautchuk, C.E. (2019) 'The Piracuru Net: Artefact, Animism and the Technical Object.' *Journal of Material Culture* 24(2): 176–193.

Savage, M., E. Ruppert, and J. Law (2010) 'Digital Devices: Nine Theses.' CRESC Working Paper Series. Manchester: University of Manchester.

Schacter, R. (2019) 'Curating the Intermural: Graffiti in the Museum 2008–2018.' In R. Sansi (ed.) *The Anthropologist as Curator*. Abingdon and New York: Bloomsbury.

Schacter, R. and C. Balaguer (2017) 'The Tears of the Hip-Hoptivist or The Rock and the Hard Place: Social Practice in the Philippines.' *World Art* 7(2): 253–282.

Schaffer, S., J. Tresch, and P. Gagliardi (eds) (2017) *Aesthetics of Universal Knowledge*. London: Palgrave Macmillan.

Schatzberg, E. (2018) *Technology: Critical History of a Concept*. Chicago: University of Chicago Press.

Schilder, P. (1935[1923]) *The Image and Appearance of the Human Body*. London: Kegan Paul, Trench, and Trubner.

Schlanger, N. (2004) 'Suivre les gestes, éclat par éclat' – la chaîne opératoire d'André Leroi-Gourhan. In F. Audouze and N. Schlanger (eds) *Autour de l'homme: contexte et actualité d'André Leroi-Gourhan*. Antibes: Editions APDCA, 127–148.

Schroeder, R. (2010) *Being There Together: Social Interaction in Shared Virtual Environments*. New York and Oxford: Oxford University Press.

Schüll, N.D. (2012) *Addiction by Design: Machine Gambling in Las Vegas*. Princeton, NJ and Oxford: Princeton University Press.

Schwartz-Cowen, R. (1985) 'How the Refrigerator Got Its Hum.' In D. Mackensie and J. Wajcman (eds) *The Social Shaping of Technology*. Oxford: Oxford University Press.

Seaver, N. (2015a) 'Bastard Algebra.' In T. Boellstorff and B. Maurer (eds) *Data, Now, Bigger and Better*! Chicago: Prickly Paradigm Press.

Seaver, N. (2015b) 'The Nice Thing about Context Is That Everyone Has It.' *Media, Culture and Society* 37(7): 1101–1109.
Seaver, N. (2017) 'Algorithms as Culture: Some Tactics for the Ethnography of Algorithmic Systems.' *Big Data and Society* 4(2): 1–12.
Seaver, N. (2019) 'Captivating Algorithms: Recommender Systems as Traps.' *Journal of Material Culture* 24(4): 41–436.
Seizer, S. (2002) 'Offstage with Special Drama Actresses in Tamilnadu: Roadwork.' In D.P. Mines and S. Lamb (eds) *Everyday Life in South Asia*, 116–131. Bloomington: Indiana University Press.
Sen, A. (1992) *Inequality Re-examined*. Oxford: Oxford University Press.
Sheller, M. (2010) 'Foreword.' In B. Fincham et al. (eds) *Mobile Methodologies*. New York: Palgrave Macmillan.
Shelley, P.B. (1840). 'A Defence of Poetry.' In P.B. Shelley (ed) *Essays, Letters and Abroad, Translations and Fragments*. London: Edward Moxon.
Siddiqi, A. (2010) *The Red Rocket's Glare: Spaceflight and the Soviet Imagination 1857–1957*. Cambridge: Cambridge University Press.
Siegert, B. (2015) *Cultural Techniques: Grids, Filters, Doors, and Other Articulations of the Real*. Translated by G. Winthrop-Young. New York: Fordham University Press.
Sigaut, F. (2003) 'La formule de Mauss.' *Techniques and Culture*, 40: 153–168.
Sigaut, F. (2012) *Comment* Homo *devint* faber. Paris: CNRS Editions.
Simmel, G. (1971[1903]) 'The Metropolis and Mental Life,' trans. E. Shils. In D. Levine (ed.) Georg Simmel on Individuality and Social Forms. Chicago: University of Chicago Press, 342–339. Originally published as 'Die Grosstadt und das Geisstesleben' in *Die Grosstadt. Jahrbuch der Gehe-Stiftung* 9.
Simmel, G. (1950) 'The Metropolis and Mental Life.' In K.H. Wolff (ed.) *The Sociology of Georg Simmel*. Glencoe, IL: The Free Press, 409–424.
Simondon, G. (1964) *L'individuation à la lumière des notions de forme et d'information*. Paris: Presses universitaires de France.
Simondon, G. (2005) *L'Invention dans les techniques: Cours et conferences*, edited by Jean-Yves Château. Paris: Seuil.
Simondon, G. (2017[1958]) *On the Mode of Existence of Technical Objects*. Minneapolis: Univocal Publishing.
Singer, M. (1980) 'Signs of the Self: An Exploration in Semiotic Anthropology.' *American Anthropologist* 82(3): 485–507.
Sloterdijk, P. (2016) *Foams*. MIT Press.
Smith, R. (2013). 'Designing Heritage for a Digital Culture.' In W. Gunn, T. Otto, and R. Smith (eds) *Design Anthropology. A Different Style of Knowing*, 117–139. London, Bloomsbury.
Sørensen, T.F. (2016) 'Hammers and nails. A response to Lindstrøm and to Olsen and Witmore.' *Archaeological Dialogues*, 23(1): 115–127.
Søresen, T.F. (2018) 'Agency (Again). A Response to Lindstrøm and Ribeiro.' *Archaeology Dialogues* 25(1): 95–101.
Souriau, É. (2015[1943]) *The Different Modes of Existence*. Minneapolis, MN: University of Minnesota Press.
Sprague, S. (2003) 'Yoruba Photography: How the Yoruba See Themselves.' In C. Pinney and N. Thomas (eds) *Photography's Other Histories*. London and Durham, NC: Duke University Press.
Spyer, J. (2017) *Social Media in Emergent Brazil*. London: UCL Press.

Spyer, P. and M.M. Steedly (2013) *Images That Move*. Santa Fe: School for Advanced Research Press.

Stallybrass, P. (1998) 'Marx's Coat.' In P. Spyer (ed.) *Border Fetishisms: Material Objects in Unstable Spaces*. London: Routledge.

Star, L.S. and K. Ruhleder (1996) 'Steps Toward an Ecology of Infrastructure: Design and Access for Large Information Spaces.' *Information Systems Research* 7(1).

Starosielski, N. (2015) *The Undersea Network*. Durham, NC: Duke University Press.

Stengers, I. (2010) *Cosmopolitics I: 1. The Science Wars, 2. The Invention of Mechanics, 3. Thermodynamics*. Minneapolis: University of Minnesota Press.

Sterne, J. (2012) *MP3: The Meaning of a Format*. Durham, NC: Duke University Press Books.

Stewart, K. (2011) 'Atmospheric Attunements.' *Environment and Planning D: Society and Space* 29(3): 445–453.

Stiegler, B. (1998) *Technics and Time, 1: The Fault of Epimetheus*. Stanford: Stanford University Press.

Stiegler, B. (2009) 'Teleologics of the Snail: The Errant Self Wired to a WiMax Network.' *Theory, Culture and Society* 26(2–3): 23–45.

Stocking, G.W. (1985) *Objects and Others: Essays on Museums and Material Culture*. Madison: University of Wisconsin Press.

Stocking, G.W. (1992) *The Ethnographer's Magic and Other Essays in the History of Anthropology*. Madison: University of Wisconsin Press.

Stocking. G.W. (1988) 'Essays on Museums and Material Culture.' In G.W. Stocking, *Objects and Others: Essays on Museums and Material Culture*. Madison: University of Wisconsin Press.

Stoller, P. (1989) *The Taste of Ethnographic Things: The Senses in Anthropology*. Philadelphia: University of Philadelphia Press.

Strassler, K. (2010) *Refracted Visions: Popular Photography and National Modernity in Java*. Durham, NC: Duke University Press.

Strathern, M. (1987a) 'Out of Context: The Persuasive Fictions of Anthropology.' *Current Anthropology* 28(3): 251–281.

Strathern, M. (1987b) 'An Awkward Relationship: The Case of Feminism and Anthropology.' *Signs* 12(2): 276–292.

Strathern, M. (1988) *The Gender of the Gift. Berkeley*. Los Angeles, and London: University of California Press.

Strathern, M. (1990) 'Artifacts of History: Events and the Interpretation of Images.' In Jukka Siikala (ed.) *Culture and History in the Pacific*. Helsinki: Finnish Anthropological Society, 25–44.

Strathern, M. (1991) *Partial Connections*. Savage, MD: Rowman and Littlefield Publishers.

Strathern, M. (1992) 'Partners and Consumer.' In *Reproducing the Future: Essays on Anthropology, Kinship and the New Reproductive Technologies*. Manchester: Manchester University Press, 118–138.

Strathern, M. (1996) 'Cutting the Network.' *Journal of the Royal Anthropological Institute* 2(3): 517–535.

Strathern, M. (2000) *Audit Cultures: Anthropological Studies in Accountability, Ethics and the Academy*. London: Routledge.

Strathern, M. (2015) 'Learning to See in Melanesia: Four Lectures Given in the Department of Social Anthropology, Cambridge University, 1993–2008.' Hau Books. Available at: https://haubooks.org/learning-to-see-in-melanesia/

Stuster, J. (2011) *Bold Endeavours: Lessons from Polar and Space Exploration*. Annapolis, MD: Naval Institute Press.

Suchman, L. (2007) *Human–Machine Reconfigurations: Plans and Situated Actions*. New York: Cambridge University Press.

Suchman, L. (2013) 'Consuming Anthropology.' In A. Barry and G. Born (eds) *Interdis ciplinarity: Reconfigurations of the Social and Natural Sciences*. London: Routledge, 141–161.

Sweeney, L. (2013) 'Discrimination in Online Ad Delivery.' Available at SSRN: https://ssrn.com/abstract=2208240 or http://dx.doi.org/10.2139/ssrn.2208240.

Sweetman, P. (1999) 'Only Skin Deep?: Tattooing, Piercing and the Transgressive Body.' In M. Aaron (ed.), *The Body's Perilous Pleasures: Dangerous Desires and Contemporary Culture*. Edinburgh: Edinburgh University Press, 165–187.

Taussig, K.S., Hoeyer, K., and Helmreich, S. (2013) 'The Anthropology of Potentiality in Biomedicine: An Introduction to Supplement 7.' *Current Anthropology*, 54(7): 3–14.

Thatcher, J., D. O'Sullivan, and D. Mahmoudi (2016) 'Data Colonialism through Accumulation by Dispossession: New Metaphors for Daily Data,' Environment and Planning D: Society and Space 34(6): 990–1006.

Thomas, N. (2010) 'The Museum as Method.' *Museum Anthropology* 33(1): 6–10.

Thompson, R.F. (1973) 'An Aesthetic of the Cool.' African Arts 7(1): 40–4, 364–367, 89–91.

Thrift, N. (2005) Knowing Capitalism. London: Sage.

Thrift, N. (2014) 'The "Sentient" City and What It May Portend.' *Big Data and Society* 1(1): 1–21.

Tilley, C. (1991) 'Materialism and an Archaeology of Dissonance.' *Scottish Archaeological Review* 8: 14–22.

Tilley, C. (1994[1974]) *A Phenomenology of Landscape: Places, Paths and Monuments*. Oxford: Berg.

Tilley, C. (1994) 'Space, Place, Landscape and Perception: Phenomenological Perspectives.' In *A Phenomenology of Landscape: Places, Paths and Monuments*. Oxford: Berg, 7–34.

Tilley, C. (1996) An Ethnography of the Neolithic: Early Prehistoric Societies in Southern Scandinavia. Cambridge: Cambridge University Press.

Tilley, C. (1999) Metaphor and Material Culture. Oxford: Blackwell.

Tilley, C. (2004) *The Materiality of Stone*. Oxford: Berg.

Tilley, C. (2006) 'Introduction: Identity, Place, Landscape and Heritage.' Journal of Material Culture 11(1–2): 7–32.

Tilley, C. (2007) 'Materiality in Materials.' *Archaeological Dialogues* 14(1): 16–20.

Tilley, C., W. Keane, S. Kuechler, M. Rowlands, and P. Spyer (eds) (2006) *Handbook of Material Culture*. London: Sage.

Tsing, A. (2005) *Friction: An Ethnography of Global Connection*. Princeton, NJ: Princeton University Press.

Turkle, S. (2012) *Alone Together: Why We Expect More from Technology and Less from Each Other*. New York: Basic Books.

Turnbull, D. (1989) *Maps are Territories: Science is an Atlas*. Chicago: University of Chicago Press.

Turner, F. (2013) *The Democratic Surround: Multimedia and American Liberalism from World War II to the Psychedelic Sixties*. Chicago: University of Chicago Press.

Ucko, P. (1969) 'Ethnography and Archeological Interpretation of Funerary Remains.' *World Archaeology* 1(2): 262–280.

Urry, J. (1972) 'Notes and Queries on Anthropology, and the Development of Field Methods in British Archaeology, 1870–1920.' *Proceedings of the Royal Anthropological Institute of Great Britain and Ireland* (1972): 45–57.

Valentine, D. (2016) 'Atmosphere: Context, Detachment, and the View from Above Earth.' *American Ethnologist* 43(3) 511–524.

Valentine, D. (2017) 'For the Machine.' *History and Anthropology* 28(3): 302–307.

Valentine, D., V. Olson, and D. Battaglia (2012) 'Extreme: Limits and Horizons in the Once and Future Cosmos.' *Anthropological Quarterly* 85(4): 1007–1026.

Vartiainen, K., and J. Enkenberg (2013) 'Learning from and with Museum Objects: Design Perspectives, Environment, and Emerging Learning System.' *Educational Technology Research and Development* 61(5): 841–862.

Varzi, R. (2006) *Warring Souls: Youth, Media and Martyrdom in Post-Revolution Iran*. Durham, NC and London: Duke University Press.

Venkatraman, S. (2017) *Social Media in South India*. London: UCL Press.

Venturini, T., M. Jacomy, A. Meunier, and B. Latour (2017) 'An Unexpected Journey: A Few Lessons from Sciences Po Médialab's Experience.' *Big Data and Society* 4(2): 1–11.

Verdon, M. (2007) 'Franz Boas: Cultural History for the Present, or Obsolete Natural History?' *The Journal of the Royal Anthropological Institute* 13(2): 433–451.

Virilio, P. (1986) *Speed and Politics*, trans. M. Polizzatti. New York: Semiotext(e).

Virilio, P. (1991) *The Aesthetics of Disappearance*, trans. P. Beitchman. New York: Semiotext(e).

Virilio, P. (2007) *The Original Accident*. Cambridge: Polity.

Vonderau, A. (2018) 'Scaling the Cloud: Making State and Infrastructure in Sweden.' *Ethnos* 84(4): 698–718.

Von Schnitzler, A. (2013) 'Traveling Technologies: Infrastructure, Ethical Regimes, and the Materiality of Politics in South Africa.' *Cultural Anthropology* 28(4): 670–693.

Wagner, R. (1975) *The Invention of Culture*. Chicago: University of Chicago Press.

Wagner, R. (1986) *Symbols That Stand for Themselves*. Chicago: University of Chicago Press.

Wagner, R. (2001). *An Anthropology of the Subject: Holographic Worldview in New Guinea and Its Meaning and Significance for the World of Anthropology*. Berkeley: University of California Press.

Walford, A. (2012) 'Data Moves: Taking Amazonian Climate Science Seriously.' *Cambridge Anthropology* 30(2): 101–117.

Walford, A. (2015) 'Double Standards: Examples and Exceptions in Scientific Metrological Practices in Brazil.' *Journal of the Royal Anthropological Institute* 21(S1): 64–77.

Walford, A. (2017) 'Raw Data: Making Relations Matter.' *Social Analysis* 61(2): 65–80.

Walls, M. (2015) 'Making as a Didactic Process: Situated Cognition and the *chaîne opératoire*.' *Quaternary International*, 405: 21–30.

Wallace, A. (1978) *Rockdale: The Growth of an American Village in the Early Industrial Revolution*. New York: Alfred Knopf and Random House.

Wallerstein, I. (1991) *Report on an Intellectual Project: The Fernand Braudel Center, 1976–1991*. Binghamton, NY: Fernand Braudel Center.

Walsh, J., and A. Gorman (2020) 'A Methodology for Research in Space Archaeology: The International Space Station Archaeological Project.' *SocArXiv*. 8 August 2020. doi:10.31235/osf.io/qtcxe.

Walton, S. (2015) 'Re-envisioning Iran Online: Photoblogs and the Ethnographic "Digital-Visual Moment."' Special Issue: Critical Histories of Photography in the Middle East. *Middle East Journal of Culture and Communication* (MEJCC): 8(2–3): 398–418.

Walton, S. (2017) '"Being ~~There~~ Where?" Designing Digital-Visual Methods for Moving With/In Iran.' In A. Elliot, R. Norum, and N.B. Salazar (eds) *Methodologies of Mobility: Ethnography and Experiment*. Oxford: Berghahn Books.

Wang, X. (2016) *Social Media in Industrial China*. London: UCL Press.

Warnier, J.P. (2001) 'A Praxeological Approach to Subjectivation in a Material World.' *Journal of Material Culture* 6(1): 5–24.

Warnier, J.P. (2007) *The Pot-King: The Body and Technologies of Power*. Leiden and Boston: Brill.

Warwick, K. (2003) 'Cyborg Morals, Cyborg Values, Cyborg Ethics.' *Ethics and Information Technology* 5(3): 131–137.

Warwick, K. (2004) *I, Cyborg*. Chicago: University of Illinois Press.

Weibel, P. and B. Latour (2007) 'Experimenting with Representation: Iconoclash and Making Things Public.' In S, Macdonald and P. Basu (eds) *Exhibition Experiments*. Oxford: Blackwell.

Were, G. (2010) 'Re-engaging the University Museum: Knowledge, Collections and Communities at University College London.' *Museum Management and Curatorship* 25(3): 291–304.

West, K. (2013) *The Garments of Salvation: Orthodox Christian Liturgical Vesture*. Yonkers, NY: St Vladimir's Seminary Press.

White, D. (2018) 'Contemplating the Robotic Embrace: Introspection for Affective Anthropology.' *More-than-Human Worlds: A NatureCulture Blog Series*. [Online] Retrieved from www.natcult.net/contemplating-the-robotic-embrace/.

Whitehead, N. and M. Wesch (eds) (2012) *Human No More: Digital Subjectivities, Unhuman Subjects, and the End of Anthropology*. Boulder: University Press of Colorado.

Willerslev, R. (2011) 'Frazer Strikes Back from the Armchair: A New Search for the Animist Soul.' *Journal of the Royal Anthropological Institute* 17(3): 504–526.

Winner, L. (1985[1977]) 'Citizen Virtues in a Technological Order.' *Inquiry* 35: 341–361.

Winner, L. (1986) 'Do Artefacts Have Politics?' In *The Whale and the Reactor: A Search for Limits in an Age of High Technology*. Chicago: University of Chicago Press.

Witmore, C. (2009) 'The Realities of the Past: Archaeology, Object-orientations, Pragmatology.' *Modern Materials. Proceedings from the Contemporary and Historical Archaeology in Theory Conference*, 25–36.

Witmore, C. (2014) 'Archaeology and the New Materialism.' *Journal of Contemporary Archaeology* 1(2): 203–246.

Wittkower, R. (1977) 'East and West: The Problem of Cultural Exchange.' In *Allegory and the Migration of Symbols*. London: Thames and Hudson, 9–14.

Wood, A. (1954) *Thomas Young, Natural Philosopher 1773–1829*. Cambridge: Cambridge University Press.

Young, G. (2012) *The Russian Cosmists: The Esoteric Futurism of Nikolai Fedorov and His Followers*. Oxford: Oxford University Press.

Young, M.Y. (2000) 'The Careless Collector: Malinowski and the Antiquarians.' In M. O'Hanlon and R.L. Welsch (eds.) *Hunting the Gatherer: Ethnographic Collectors, Agents and Agency in Melanesia, 1870s–1930s.* Oxford: Berghahn,
Zabusky, S. (1995) *Launching Europe*. Princeton: Princeton University Press.
Zhao, S. (2003) 'Toward a Taxonomy of Copresence.' *Presence* 12(5): 445–455.
Zuboff, S. (2018) *The Age of Surveillance Capitalism: The Fight for the Human Future at the New Frontier of Power*. New York: Public Affairs.

Index

For Product Safety Concerns and Information please contact our EU representative GPSR@taylorandfrancis.com Taylor & Francis Verlag GmbH, Kaufingerstraße 24, 80331 München, Germany

Batch number: 10725966

Printed by Printforce, the Netherlands